MULTILINGUAL COMPENDIUM OF PLANT DISEASES (English)

COMPENDIUM POLYGLOTTE DES MALADIES DES PLANTES (French)

موجز بعدة لغات عن أمراض النبـات (Arabic)

COMPENDIO DE ENFERMEDADES DE LAS PLANTAS EN VARIOS IDIOMAS (Spanish)

МНОГОЯЗЫЧНЫЙ СПРАВОЧНИК-СЛОВАРЬ ПО БОЛЕЗНЯМ РАСТЕНИЙ (Russian)

COMPENDIO POLIGLOTA DAS DOENCAS DAS PLANTAS (Portuguese)

COMPENDIO MULTILINGUE DELLE MALATTIE DELLE PIANTE (Italian)

خلاصه لغات گوناگون امراض نباتات (Persian)

FLERSPROGLIGT KOMPEN ... ME (Danish)

FLERSPRAKLIG HANDBOK OM PLANTESYKDOMMER (Norwegian)

FLERSPRAKIG HANDBOK OM VAXTSJUKDOMAR (Swedish)

МНОГОЕЗИКОВ НАРЪЧНИК ПО БОЛЕСТИ НА РАСТЕНИЯТА (Bulgarian)

MEHRSPRACHIGES HANDBUCH DER PFLANZENKRANKHEITEN (German)

MEERTALIGE SAMENVATTENDE LIJST VAN PLANTENZIEKTEN (Dutch)

အပင်ရောဂါ ဘာသာစုံကျမ်း (Burmese)

NOVEHYI BETEGSEGEK TOBBNYELVU SZOFARA (Hungarian)

สารานุกรมโรคพืชภาษาต่างๆ (Thai)

विरुवाको रोगहरूको बहुभाषीय संकलन (Nepali)

पौधों के रोगों का बहुभाषा संग्रह (Hindi)

多國語言植物病蟲害簡述 (Chinese)

ÇEŞİTLİ DİLLERDE BİTKİ HASTALIKLARI ÖZETİ (Turkish)

各国語による植物疫病概要 (Japanese)

DANH SÁCH NHỮNG BỆNH CỦA CÂY CỎ VIẾT BẰNG NHIỀU THỨ TIẾNG (Vietnamese)

COMPENDIO MULTILINGUAL DEL NOMINES POPULAR DE MALADIAS BOTANIC (Interlingua)

Copyright © 1976, by
The American Phytopathological Society

ISBN: 0-89054-018-7
Library of Congress Catalog Card Number: 75-46392

The American Phytopathological Society
3340 Pilot Knob Road, St. Paul, Minnesota 55121

Printed in the United States of America

CONTENTS

INTRODUCTION

The purpose of this <u>Multilingual Compendium of Plant Diseases</u> is to help improve communication among scientists throughout the world, especially those who are concerned with crop production and crop protection. Its ultimate aim is to aid in increasing food production, primarily in the developing countries and to reduce the confusion and misunderstanding caused by the multiplicity of common names of plant diseases. We wished to produce a useful publication which will achieve these three major objectives. It was recognized that the cooperation of many scientists would be needed and that, regardless of how carefully and competently the work might be done, it would never be entirely accurate or complete. However, in order for the compendium to be as comprehensive and useful as possible, letters were sent to approximately one hundred and twenty-five plant pathologists throughout the world soliciting help in supplying lists of their most important plant diseases, including the scientific and common names of each. The response to these letters was gratifying. These lists, in addition to furnishing the common names used in the many languages, provided information as to the relative importance of crops for each of the countries. The criteria utilized in selecting the crops and diseases were the economic importance of the crop on a worldwide basis, with emphasis on food production, the potential of the disease to cause loss, and the availability of immediate controls such as the safe and efficient use of fungicides. Only diseases caused by fungi and bacteria have been included, since from the standpoint of common names the diseases caused by viruses and nematodes are not available in sufficient numbers to justify their inclusion. It is hoped that when knowledge has developed to provide such common names these will be included in future revisions.

The following example represents a shortened version of the format which will be followed for each disease. In order to explain fully the complex organization of the material required by the nomenclature of both pathogens and hosts and the multiplicity of languages involved, we have numbered the variables and will clarify each in the example. See page 16 for the complete version.

I.   The individual sections (a) through (f), are defined as follows:

    (a)   The Latin name of the host listed first is the one that we have selected to use in connection with the association between pathogen and host.

    (b)   These are the Latin names of the hosts that are also considered to be of importance in connection with this pathogen. When a disease which needs an alternate host in order to complete its life cycle occurs, such as Stem Rust of wheat, we have selected the economically important host's symptoms to describe.

    (c)   Indicates the symbol used to mean "in association with."

    (d)   Latin names of the pathogen (disease-producing organism).

    (e)   Represents other names which have been used as synonyms for the pathogen. We are defining "synonym" as either a word accepted by some scientists as indicating the same pathogen or, in some cases, to be either the sexual or conidial stage of the fungus, whichever is opposite the first. These may be words gleaned either from the literature or from the lists submitted to us by the cooperating pathologists.

    (f)   Indicates the symbol used to mean "is called."

         (a)   (b)   (c)

I.   <u>Avena</u> <u>sativa</u> (<u>Triticum</u> <u>aestivum</u>, <u>Zea</u> <u>mays</u> and other Gramineae) +

         (d)   (e)   (f)

<u>Gibberella</u> <u>zeae</u> (Schw.) Petch (<u>Gibberella</u> <u>saubinetti</u> (Mont.) Sacc.) =

This example indicates that the host associated with the pathogen results in a biological phenomenon which is known by the following common disease names:

II.   ENGLISH   Head Mold, Scab, Seedling Blight, Head Blight
      FRENCH   Flétrissement des plantules des céréales, Fusariose, Maladie d'enivrement
      DUTCH   Kiemschimmel
      GERMAN   Fusariose, Tammelkrankheit

II. As is well known, common disease names are assigned to the resulting conditions caused by the many combinations which exist between plant pathogens and their hosts. In our descriptions we have sometimes used the terms "disease" and "pathogen" (fungus or bacterium) as synonyms. We are aware that this is not scientifically correct; however, we chose to do this for the sake of brevity and clarity for all readers. Most of the disease situations that we are dealing with fall within one of the following nine catagories:

1. One disease name identifying the association of one host species and one pathogen species.
2. One disease name identifying the association of one host species and two or more pathogen species.
3. One disease name identifying the association of two or more host species and one pathogen species.
4. One disease name identifying the association of two or more host species and two or more pathogen species.
5. Two or more disease names identifying the association between one host species and one pathogen species.
6. Two or more disease names identifying the association between one host species and two or more pathogen species.
7. Two or more disease names identifying the associations between two or more host species and one pathogen species.
8. Two or more disease names identifying the associations between two or more host species and two or more pathogen species.
9. In addition, on some occasions there are disease situations which involve two or more of these combinations.

It is apparent that the many common plant disease names have come about as a result of the different bases used in assigning them. An illustration of this confusion would be for a disease caused by the conidial stage of a fungus to be called "damping-off" and for the sexual stage to be called "blight." Furthermore, our observations after dealing with these common names in many languages have made it clear that the majority of common names have been derived from two major sources. Either they are translations of the common names from the English language or translations of the Latin names of the causal organism. Thus the confusion is perpetuated.

We have selected twenty-one languages for the section dealing with common names. These are the ones for which we have information in sufficient quantity to justify their use. We do not mean to imply that the names have been listed in order of frequency of use, but merely that they are all in the literature. Neither do we intend for the reader to believe that these are necessarily all the names that have been used in association with the disease.

III. The earliest and most conspicuous symptom.........................stem blight.
IV. Le prime e le plus remarcabile symptoma ....plaga de stirpes.
V. Le symptôme le plus précoce .... et des tiges.
VI. Los más conspicuous .... tizón de tallo.

III - VI. If all the possible diseases resulting from combinations of hosts and pathogens which are mentioned in this compendium were described, the number would exceed 4000. Therefore, in the case of multiple hosts, we have chosen the host which would provide the most characteristic symptoms. This process has limited the number of diseases described to 325 while still providing a general concept (description of symptoms) of the disease which may also apply to the other possible combinations.

Most plant diseases indicate a biological phenomenon with fairly well-defined limits. Our approach in describing what happens when a pathogen is associated with a host is perhaps unique because it is based upon the concept of the disease rather than signs. This is not unlike the results obtained from time-lapse photography as opposed to a single close-up snapshot. Utilizing the conceptual approach based upon this phenomenon, we have attempted to describe each disease in such a manner that those interested in crop protection and production, regardless of their native language, will think of the same disease when the various common names are used in oral or written communication. The descriptions were written in non-technical terms for two reasons: (1) the terminology used will permit the reader, no matter what his scientific training, to visualize what these diseases really look like and (2) ease in translating them.

In contrast to the preceding section pertaining to common names which were largely contributed by pathologists in the various countries throughout the world, for the section on disease concepts (descriptions) we have selected only four languages -- English, French, Spanish and Interlingua. It is generally accepted that approximately 90% of the world population has a rudimentary understanding of English, French or Spanish. For those who may not be familiar with Interlingua, it is a language in which the vocabulary is derived from Latin, Italian, Spanish, Portuguese, French, English, German and Russian. It embodies all the word material that the languages of the western world have typically in common. The grammar is greatly simplified. Many will find the Interlingua vocabulary familiar to them because the words are, for the most part, the common roots of English and the Romance languages.

There is every reason to believe that there will be an additional number of readers who will be able to read and understand Interlingua. Our experience in the use of this language in connection with material published in the Plant Disease Reporter showed us that many plant pathologists throughout the world, although they had not previously studied Ingerlingua, found it easy to comprehend. It is conceivable that Interlingua could play an important role in providing a universal language for common names of plant diseases in the same manner that Latin serves as a standard for scientific names. Because we believe the language can provide a useful tool in communication among scientists, we have included some helpful material on pages 454 and 455.

VII. DISTRIBUTION: Af., A., Aust., Eur., N.A., C.A., S.A.

In this section the distribution is listed by continent or definite geographical area. For the most part this information was taken from "Distribution Maps of Plant Diseases" published by the Commonwealth Mycological Institute of England. On the reverse side of these maps the distribution by individual countries within the continents, with supporting documentation, can be found. The following abbreviations will be standard throughout the compendium:

Af.	(Africa)	N. A.	(North America)
A.	(Asia)	S. A.	(South America)
Aust.	(Australasia)	C. A.	(Central America)
Eur.	(Europe)	W. I.	(West Indies)

It is believed that the close-up color illustrations of the most typical symptoms in combination with the general symptomotology presented in the descriptions will provide excellent diagnostic tools. The number on each illustration refers to the corresponding page on which the pertinent information for the disease can be found. The missing photographs are the result of our inability to obtain suitable transparencies. Generally, either root diseases which have symptoms that are difficult to photograph or diseases which, although of major importance in some geographical areas, were not possible for us to secure account for the omissions.

Le scopo de iste <u>Compendio Multilingual del Nomines Popular de Maladias Botanic</u> es ameliorar communication inter scientistas per tote le mundo, specialmente inter illos qui se occupa del production e del protection de plantas alimentari. Le objectivo ulterior es fomentar le augmentation del productos alimentari, principalmente in le paises in via de developpamento. Non existe actualmente un tractato complete le qual elimina le confusion e le conceptos erronee que es le resultato del multiplicitate del nomines popular de maladias botanic. Ergo, il esseva nostre intention editar un texto utile que va realisar iste tres objectivos importante. On ha recognite que le cooperation de multe scientistas esserea necessari e que, le attention meticulose e le aptitudes capabile del collaboratores nonobstante, le obra non poterea esser totalmente correcte o complete. Malgrado iste limitation, al fin que le <u>Compendio</u> sia tanto complete e utile como possibile, on ha inviate litteras a circa 125 pathologistas de plantas per tote le mundo, demandante lor adjuta in le compilation de listas del plus importante maladias de plantas -- e le nomines scientific e le nomines popular. Le responsas al litteras esseva agradabile. Iste listas forniva le nomines popular actual in le multe linguas e, in plus, illos indicava le importantia relative de plantas pro cata pais. In le selection de plantas e de maladias, nos ha utilisate como criterios: le importantia economic de un planta, ab un vista mundial, specialmente in relation a nutrition; le capacitate del maladia pro destruction; e le disponibilitate immediate de medios de controlo como fungicidas inoffensive e efficace. Solmente le maladias que es causate per fungos e bacterios esseva hic incluse, perque nomines popular generalmente non existe pro le viruses e nematodos. On spera que le cognoscentia general essera avantiate in le futuro de maniera que nomines popular de iste maladias essera divulgate e incluse in revisiones veniente.

Le exemplo sequente presenta un exposition abbreviate del formula pro cata maladia. Pro un explication exacte del organisation intricate de detalios causate per le nomenclatura e de pathogenos e de plantas-hospites e per le mutliplicitate de linguas, nos ha enumerate le detalios variable. Un clarification seque.

Le articulos (a) usque a (f) significa:
(a) le planta-hospite cuje nomine latin se trova in prime position esseva selecte pro illustration del association inter pathogeno e hospite.
(b) le nomines latin de altere hospites importante in connexion con le mesme pathogeno. Si un maladia, como (fer)ruggine nigre (linear) in Gramineae, require un hospite alternative pro su cyclo vital, on ha selecte describer le symptomas del hospite con le plus grande valor economic.
(c) "in association con."
(d) nomine latin del pathogeno (le organismo que causa le maladia).
(e) altere nomines acceptate como synonymos pro le pathogeno. Le interpretation de "synonymo" es un vocabulo o acceptate per qualque scientistas pro le mesme pathogeno o, a vices, le nomine del stadio sexual o conidial del fungo (qualcunque es opposite al prime nomine). Iste synonymos esseva extracte o ab le litteratura o ab le listas inviate per le pathologistas collaborante.
(f) "se nomina."

    (a)   (b)   (c)
I. <u>Avena</u> <u>sativa</u> (<u>Triticum</u> <u>aestivum</u>, <u>Zea</u> <u>mays</u> and other Gramineae) +

    (d)   (e)   (f)
<u>Gibberella</u> <u>zeae</u> (Schw.) Petch (<u>Gibberella</u> <u>saubinetti</u> (Mont.) Sacc.) =

Iste exemplo indica que le hospite in association con le pathogeno presenta un phenomeno biologic le qual se nomina popularmente:

II. ENGLISH   Head Mold, Scab, Seedling Blight, Head Blight
    FRENCH   Flétrissement des plantules des céréales, Fusariose, Maladie d'enivrement
    DUTCH   Kiemschimmel
    GERMAN   Fusariose, Taumelkrankheit
Como on sape, nomines popular describe le conditiones consecutive al multe combinationes de pathogenos e los hospites. In le descriptiones nos ha a vices tractate le vocabulos

"maladia" e "pathogeno" (fungo o bacterio) como synonymos. Es ver que iste formulation non es scientificamente e technicamente correcte, ma nos ha lo preferite a causa de brevitate e claritate pro nostre lectores. Le majoritate del conditiones pathologic hic tractate es incluse in nove (9) categorias:

    (1) Nomine de un maladia que significa le association de un hospite con un pathogeno.

    (2) Nomine de un maladia que significa le association de un hospite con duo o plure pathogenos.

    (3) Nomine de un maladia que significa le association de duo o plure hospites con un pathogeno.

    (4) Nomine de un maladia que significa le association de duo o plure hospites con duo o plure pathogenos.

    (5) Duo o plure nomines de maladias que significa le association inter un hospite e un pathogeno.

    (6) Duo o plure nomines de maladias que significa le association inter un hospite e duo o plure pathogenos.

    (7) Duo o plure nomines de maladias que significa le associationes inter duo o plure hospites e un pathogeno.

    (8) Duo o plure nomines de maladias que significa le associationes inter duo o plure hospites e duo o plure pathogenos.

    (9) In plus, a vices il existe conditiones pathologic que es combinationes del supracitate conditiones.

Il es evidente que le multe nomines popular del maladias botanic es le resultato de multe punctos de vista. Per exemplo, le stadio conidial de un fungo pote nominar se un "putrefaction," ma le stadio sexual pote nominar se un "necrosis." In plus, post examination del multe linguas, nos ha constatate que le nomines popular se deriva, generalmente, ab duo origines principal. Illos es traductiones o del nomines popular anglese o del nomines latin pro le organismos pathogenic. Confusion ergo es inevitabile.

Nos ha selecte 21 linguas pro le section de nomines popular. Pro iste linguas nos ha evidentia sufficiente. Non es a supponer que le ordine de nostre listas indica le frequentia actual, ma solmente que tote le nomines occurre in le litteratura. In plus, le lector non debe supponer que tote le nomines associate con un maladia se trova ici.

III. The earliest and most conspicuous symptom...................................blight.
IV. Le prime e le plus remarcabile symptoma......................... plaga de stirpes.
V. Le symptome le plus precoce... et des tiges.
VI. Los más conspicuos.. tízon de tallo.

Si on habeva selecte describer tote le maladias consecutive a varie combinationes del hospites e del pathogenos citate in iste compendio, lor numero haberea essite super 4,000. Ergo, in le caso de plure hospites, nos ha selecte le uno cuje symptomas es le plus characteristic. In tal maniera, on ha limitate le numero de maladias a 325, ma al mesme tempore on forni notiones general (descriptiones de symptomas) del maladia le quales es descriptive del altere combinationes potential.

Le majoritate del maladias botanic presenta un phenomeno biologic con limites assatis definite. Nostre methodo in le description de lo que resulta ab le association de un pathogeno e un hospite es forsan unic, perque su fundamento es le concepto del maladia e non del symptomas. Iste methodo forni resultatos similar a photographia per un intervallo de tempore, in contradistinction a un sol photo instantanee. Alora, nos ha utilisate iste methodo conceptual comenciante ab le phenomeno biologic; nos ha tentate describer cata maladia in tal maniera que illes qui se occupa del protection e production de plantas, sin consideration de lor linguas matre, va precisar le mesme maladia quando on utilisa le multe nomines popular o in communication scripte o oral. On ha scripte le descriptiones in terminologia nontechnic per duo rationes: (1) iste terminologia permitte le lector, sin consideration de su education scientific, imaginar le actual aspectos visual de iste maladias; (2) iste terminologia facilita le traduction.

In contradistinction al parte precedente, que tracta le nomines popular le quales esseva de grande mesura le contributiones de pathologistas in multe paises per tote le mundo, in le parte tractante conceptos (descriptiones) del maladias on ha selecte solmente 4 linguas -- anglese, francese, espaniol, e interlingua. On crede que circa

90% del homines in le mundo pote comprehender, plus o inus, o anglese o francese o espaniol. Interlingua es le lingua scientific international cuje vocabulario se deriva ab latino, italiano, espaniol, portuguese, francese, anglese, germano, e russo. Illo incorpora tote le materiales lexical le quales es commun al linguas del occidente. Le grammatica es multo simplificate. Le majoritate del lectores va trovar le vocabulario interlingual familiar perque le vocabulos es typicamente le radices commun de anglese e del linguas romanic.

On crede que multe lectores additional potera comprehender e leger le texto interlingual. Per nostre experientia con le usage de Interlingua in le Plant Disease Reporter, nos ha discoperte que multe pathologistas de plantas per tote le mundo ha trovate Interlingua facile a comprehender, ben que illos non ha estudiate lo. On pote imaginar que il es possibile que Interlingua functiona como un lingua universal pro le nomines popular de maladias botanic de mesme maniera que latino es le standard pro le nomines scientific. Alora, perque on crede que le lingua es un instrumento utile in communication inter scientistas, on ha includite alicun commentos grammatical in le paginas 454 e 455.

VII. DISTRIBUTION: Af., A., Aust., Eur., N.A., C.A., S.A.

In iste portion le distribution es citate per continente o per un area geographic definite. Generalmente, le information se deriva ab "Distribution Maps of Plant Diseases" publicate per le Commonwealth Mycological Institute of England. Al reverso de iste mappas se trova distribution per varie paises in cata continente, con documentation corroborante. Le abbreviationes sequente occurre in le compendio:

Af.	(Africa)	N.A.	(America del Nord)
A.	(Asia)	S.A.	(America del Sud)
Aust.	(Austral-asia)	C.A.	(America Central)
Eur.	(Europa)	W.I.	(Antilles)

On crede que le detaliate photographias in color del symptomas le plus typic in combination con le symptomatologia general offerte in le descriptiones va fornir excellente instrumentos diagnostic. Le numero in cata illustration corresponde al pagina in le qual se trova information pertinente al maladia. Photographias absente occurre quando nos esseva impotente obtener diapositivas satisfactori. Le omissiones generalmente esseva causate per le facto que maladias radical es difficile a photographar e, a vices, photographias de maladias que es importante in varie areas geographic non esseva obtenibile.

INTRODUCTION

Ce Compendium Polyglotte des Maladies des Plantes a pour but de faciliter les communications entre les hommes de science de par le monde, spécialement ceux qui oeuvrent dans la production et la protection des plantes. Son objectif ultime est de promouvoir la production des vivres, surtout dans les pays en voie de développement. Il ne se trouve présentement aucun traité d'ensemble susceptible d'atténuer les confusions et malentendus causés par la multiplicité des noms populaires des maladies des plantes. C'est pour combler cette lacune et atteindre ces objectifs que nous avons voulu publier ce compendium. Nous nous sommes rendu compte que la collaboration de nombreux hommes de science s'imposerait et que, quels que soient les soins et la compétence mis à contribution, cette oeuvre ne serait jamais tout à fait à point ni achevée. Néanmoins, afin de rendre ce recueil aussi pratique et complet que possible, nous avons écrit à quelque cent vingt-cinq phytopathologistes de par le monde et les avons priés de nous fournir chacun une liste des plus importantes maladies des plantes de leur pays, avec, pour chacune, le nom scientifique et le nom populaire. Les réponses à nos lettres ont été très encourageantes. En plus de nous faire connaître les noms populaires employés dans les diverses langues, ces listes nous ont renseigné sur l'importance relative des cultures dans chaque pays. Pour fixer notre choix des cultures et des maladies, nous avons considéré l'importance économique de la culture à une échelle mondiale, en mettant l'accent sur la production des vivres, sur la possibilité pour la maladie de causer des pertes et sur la disponibilité de moyens immédiats de lutte, comme le recours à des fongicides efficaces et sans danger. Nous n'avons inclus que les maladies causées par les champignons et les bactéries, parce que, en ce qui concerne les noms populaires, on ne peut réunir un assez grand nombre de maladies causées par les virus et les nématodes pour que ça vaille la peine de les inclure. Il est à espérer que lorsque l'état de nos connaissances aura suffisamment évolué pour fournir plus de noms populaires, il y aura lieu de les inclure dans des éditions revisées.

L'exemple suivant représente une version abrégée du format qui sera adopté pour chaque maladie. Pour bien expliquer l'organisation complexe des matériaux requis par la nomenclature des pathogènes et des hôtes et par la multiplicité des langues impliquées, nous avons numéroté les variables et nous expliquerons chacune dans l'exemple qui suit (voir page 16 pour la version complète):

Les sections (a) à (f) sont ainsi agencées:
- (a) Le premier nom latin sur la liste est celui que nous avons choisi pour marquer le rapport entre le pathogène et l'hôte.
- (b) Ici figurent les noms latins des hôtes jugés également importants en rapport avec le pathogène. Dans le cas d'une maladie dont l'évolution nécessite un hôte complémentaire, comme la rouille de la tige du blé nous avons décrit les symptômes sur l'hôte de plus grande importance économique.
- (c) C'est le symbole employé pour désigner "en association avec".
- (d) Noms latins du pathogène (organisme causant la maladie).
- (e) Représente les autres noms du pathogène employés comme synonymes. Nous entendons par "synonyme" un mot qui, selon certains hommes de science, désigne le même pathogène, ou, dans certains cas, la phase sexuée ou la phase condienne d'un même champignon. Ces termes peuvent avoir été pigés dans des documents ou avoir été tirés de listes soumises par les phytopathologistes coopérants.
- (f) C'est le symbole employé pour signifier "se nomme".

I.   (a)   (b)   (c)

Avena sativa (Triticum aestivum, Zea mays and other Gramineae)   +

             (d)   (e)   (f)

Gibberella zeae (Schw.) Petch (Gibberella saubinetti (Mont.) Sacc.)   =

Cet exemple montre que l'association de l'hôte et du pathogène résulte en un phénomène biologique connu sous les noms populaires des maladies qui suivent.

II. ENGLISH   Head Mold, Scab, Seedling Blight, Head Blight
     FRENCH   Flétrissement des plantules des céréales, Fusariose, Maladie d'enivrement
     DUTCH   Kiemschimmel
     GERMAN   Fusariose, Tammelkrankheit

Comme on sait, les noms populaires des maladies sont assignés aux conditions qui dérivent des combinaisons multiples entre les phytopathogènes et leurs hôtes. Dans les descriptions, les termes "maladie" et "pathogène" (champignon ou bactérie) sont parfois employés comme synonymes. Nous savons que cela est inexact au point de vue scientifique, mais nous avons pensé que cela pourrait contribuer à rendre le texte plus bref et plus clair. La plupart des cas de maladie inclus tombent dans l'une des neuf catégories suivantes:

1. Le nom d'une maladie désignant l'association d'un hôte et d'un pathogène.
2. Le nom d'une maladie désignant l'association d'un hôte et de deux pathogènes ou plus.
3. Le nom d'une maladie désignant l'association de deux hôtes ou plus et d'un pathogène.
4. Le nom d'une maladie désignant l'association de deux hôtes ou plus et de deux pathogènes ou plus..
5. Deux noms de maladie ou plus désignant l'association entre un hôte et un pathogène.
6. Deux noms de maladies ou plus désignant l'association entre un hôte et deux pathogènes ou plus.
7. Deux noms de maladies ou plus désignant l'association entre deux hôtes ou plus et un pathogène.
8. Deux noms de maladies ou plus désignant l'association entre deux hôtes ou plus et deux pathogènes ou plus.
9. De plus, il peut y avoir des cas de maladie impliquant deux de ces combinaisons ou plus.

Il est clair que les nombreux noms populaires de maladies des plantes tirent leur origine des différentes données sur lesquelles on s'est fondé pour les attribuer. Pour illustrer la confusion qui règne, signalons le cas d'une maladie qui s'appelle "fonte" lorsqu'elle est causée par la phase conidienne d'un champignon, et "brûlure" s'il s'agit de la phase sexuée du même organisme. Bien plus, après nous être familiarisé avec les noms populaires en plusieurs langues, il nous a paru clair que la majorité de ces noms sont tirés de deux sources principales. Ce sont le plus souvent des traductions des noms populaires anglais des maladies ou des noms latins du pathogène responsable, de sorte que la confusion reste complète.

Nous avons inclus vingt et une langues dans la section qui traite des noms populaires. Ce sont les langues dans lesquelles nous avons pu recueillir une somme suffisante de renseignements. Cela ne signifie pas que les noms ont été inscrits par ordre de fréquence d'emploi, mais seulement qu'on peut les retrouver tous dans la documentation. Nous ne voulons pas non plus que le lecteur soit porté à croire que ce sont là nécessairement tous les noms employés en rapport avec la maladie.

III. The earliest and most conspicuous symptom . . . . stem blight.
IV. Le prime e le plus remarcabile symptoma . . . . plaga de stirpes.
V. Le symptômé le plus précoce . . . . et des tiges.
VI. Los más conspicuous . . . . tízon de tallo.

S'il avait fallu décrire toutes les maladies qui peuvent résulter des relations entre les hôtes et les pathogènes, il y en aurait eu plus de quatre mille. C'est pourquoi, lorsqu'il y a plusieurs hôtes, nous avons retenu celui qui présente les symptômes les plus caractéristiques. Cette façon de procéder nous a permis de limiter à 325 le nombre des maladies décrites, tout en donnant de la maladie une conception générale (description des symptômes) qui peut s'appliquer aussi à d'autres combinaisons possibles.

La majorité des maladies des plantes relèvent d'un phénomène biologique assez bien délimité. Notre façon d'aborder le problème, lorsqu'il s'agit de décrire ce qui se produit quand un pathogène est associé à un hôte, est peut-être unique du fait qu'elle se fonde sur la conception de la maladie plutôt que sur les signes. Cela rappelle ce qu'on obtient avec la chronophotographie au lieu d'un seul instantané de plan rapproché. En appliquant la méthode conceptuelle basée sur ce phénomène, nous nous sommes efforcé de décrire chaque maladie de façon que les producteurs et les protecteurs des plantes, quelle que soit leur langue maternelle, aient dans l'esprit la même maladie, malgré la diversité de ses appellations dans la langue érite ou parlée. Nous avons évité les termes techniques pour deux raisons: (1) la terminologie employée permettra au lecteur, quelle que soit sa formation scientifique, de se faire une idée juste de ces maladies et (2) nous avons voulu rendre la traduction plus facile.

Par comparaison à la section précédente relative aux noms populaires qui nous sont venus surtout de phytopathologistes de divers pays du monde, nous n'avons retenu, dans cette section des concepts de maladies (descriptions), que quatre langues; soit l'anglais, le français, l'espagnol et l'interlingua. On s'accorde généralement à croire qu'environ 90% des habitants de notre monde peuvent comprendre assez bien l'anglais, le français ou l'espagnol. Il faut savoir que l'interlingua est une langue dont le vocabulaire est dérivé du latin, de l'italien, de l'espagnol, du portuguais, du français, de l'anglais, de l'allemand et du russe. Il comporte tous les vocables que les langues occidentales ont typiquement en commun. La grammaire est très simplifiée. Plusieurs s'apercevront que le vocabulaire de l'interlingua leur est fondamentalement familier parce qu'en général les mots sont des racines communes à l'anglais et aux langues romanes.

Il y a toutes les raisons de croire qu'un nombre encore plus grand de lecteurs pourront lire et comprendre l'interlingua. Notre expérience de l'emploi de ce langage dans des articles parus dans le "Plant Disease Reporter" nous a montré que plusieurs phytopathologistes, qui pourtant n'avaient jamais étudié l'interlingua, l'ont trouvé facile à comprendre. Il n'est pas impossible que l'interlingua soit appellé à jouer un rôle important en servant de langue universelle pour désigner les noms populaires des maladies des plantes, de la même façon que le latin sert de standard pour les noms scientifiques. Parce que nous croyons que ce langage peut fournir un moyen utile de communication aux hommes de science, nous avons inclus quelques indications utiles à les pages 454 et 455.

VII. DISTRIBUTION: Af., A., Aust., Eur., N.A., C.A., S.A.

Dans cette section, la distribution est inscrite par continent ou aire géographique définie. La plupart des renseignements ont été puisés dans "Distribution Maps of Plant Diseases" publié par le "Commonwealth Mycological Institute of England". On peut trouver, au dos des cartes, la distribution, avec commentaires à l'appui, dans chaque pays à l'intérieur des continents. Les abréviations suivantes seront employées dans tout le Compendium:

Af. (Afrique)	N.A. (Amérique du Nord)
A. (Asie)	S.A. (Amérique du Sud)
Aust. (Australie)	C.A. (Amérique Centrale)
Eur. (Europe)	W.I. (Antilles)

Nous croyons que les illustrations de plan rapproché en couleurs des symptômes les plus typiques, de concert avec la symptomatologie présentée dans les descriptions, constitueront d'excellents outils de diagnostic. Le numéro de chaque illustration se rapporte à la page correspondante dans laquelle on peut trouver les renseignements propres à la maladie en question. Nous n'avons pas toujours obtenu des diapositives qui permettent de voir tous les illustrations. En général, les maladies des racines, dont les symptômes sont difficiles à photographier, ou les maladies, si importantes soient-elles dans leur aire de distribution, dont nous n'avons pu obtenir d'échantillons, n'ont pu figurer dans le Compendium.

INTRODUCCION

El propósito de este <u>Compendio de Enfermedades de las Plantas en Varios Idiomas</u>, es ayudar a mejorar la comunicación entre los científicos a través del mundo, especialmente entre aquéllos que están involucrados en la producción y protección de los cultivos. Su objetivo esencial es ayudar a aumentar la producción de alimentos, especialmente en los países en desarrollo. Actualmente no hay ningún tratado comprensible disponible para reducir la confusión y mal entendido causado por los múltiples nombres comunes de las enfermedades de las plantas. Por lo tanto, ha sido nuestro deseo producir una publicación útil que cumpla con estos tres objetivos principales. Se reconoció que se necesitaba la cooperación de muchos científicos y que, a pesar del cuidado y esmero que se haya puesto en la realización del trabajo, nunca sería totalmente exacto o completo. Sin embargo, para lograr que el compendio fuera comprensible y útil, se enviaron cartas a aproximadamente ciento veinticinco fitopatólogos en todo el mundo, solicitándoles su ayuda para que proporcionaran una lista de las enfermedades más importantes en su área, incluyendo los nombres científicos y comunes de cada una. La respuesta a estas cartas fue compensatoria. Estas listas, además de proporcionar los nombres comunes usados en muchos idiomas, también suministraron información respecto a la importancia relativa de los cultivos en cada uno de los países. El criterio que se usó al seleccionar los cultivos y enfermedades fue para determinar la importancia economía del cultivo en bases mundiales, con énfasis en la producción de alimentos, la potencialidad de la enfermedad para causar pérdidas y la disponibilidad de controles immediatos tales como el uso eficiente y seguro de los fungicidas. Sólo enfermedades causadas por hongos y bacterias han sido incluídas, ya que desde el punto de vista de nombres comunes, las enfermedades causadas por virus y nemátodos no están disponibles en cantidades suficientes como para justificar el incluirlas. Se espera que al contar con mayores conocimientos para proporcionar tales nombres comunes, éstos serán incluídos en futuras revisiones.

El siguiente ejemplo representa una versión corta del formato que se seguirá en cada enfermedad. Para poder explicar completamente la compleja organización del material requerido por la nomenclatura, tanto de patógenos como de hospederas y la multiplicidad del lenguaje involucrado, hemos enumerado las variables y las aclararemos cada una en el ejemplo. Ver la versión completa en la página 16.

Las secciones individuales (a) hasta (f), se definen como sigue:
(a) El nombre latino de la hospedera que se menciona primero es el que hemos seleccionado en relación con la asociación entre patógeno y hospedera.
(b) Estos son los nombres latinos de las hospederas que también se consideran de importancia en relación con este patógeno. Cuando aparece una enfermedad que necesita una hospedera alterna para completar su ciclo de vida, tal como la Roya del Tallo del trigo, hemos considerado describir los síntomas económicamente importantes de la hospedera.
(c) Indica el símbolo usado para significar "en asocio con".
(d) Nombres latinos del patógeno (organismo causante de la enfermedad).
(e) Representa otros nombres que han sido usados como sinónimos para el patógeno. Estamos definiendo "sinónimo" ya sea como una palabra aceptada por algunos científicos como indicadora del mismo patógeno o, en algunos casos, cuando sea el estado sexual o el conidial del hongo, el que sea opuesto al primero. Estas pueden ser palabras recogidas ya sea de la literatura o de las listas que nos entregaron los fitopatólogos que han cooperado.
(f) Indica el símbolo usado para significar "es llamado".

(a)   (b)   (c)

I. <u>Avena sativa</u> (<u>Triticum aestivum</u>, <u>Zea mays</u> y otras Gramineae) +

(d)   (e)   (f)

<u>Gibberella zeae</u> (Schw.) Petch (<u>Gibberella saubinetti</u> (Mont.) Sacc.) =

Este ejemplo indica que la hospedera asociada con el patógeno resulta en un fenómeno biológico, el cual es conocido por los siguientes nombres de enfermedades:
II. ENGLISH Head Mold, Scab, Seedling Blight, Head Blight
    FRENCH   Fletréssement des plantules des céréales, Fusariose, Maladie d'enivrement

DUTCH   Kiemschimmel
GERMAN   Fusariose, Tammelkrankheit

Como es bien sabido, los nombres comunes de las enfermedades se asignan a las condiciones resultantes causadas por las muchas combinaciones que existen entre los patógenos vegetales y sus hospederas.   En nuestras descripciones algunas veces hemos usado los términos "enfermedad" y "patógeno" (hongo o bacteria) como sinónimos. Estamos conscientes de que esto no es científicamente correcto; sin embargo, lo hemos hecho así para abreviar y dar mayor facilidad a su lectura. Muchas de las situaciones de las enfermedades a las que nos referimos caben dentro de alguna de las siguientes categorías:

1. Un nombre de enfermedad significa la asociación entre una especie de hospedera y otra de patógeno.
2. Un nombre de enfermedad identificando la asociación de una especie de hospedera y dos o más especies patógenas.
3. Un nombre de enfermedad identificando la asociación de dos o más especies de hospederas y una especie patogéna.
4. Un nombre de enfermedad identificando la asociación de dos o más especies de hospederas y dos o más especies patógenas.
5. Dos o más nombres de enfermedad identificando la asociación entre una especie de hospedera y una especie de patógeno.
6. Dos o más nombres de enfermedades identificando la asociación entre una especie de hospedera y dos o más especies patógenas.
7. Dos o más nombres de enfermedades identificando la asociación entre dos o más especies de hospederas y una especie de patógeno.
8. Dos o más nombres de enfermedad identificando la asociación entre dos o más especies de hospedera y dos o más especies de patógenos.
9. Además, en algunas ocasiones existen situaciones de enfermedades que involucran dos o más de estas combinaciones.

Es aparente que la gran cantidad de nombres comunes que ha surgido como resultado de las .iferentes bases usadas al asignarlos. Una ilustración de esta confusión existente sería la enfermedad causada por el estado conidial del hongo llamada "mal del talluelo" y que en el estado sexual se le llama "tizón". Además, nuestras observaciones, después de tratar con estos nombres en muchos idiomas, han aclarado que la mayoría de nombres comunes se ha derivado de dos fuentes principales: ya sea que se hayan traducido de los nombres comunes del idioma inglés o traducciones del nombre en latín del organismo causal. Así se perpetúa la confusión.

Nosotros hemos seleccionado veintiún idiomas para la sección que contiene los nombres comunes. Estos son aquéllos sobre los cuales tenemos suficiente cantidad de información para justificar su uso. No queremos decir que los nombres se hayan mencionado en orden o frecuencia de uso, sino solamente que todos ellos están en la literatura. Tampoco intentamos que el lector area que éstos son necesariamente todos los nombres que han sido usados en asociación con la enfermedad.

III. The earliest and most conspicuous symptoms .... stem blight.
IV. Le prime e le plus remarcabile symptoma......................... plaga de stirpes.
V. Le symptôme le plus précoce .... et des tiges.
VI. Los más conspicuous... tizón de tallo.

Si todas las posibles enfermedades resultantes de las combinaciones de hospederas y patógenos que se mencionan en este compendio fueran descritas, el número excedería a los 4000. Por lo tanto, en el caso de hospederas múltiples, hemos escogido la hospedera que muestra los síntomas más característicos. Este proceso ha limitado el número de enfermedades descritas a 325, las que al proporcionar un concepto general (descripción de síntomas) de la enfermedad, pueden también alicarse a otras posibles combinaciones.

La mayoría de las enfermedades de las plantas indican un fenómeno biológico con límites bastante bien definidos. Nuestro enfoque al describir lo que pasa cuando un patógeno está asociado con una hospedera es talvez único, debido a que ha sido basado en el concepto de la enfermedad y no en los síntomas. Esto no difiere de los resultados obtenidos de la fotografía de lapso de tiempo comparada con una sola fotografía de cerca. Usando el enfoque conceptual basado en este fenómeno, hemos tratado de describir cada enfermedad de tal manera que aquéllos que están interesados en la protección y producción de cultivos, sin importar su idioma nativo, piensen en la misma enfermedad cuando se usen los diferentes nombres comunes en la comunicación oral o escrita. Las descripciones han sido escritas en términos no técnicos por dos razones: (1) la terminología empleada permitirá al lector, a pesar de su entrenamiento científico, visualizar cómo lucen en realidad estas enfermedades y (2) se trató en lo posible, de facilitar su traducción.

En contraste con la sección anterior relacionada con los nombres comunes que fueron proporcionados en su mayoría por fitopatólogos de diferentes países del mundo, para esta sección de conceptos de enfermedades (descripciones) hemos seleccionado solamente cuatro idiomas: inglés, francés, español e interlengua. Es aceptado en general, que aproximadamente 90% de la población mundial tiene un conocimiento general del inglés, francés o español. Para aquéllos que no están familiarizados con la interlengua, este es un lenguaje en el que el vocabulario se deriva del latín, italiano, español, portugués, francés, inglés, alemán y ruso. Abarca todo el material mundial que los idiomas del mundo occidental tienen típicamente en común. La gramática es bastante simple. Muchos encontrarán el vocabulario del interlengua básicamente familiar porque las palabras son generalmente las raíces comunes del inglés y de las lenguas romance.

Existen muchas razones para creer que habrá una substancial cantidad adicional de lectores que podrán leer y comprender el interlengua. Nuestra experiencia en el uso de este idioma en relación con material publicado en el Plant Disease Reporter nos mostró que muchos fitopatólogos del mundo, aunque no habían estudiado el interlengua previamente, lo encontraron fácil de entender. Se considera que el interlengua podría jugar un importante papel proporcionando un lenguaje universal para los nombres comunes de enfermedades de las plantas, de la misma manera que el latín sirve como un standard para los nombres científicos. Debido a que creemos que el idioma puede constituir una herramienta útil en la comunicación entre científicos, hemos incluido algún material de utilidad en las páginas 454 y 455.

VII. DISTRIBUCION: Af., A., Aust., Eur., N.A., C.A., S.A.

En esta sección, la distribución se menciona por continente o área geográfica definida. La mayor parte de esta información se tomó de los "Mapas de Distribución de Enfermedades de las Plantas" publicados por el Commonwealth Mycological Institute de Inglaterra. Al reverso de estos mapas puede encontrarse la distribución por países individuales dentro de los continentes, con información al respecto. Las siguientes abreviaturas serán usadas uniformemente durante todo el compendio:

Af.	(Africa)	N.A.	(Norte América)
A.	(Asia)	S.A.	(Sur América)
Aust.	(Australasia)	C.A.	(Centro América)
Eur.	(Europa)	W.I.	(Antillas)

Creemos que las fotografías de cerca y a colores, de los síntomas más típicos en combinación con la sintomatología general presentada en las descripciones, proporcionará excelentes medios para el diagnóstico. El número de cada ilustración se refiere a la página correspondiente en la cual puede encontrarse la información pertinente a la enfermedad. Los números faltantes han resultado de la imposibilidad de obtener las transparencias adecuadas. En general, las omisiones son de caso y sea de enfermedades radiculares que tienen síntomas difíciles de fotografiar o enfermedades que, aunque de mayor importancia en algunas áreas geográficas, nos fue imposible obtener.

ACKNOWLEDGEMENTS

Obviously, it is impossible to list all of the sources of information used during the preparation of this compendium which, because of its multilingual nature, is rather complex. However, we shall attempt to identify the principal ones.

The following contributed common names of plant diseases occurring in their countries:

AFRICA:
  EGYPT:
     Dr. M. K. Abo-El-Dahab   Plant Pathology Department
     Dr. M. A. El-Goorani   Faculty of Agriculture
                                          Alexandria University
                                          Alexandria, Egypt

  KENYA:
     Dr. Walter Kaiser   E. A. A. F. R. O.
                                          P. O. Box 30148
                                        Nairobi, Kenya

  MALAWI:
     Mr. Moid A. Siddiqi   Ministry of Agriculture
                                          Agricultural Research Station
                                        Bvumbwe, P. O. Box 5748
                                        Limbe, Malawi

  REPUBLIC of SOUTH AFRICA
     Mr. S. W. Baard   Faculty for Agriculture
                                          U. O. F. S.
                                        Private Bage X0576
                                        Bloemfontein
                                          Republic of South Africa

  SIERRA LEONE:
     Dr. R. C. Kapooria   Njala University College
                                          University of Sierra Leone
                                        Private Mail Bag
                                        Freetown, Sierra Leone

ASIA:
  BURMA:
     Dr. Maung Mya Thaung   Department of Plant Pathology
                                          Institute of Agriculture
                                        Mandalay, Burma

  CEYLON:
     Dr. O. S. Peries   Rubber Research Institute
                                          Agalawatta, Ceylon

  CHINA:
     Dr. I. J. Hodgkiss   Department of Botany
                                          The University of Hong Kong
                                        Hong Kong, China

  INDIA:
     Dr. V. V. Chenulu   Division of Mycology and Plant Pathology
     Dr. Ashok Gaur   Indian Agricultural Research Institute
     Dr. Ram Nath   New Delhi 12, India
     Dr. Shamsher Singh

INDONESIA:
    Dr. W. Y. H. Peregrine   Department of Agriculture
                                              Pengarah Pertanian
                                              Brunei, Borneo, Indonesia

IRAN:
    Dr. Dariush Danesh   Karaj College of Agriculture
                                              Karaj, Tehran,
                                              Iran

    Dr. Nasser Zalpoor   Plant Pest Diseases Research Institute
                                              P. O. Box 3178
                                            Evin, Tehran
                                            Iran

IRAQ:
    Dr. Mahdi M. Al-Shukri   Department of Botany
                                              College of Science
                                            University of Baghdad
                                            Adhamea, Iraq, West Asia

JAPAN:
    Dr. Hajame Kato   The Phytopathological Society of Japan
    Dr. N. Nishihara   Shokubo Building
    Dr. K. Yora   1-43-11 Komagome, Toshima-ku,
                                              Tokyo, Japan

    Dr. Hiroshi Tochihara   Institute for Plant Virus Research
                                              CHIBA-shi
                                            Chiba Prefecture, Japan

KOREA:
    Dr. Hoo Sup Chung   Department of Agricultrual Botany
                                              College of Agriculture, Seoul
                                            National University
                                            Suwon, Korea

    Dr. H. Arthur Lamey   c/o United Nations Development Program
                                              C. P. O. Box 143
                                            Seoul, Korea

MALAYSIA:
    Dr. K. H. Chee   Pathology Division RRI
                                            P. O. Box 150
                                            Kuala Lumpur
                                            Malaysia, S. E. Asia

NEPAL:
    Miss Nisha Pokharel   Department of Botany
    Dr. S. C. Singh   Tribhuvan University
                                            Kirtipur, Kathmandu
                                            Nepal

THE PHILLIPINES:
    Mrs. Erlinda P. Rillo   Guinobatan Experiment Station
                                            Guinobatan, Albay H-104
                                            The Phillipines

REPUBLIC of CHINA:
Dr. Chii Chhan Chen

Department of Plant Pathology
College of Agriculture
Taiwan University
Taiwan, Republic of China

Dr. Lung Chi Wu

Phytopathological Laboratory
National Taiwan University
Taipei, Taiwan
Republic of China

Dr. Charles Y. Yang

Department of Plant Pathology
Asian Vegetable Research and
   Development Center
P. O. Box 42, Shanhua Tainan,
Taiwan (741)
Republic of China

THAILAND:
Dr. Givatgong Piya

Technical Division
Kasetsart University
Bangkhen, Bangkok
Thailand

Dr. Udom Pupipat

Department of Entomology and
   Plant Pathology
Kasetsart University
Bangkhen, Bangkok
Thailand

Dr. Bobby Renfro

G. P. O. Box 2453, The Rockefeller Foundation
Bangkok, Thailand

TURKEY:
Prof. Dr. Mithat Özsan

A. U. Zirrat Fakultesi
Ankara, Turkey

VIET NAM:
Dr. Thái-Cong Tụng

Ministry of Land Reform and Agriculture
Institute of Research
P. O. Box 232
Saigon, Viet Nam

AUSTRALASIA:
AUSTRALIA:
Dr. L. W. Burgess

Department of Agricultural Botany
University of Sydney
New South Wales 2116
Australia

Mr. John Walker

New South Wales Department of Agriculture
Plant Pathology Branch
B. C. R. I. Rydalmere
New South Wales, Australia

EUROPE:
AUSTRIA:
Dr. Gerald Neider

Bundesanstalt für Planzenschuta
Landw. Bakteriologische Versuchsanstalt
Postface 154, 1021 Wien
Österreich

BELGIUM:
Dr. J. Bernard
        Station de Zoologie Appliquée
           de L'Etat
        8, Av. Maréchal
        5800 Gembloux
        Belgique

BULGARIA:
Dr. Tz. Hinkovsky
        General Scientific Secretary
        Agricultural Academy "Georgy Dimitrov"
        Bulgaria

CYPRUS:
Dr. N. Achillides
        Department of Agriculture
        Nicosia, Cyprus

DENMARK:
Dr. H. Rønde Kristensen
        Statens Platenpatologiske Forsog
        Lottenborgvej 2
        Lynbgy, Denmark

Dr. Paul Neergaard
        Fropatologisk Institute
        Ryvangs Alle 78
        2900 Hllerup,
        Denmark

ENGLAND:
Dr. D. A. Doling
        Crop Science Department
        The Lord Rank Research Center
        Lincoln Road, High Wycombe
        Bucks. England

FINLAND:
Dr. Eeva Tapio
        Department of Plant Pathology
        Agricultural Research Center
        Box 18
        S. F. - 01301
        Tikhurila, Finland

FRANCE:
Prof. Dr. G. Viennot-Bourgin
        INRA, Route de St. Cyr 78
        Versailles, France

GERMANY:
Prof. Dr. F. Grossmann
        Lehrstuhl für Phytopathologie
           und Pflanzenschutz
        Universität Hohenheim
        7000 Stuttgart 70
        Deutschland

Dr. G. Schmidt
        Biologische Bundesanstalt für Land-
           und Forstwirtschaft
        1 Berlin 33 Dahlem, den
        Königin-Luise-Strasse 19
        Berlin-West, Deutschland

HUNGARY:
Dr. Tibor Jermy
        Hungarian Research Institute
           for Plant Protection
        Herman Ottostr 15
        Budapest 11, Hungary

NORWAY:
  Dr. Tor Munthe

Norwegian Plant Protection Institute
Division of Plant Pathology
1432 Vollebekk, Norway

SWITZERLAND:
  Dr. H. Zogg

Eidg. Landwirtschaftliche Versuchsanstalt
Reckenholzstrasse 191-211
8046 Zürich, Switzerland

NORTH AMERICA:
  CANADA:
    Dr. Lorne Callbeck

Research Station
Canadian Department of Agriculture
Box 1210, Charlottetown
Prince Edward Island, Canada

    Dr. Denis Lachance

Centre de Recherche forestère
  des Laurantides
1080 Route du Vallon
C. P. 3800
Ste-Foy, Québec, 10, Canada

  MEXICO:
    Dr. Jorge Galindo A.

Colegio de Postgraduados
Escuela Nacional de Agricultura
Chapingo, Mexico

  UNITED STATES of AMERICA:
    Dr. M. A. Tabatabai

Department of Agronomy
Iowa State University
Ames, Iowa

CENTRAL AMERICA and WEST INDIES:
  COSTA RICA:
    Dr. Luis C. González

Universidad de Costa Rica
Costa Rica, Central America

  GUATEMALA:
    Dr. Eugenio Schieber

P. O. Box 226
Antigua, Guatemala
Central America

  JAMAICA:
    Dr. Caroll Henry

Ministry of Agriculture and Fisheries
P. O. Box 480
Hope, Kingston, Jamaica

    Dr. Reginald E. Pierre

Faculty of Agriculture
University of the West Indies
Mona, Kingston 7, Jamaica

SOUTH AMERICA:
  ARGENTINA:
    Dr. J. M. Feldman
    Dr. Olga Gracia

Estación Experimental Agropecuaria
La Consulta
Casilla de Correo 8
La Consulta, Mendoza, Argentina

    Drs. Abel and Maria Sarasola

Instituto de Patologia Vegetal
Castelar, Buenos Aires, Argentina

BRAZIL:
    Dr. Agelisau A. Bitancourt

Instituto Biologico
Caixa Postal 7119
Saõ Paulo, Brazil

    Dr. Paul S. Lehman

USAID/ Brasilia
Agency for International Development
Washington, D. C. 20523
U. S. A.

    Dr. J. P. Costa Neto
    Mr. Caio Machado

Departmento de Fitotecnia
Setor de Fitopatologia
Faculdade de Agronomia-CxP. 776
Universidade Federal do R. S.
90,000-Porto Alegre-RS.
Brazil

CHILE:
    Dr. Claudio Vergara C.

Departemento de Investigaciones Agricolas
Instituto de Investigaciones Agropecuarias
La Platina, Santiago, Chile

COLUMBIA:
    Dr. Guillermo E. Galvez

CIAT
Apartado aereo 6713
Cali, Columbia

ECUADOR:
    Ing. Cristobal Barba D.
    Ing. Ramiro Eguiguren C.
    Ing. Hugo Vallejo M.
    Ing. César Wamdenberg S.

Ministerio de Agricultura y Ganaderia
Departamento de Fitopatologia
Quito, Ecuador

PARAGUAY:
    Ing. Agr. M. S. Delio Sanchez
    Ing. Agr. Pedro Seall

Facultad de Agronomia y Veterinaria
Universidad Nacional de Asuncion
San Lorenzo, Paraguay

PERU:
    Ing. Consuelo Bazan de Segura

Instituto de Ciencias Forestales
Universidad Agraria La Molina
Apartado 456
Lima, Peru

    Dr. E. N. Fernandez-Northcote

Departemento de Sanidad Vegetal
Sección Fitopatología
Universidad Nacional Agraria
Aptdo. 456, La Molina
Lima, Peru

SURINAM:
    Dr. F. A. del Prado

Agricultural Experiment Station
P. O. Box 160, Paramaribo
Surinam, South America

URAGUAY:
    Sra. Ing. Agr. Lucía Koch de Brotos

Dirección de Investigaciones en
    Filopatología
Abda. Millán 4715
Montevideo, Uraguay

We also wish to acknowledge the following persons who translated the plant disease descriptions into either Spanish, Interlingua or French:

1) The Spanish translations were made by various members of the Caribbean Division of the American Phytopathological Society under the supervision of Mr. Benjamen Waite and Dr. José Nolla.

2) Interlingua translations were prepared by Dr. Frank Esterhill, Director of the Interlingua Institute (P. O. Box 126 Canal Station-New York, New York 10013) and reviewed for correctness and international character by members of the Union Mundial pro Interlingua.

3) French translations were made by Henri Généreux, Louis-Joseph Coulombe, Denis Lachance, Pierre Duval, Péal Pelletier, Omer Beaudoin, Roland Loiselle, Lionel Dessureaux, Claude Aubé, Camillien Gagnon, Jean-Marc Girard and Richard Cayouette, members of the Société de Protection des Plantes du Québec under the guidance and coordination of Albert Payette, Jacquelin Santerre and Pierre Thibodeau.

Appreciation is expressed to the three agencies, namely the American Phytopathological Society, United States Agency for International Development and the Agricultural Research Service, U. S. Department of Agriculture which cooperated in sponsoring the project. Special mention should be made of the contributions of W. Q. Loegering who, as chairman of the International Cooperation Committee of APS in 1962, recognized the need for better communication among the world's plant pathologists and suggested that a type of plant disease thesaurus might be helpful; T. van der Zwet, chairman in 1971, who provided effort and help in initiating the program; Charles Yang, for his exceptional cooperation; Mr. Baburam Sharma of the Royal Nepalese Embassy and Dr. Lekh Batra of ARS, USDA for their generous help with the languages based on Sanskrit; and to all of the plant pathologists who permitted us to reproduce their color transparencies.

Applied Plant Pathology Laboratory
Plant Protection Institute
Agricultural Research Service
United States Department of Agriculture
Beltsville, Maryland

Paul R. Miller
Hazel L. Pollard

Multilingual Compendium
of Plant Diseases

<u>Allium</u> spp. + <u>Alternaria porri</u> (Ell.) Cif. =

ENGLISH	Purple Blotch
FRENCH	Tache Pourpre, Alternariose
SPANISH	Mancha púrpura, Mancha rosada
PORTUGUESE	Queima das fôlhas, Manchas púrpuras, Môfo prêto, Crestamento, Pinta
ITALIAN	
RUSSIAN	альтернариоз
SCANDINAVIAN	
GERMAN	Purpurfleckigkeit
DUTCH	
HUNGARIAN	
BULGARIAN	
TURKISH	
ARABIC	El-lafhaa El Kormosia القرمزية اللفحـه
PERSIAN	
BURMESE	Tauk-te-mee roga ၊ အော် လတ ၁ ၄န ရုံး ဃ၁း ေၯ ၍င
THAI	Bai-Mhai ใบจุดสีม่วง
NEPALI	
HINDI	
VIETNAMESE	Chấy-lá
CHINESE	Yáng tsung dzỹy ḫanbìng 洋葱紫斑病
JAPANESE	Kokuhan-byô 黒斑症

ENGLISH:

The first symptoms are small white lesions on the leaves; these under wet conditions develop into elliptical purplish areas, spreading to several centimeters long and with a yellowish border. After 3-4 weeks the leaves collapse and infection can spread to the bulb, causing a deep yellow and reddish watery rot. The scales become desiccated and dark. Bulbs may be small or fail to develop.

INTERLINGUA:

Le prime symptomas es minute lesiones blanc super le folios. Iste lesiones, sub conditiones humide, se transforma in areas elliptic purpuree, le quales pote extender se a plure centimetros de longitude e ha un margine jalnette. Post 3 o 4 septimanas le folios suffre collapso e le infection pote extender se al bulbo, causante un brillante putrefaction de color jalne e rubiette. Le scalias se desicca e deveni obscur. Le bulbos pote o restar micre o non evolver.

FRENCH:

Les premiers symptômes sont de petites lésions blanches sur les feuilles; à l'humidité, ces lésions deviennent des plages elliptiques pourprées pouvant atteindre plusieurs centimètres de longueur, à bords jaunâtres. Au bout de trois ou quatre semaines, les feuilles s'affaissent et l'infection peut se propager au bulbe, où elle produit une pourriture aqueuse jaune foncé et rougeâtre. Les pelures se dessèchent et noircissent. Les bulbes restent petits ou ne se forment pas.

SPANISH:

Los primeros síntomas de la enfermedad son pequeñas lesiones blancas sobre las hojas; estas lesiones, bajo condiciones húmedas, se desarrollan en áreas elípticas de color púrpura, que llegan a alcanzar varios centímetros de largo y tienen la orilla amarillenta. Después de unas 3-4 semanas las hojas sufren un colapso y la infección puede llegar hasta el bulbo, causando una pudrición de color amarillo intenso y rojizo. Los cascos del bulbo se disecan y obscurecen. Los bulbos pueden ser pequeños o no llegan a desarrollarse.

DISTRIBUTION: Af., A., Aust., Eur., N.A., C.A., S.A.

<u>Allium</u> spp. + <u>Botrytis</u> <u>allii</u> Munn =

ENGLISH	Neck Rot, Gray-Mold Rot, Leaf Blight
FRENCH	Pourriture des Oignons, Pourriture Grise
SPANISH	Podredumbre del cuello, Moho gris, Pudrición gris, Tizón de la hoja
PORTUGUESE	Queima das pontas, Podridão do colo
ITALIAN	Muffa grigia, Marciume delle cipolle, Muffa
RUSSIAN	Гниль шейковая
SCANDINAVIAN	Løkgråskimmel, Løg-graskimmel, Drueskimmel
GERMAN	Zwiebelfäule, Grauschimmelfäule (Halsfäule), Botrytis-Fäule, Fusarium-Zwiebel-kuchenfäule
DUTCH	Koprot, Bodemrot
HUNGARIAN	
BULGARIAN	Шийно гниене
TURKISH	Soğan kurşuni küfü
ARABIC	Affan غـن
PERSIAN	Pōōsedẹgeyẹ khăkẹstáreyẹ pēyăz یوسیدنی خاکستری پیاز
BURMESE	
THAI	
NEPALI	
HINDI	
VIETNAMESE	
CHINESE	
JAPANESE	Haiirohuhai-byô 灰色腐敗病

Infection usually begins at the neck where affected scale tissue softens, becomes sunken and spongy and appears dried out. When cut across, the bulb scales appear cooked and the diseased tissue is water-soaked and brownish. Affected bulbs dry out and later become mummified.

Le infection usualmente comencia al collo, in qual loco le scaliose texito afflicte se emolli, deveni submerse e spongiose, e appare esser desiccate. Iste scalias del bulbo, quando on los divide transversemente, appare esser cocinate e le texito morbide es saturate de aqua e brunette. Bulbos afflicte se desicca e plus tarde se mumifica.

L'infection débute généralement au collet où les tissus écailleux s'amollissent, deviennent déprimés et spongieux et paraissent desséchés. Une fois l'oignon coupé, les pelures du bulbe ont l'air d'être cuites et les tissus malades paraissent imbibés et brunâtres. Les bulbes atteints se dessèchent et puis se momifient.

La infección generalmente comienza en el cuello, donde los tejidos afectados de los cascos del bulbo se hunden, se vuelven esponjosos y con apariencia seca. Cuando se corta el bulbo, los cascos aparecen cocidos y el tejido enfermo está empapado de agua y es de color cafesoso. Los bulbos afectados se secan y más tarde se momifican.

DISTRIBUTION: Af., A., Aust., Eur., N.A., S.A.

Allium cepa + Colletotrichum circinans (Berk.) Vogl. =

ENGLISH	Smudge
FRENCH	Anthracnose, Barbouillage
SPANISH	Antracnosis, Tizne
PORTUGUESE	Podridaõ, Podridaõ sêca, Podridao da base, Podridao da raiz, Antracnose Cachorro quente
ITALIAN	Marciume dei bulbi di cipolla
RUSSIAN	антракноз
SCANDINAVIAN	
GERMAN	Schmutzfleckenkrankheit, Brennfleckenkrankheit
DUTCH	Roetvlekken
HUNGARIAN	
BULGARIAN	
TURKISH	Soğan antraknozu
ARABIC	
PERSIAN	
BURMESE	
THAI	Anthracnose แอนแทรกโนส
NEPALI	
HINDI	
VIETNAMESE	Đén ở cổ cây
CHINESE	Yáng tsung tàn dzŭ bìng 洋葱炭疽病
JAPANESE	Tanso-byô 炭そ病

This disease appears at any time in the life of the plant, in storage, during transportation as well as in the field. The most common symptom is the dark green or black smudge on the bulb or neck. The black discoloration follows the veins of the scales or crosses the veins transversely, beginning a narrow zone of discoloration around the bulb. In a severe attack the fungus grows through the outer dry scales into the living tissue, causing a collapse of the fleshy scales.

Iste morbo se manifesta per tote le stadios del vita de plantas, o conservate in magasinage, o durante le transportation, o in le campo. Le symptoma le plus commun es un tinta obscur, verde o nigre, in le bulbo o in le collo. Le discoloration nigre seque o transversa le venas del scalias; e un zona restricte de discoloration comencia evolver circum le bulbo. In un attacco sever, le fungo, penetrante le exterior scalias desiccate, infecta le texito vivente e causa le collapso del scalias carnose.

Cette maladie apparaît a n'importe quel âge de la plante, en entrepôt, durant le transport, ainsi que dans le champ. Le symptôme le plus commun est un barbouillage vert foncé ou noir sur le bulbe ou le collet. Le noircissement suit les nervures des pelures ou traverse les nervures transversalement, ce qui marque le début de formation d'une bande sombre autour de bulbe. Dans les cas graves, le champignon quitte les pelures sèches de l'extérieur et envahit les tissus vivants pour produire un affaissement des pelures charnues.

Esta enfermedad puede aparecer en cualquier tiempo durante la vida de la planta, ya sea durante el almacenamiento, mientras se transporta o en el campo. El síntoma más común es el tizne de color verde obscuro o negro en el bulbo o en el cuello. Una decoloración negra aparece después en las venas de los cascos o cruza transversalmente las venas, iniciando un angosta zona de decoloración alrededor del bulbo. En un ataque severo el hongo crece a través de los cascos exteriores secos y pasa al tejido viviente, causando un colapso de los cascos carnosos.

DISTRIBUTION: A., Aust., Eur., N.A., S.A.

4

Allium cepa + Peronospora destructor (Berk.) Casp. =

ENGLISH	Downy Mildew
FRENCH	Mildiou
SPANISH	Mildiú, Mildiú felpudo, Mildiu velloso, Mildiú lanoso
PORTUGUESE	Mildio, Peronospora
ITALIAN	
RUSSIAN	пероноспороз, ложная мучнистая роса
SCANDINAVIAN	Løkbladskimmel
GERMAN	Falscher Mehltau, Zwiebelmehltau
DUTCH	
HUNGARIAN	Hagymaperonoszpóra
BULGARIAN	мана, пероноспора
TURKISH	Soğan mildiyösü
ARABIC	Bayad Zaghaby بياض زغبى
PERSIAN	Sefedáke dorooghéye péyáz سفیدک دروغی پیاز
BURMESE	
THAI	Ra-Num karng ราน้ำค้าง
NEPALI	
HINDI	
VIETNAMESE	
CHINESE	
JAPANESE	Beto-byð べと病

This pathogen may cause local infections on leaves or be systemic and infect the entire plant. Local infections usually appear as pale green, oval to elongate, slightly sunken spots on the leaves and seed stalks. Alternating bands of yellowish and green tissue may be evident. Later the whole leaf may turn dull pale green and then yellow. Affected foliage often breaks and shrivels.

Iste pathogeno pote causar infectiones local super le folios o poter esser systematic, infectante tote le planta. Le infectiones local usualmente se manifesta como pallide maculas verde, un pauc submerse, oval o elongate, super le folios e le pedunculo. Bandas alternative de texito jalnette e verde es a vices evidente. Plus tarde, tote le folio pote devenir mat e pallide, verde a alora jalne. Le foliage afflicte es frequentemente fracte e crispate.

Ce pathogène peut causer des infections locales sur les feuilles ou être systémique et infecter toute la plante. Les infections locales apparaissent généralement comme des taches légèrement déprimées, ovales à allongées, de couleur vert pâle sur les feuilles et les pédoncules. On peut voir alterner des bandes de tissue jaunâtres et verts. Par la suite, toute la feuille peut devenir vert pâle terne et puis jaunir. Souvent, les feuilles infectées se brisent et se recroquevillent.

Este patógeno puede causar infecciones locales en las hojas o ser sistémico e infectar toda la planta. Las infecciones locales generalmente aparecen como manchas de color verde pálido, ovaladas o alargadas, algo hundidas, sobre las hojas o tallos de semilla. Bandas alternas de tejido amarilloso y verde pueden ser evidentes. Más tarde toda la hoja puede volverse verde pálido y luego amarillo. El follaje afectado generalmente se quiebra y se arruga.

DISTRIBUTION: Af., A., Aust., Eur., N.A., C.A., S.A.

Allium cepa + Peronospora schleideni Ung. =

ENGLISH	Blight, Mold, Downy Mildew
FRENCH	Mildiou de l'oignon
SPANISH	Mildeo velloso, Tizón, Moho, Mildiu lanoso
PORTUGUESE	
ITALIAN	Peronospora delle cipolle, Muffa
RUSSIAN	
SCANDINAVIAN	Løgskimmel
GERMAN	Zwiebelmehltau, Falscher Mehltau der Zwiebel
DUTCH	Valsche Meeldauw
HUNGARIAN	
BULGARIAN	
TURKISH	
ARABIC	Bayad Zaghaby بياض زغبى
PERSIAN	
BURMESE	
THAI	
NEPALI	
HINDI	
VIETNAMESE	Bênh dóm phán
CHINESE	Yáng tsung lùh jìunn bìng 洋葱露菌病
JAPANESE	Beto-byð べと病

The attack in the field occurs first on a few plants during warm, damp weather when the leaf tips appear to have been hit with hot water. Under favorable weather conditions the disease spreads in epidemic fashion and the plants, when observed from a distance, show a violet tint. Within two or three days the plants lose their green color, turn pale or yellow; then they collapse. There is no increase in bulb size after a severe attack.

Le attacco occurre al initio in alicun plantas in le campo durante tempore calide e humide. In iste attacco le extremitates del folios sembla haber essite escaldate con aqua bulliente. Sub conditiones favorabile de tempore, le morbo se extende in un maniera epidemic, e le plantas, observate de longe, monstra un tinta violette. In le curso de 2 o 3 dies le plantas perde lor color verde, deveni pallide o jalne, e alora suffre collapso. Le bulbos non cresce in dimension post un attacco sever.

Dans le champ, ce sont d'abord quelques plantes qui subissent l'attaque par temps chaud et humide, alors que le bout des feuilles semble avoir été échaudé. Dans des conditions favorables de climat, la maladie devient épidémique et les plantes, vues à distance, paraissent violettes. Au bout de deux ou trois jours, les plantes perdent leur couleur verte, pâlissent ou jaunissent, après quoi elles s'affaissent. Le bulbe cesse de grossir à la suite d'une attaque grave.

El ataque en el campo ocurre primero en unas pocas plantas durante tiempo cálido y húmedo, cuando las puntas de las hojas parecen haber sido quemadas con agua caliente. Bajo condiciones favorables de tiempo, la enfermedad se disemina en forma epidémica y la planta muestra un tinte violáceo al observarla desde alguna distancia. En dos o tres días las plantas pierden su color verde, se vuelven pálidas o amarillas y luego sufren un colapso. No hay aumento en el tamaño del bulbo después de un ataque severo.

DISTRIBUTION: Af., A., Eur., S.A.

6

Allium cepa + Puccinia allii Rud. =

ENGLISH	Garlic Rust
FRENCH	Rouille de l'ail
SPANISH	Roya del ajo
PORTUGUESE	Ferrugem
ITALIAN	Ruggene dell'aglio, Nebbia dell'aglio
RUSSIAN	
SCANDINAVIAN	
GERMAN	Porreerost
DUTCH	
HUNGARIAN	
BULGARIAN	
TURKISH	Sarımsak pası
ARABIC	Sadaa صـدأ
PERSIAN	Zángē pēyāz زنگ پیاز
BURMESE	
THAI	
NEPALI	
HINDI	Ratua रतुवा
VIETAMESE	
CHINESE	Yáng tsung shǐou bǐng 洋葱銹病
JAPANESE	Sabi-byð さび病

The leaves, when heavily infected, turn yellow and die. Although a new crop of leaves may develop, the size of bulbs is reduced. The infected tissue of both stem and leaves usually show bright yellow pustules which contain the fungus rust.

Le folios, si severmente infectate, deveni jalne e mori. Ben que le folios forsan cresce un altere vice, le dimension del bulbos es diminute. Le texito infectate del stirpe e del folios usualmente monstra brillante pustulas jalne, le quales contine le ferrugine del fungo.

A la suite d'une forte attaque, les feuilles jaunissent et meurent. Même si de nouvelles feuilles peuvent se former, la grosseur des bulbes est réduite. Les tissus infectés de la tige et des feuilles montrent des pustules jaune vif qui portent la rouille.

Cuando las hojas están severamente atacadas, se vuelven amarillas y mueren. Aunque pueden desarrollarse hojas nuevas, el tamaño de los bulbos disminuye. El tejido infectado tanto del tallo como de las hojas generalmente muestra pústulas de color amarillo brillante, las cuáles contienen el hongo de la roya.

DISTRIBUTION: Af., A., Aust., Eur., N.A., S.A.

Allium cepa (A. ascalonicum, A. porrum etc.) + Sclerotium cepivorum Berk. =

ENGLISH	White Rot
FRENCH	Pourriture Blanche de l'oignon, Maladie des Sclérotes
SPANISH	Pudrición blanca
PORTUGUESE	Podridaᴏ branca
ITALIAN	Mal dello sclerozio, Marciume bianco della cipolla
RUSSIAN	гниль луковиц белая склероциальная
SCANDINAVIAN	Løkhvitråte, Hvidråd
GERMAN	Melhkrankheit der Zwiebel, Sclerotium-Weissfäule
DUTCH	Verschimmelen der Uien, Sclerotiĕn-rot
HUNGARIAN	
BULGARIAN	бяло гниене
TURKISH	
ARABIC	Affan Abiad غن ابيض
PERSIAN	
BURMESE	
THAI	
NEPALI	
HINDI	
VIETNAMESE	
CHINESE	
JAPANESE	Kurogusare-kinkaku-byð 黒腐菌核病

This disease is found usually in the field, seldom causing injury in storage. The leaves of an invaded plant decay at the base, turn yellow, wilt and fall over, the older ones being the first to collapse. The roots of such plants are badly rotted, so that the plant can be pulled up easily. The diseased plants are scattered promiscuously about the field, rather than in a continuous row.

Iste morbo se trova usualmente in le campo, rarmente faciente mal a plantas conservate in magasinage. Le folios de un planta invase se corrumpe al pede, deveni jalne, marcesce, e se·inclina, le plus vetules suffrente collapso ante le alteres. Le radices de tal plantas es severmente putrefacte de maniera que le plantas es facilemente extracte. Le plantas morbide es disperse confusemente per tote le campo plus tosto que situate in un linea continue.

Cette maladie se rencontre d'ordinaire dans le champ et ne cause que rarement des dommages en entrepôt. Les feuilles d'une plante envahie dépérissent à la base, jaunissent, se flétrissent et tombent, les plus vieilles s'affaissant les premières. Les racines de ces plantes sont pourries au point que la plante est facile à arracher. Les plantes malades sont dispersées au hasard par tout le champ au lieu de former une rangée continue.

Esta enfermedad se encuentra generalmente en el campo; casi nunca causa daños en el almacenamiento. Las hojas de una planta atacada se pudren en la base, se vuelven amarillas, se marchitan y caen, siendo las más viejas las primeras en sufrir el colapso. Las raíces de tales plantas están bastante podridas, por lo que la planta puede arrancarse fácilmente. Las plantas enfermas están promíscuosamente diseminadas en todo el campo, en vez de en una fila continua.

DISTRIBUTION: Af., A., Aust., Eur., N.A., S.A.

8

Allium cepa (Allium spp.) + Urocystis cepulae Frost =

ENGLISH	Smut
FRENCH	Charbon des feuilles d'oignon
SPANISH	Carbón desnudo, Tizón, Carbón
PORTUGUESE	
ITALIAN	Carbone della cipolla
RUSSIAN	
SCANDINAVIAN	
GERMAN	Zwiebelbrand
DUTCH	Zwartmop, Uienbrand
HUNGARIAN	
BULGARIAN	
TURKISH	Soğan sürmesi
ARABIC	Taphahom تفـحم
PERSIAN	
BURMESE	
THAI	Khamouw-Dum เขมาคำ
NEPALI	
HINDI	Kand कन्द
VIETNAMESE	
CHINESE	
JAPANESE	Kuroho-byô 黒穂病

Dark, slightly thickened streaks develop in the first leaves to appear on the seedlings. Similar brown to black streaks occur in the scales or leaves as they develop. Affected leaves soon become swollen, curled downward, and distorted. Affected plants are usually stunted. Infected plants may survive in a weakened condition until mid-season or harvest, but usually die within a month after the seedling emerges.

Venas de color obscur, un pauc inspissate, evolve in le prime folios que se manifesta in le plantas juvene. Venas similar de color brun o nigre se trova in le scalias o le folios emergente. Folios attaccate rapidemente deveni tumide, curvate a basso, e distorte. Le crescimento de plantas afflicte es usualmente obstructe. Plantas infectate pote superviver in un condition infirme usque al medie-saison o al rendimento, ma generalmente illos mori intra un mense depost que le planta juvene emerge.

Des stries foncées, assez épaisses, apparaissent sur les toutes premières feuilles de la plantule. Des stries semblables, brunes ou noires, se présentent sur les pelures ou feuilles à mesure que celles-ci se forment. Le feuilles atteintes ne tardent pas à se gonfler, à s'enrouler par en dessous et à se distordre. Les plantes atteintes sont ordinairement rabougries. Les plantes malades peuvent survivre tant bien que mal jusqu'à la mi-saison ou jusqu'à la récolte, mais le plus souvent, elles meurent moins d'un mois après leur sortie de terre.

En las primeras hojas que aparecen en los semilleros se desarrollan vetas obscuras, algo gruesas. Vetas similares de color café o negro aparecen en los cascos o en las hojas al irse desarrollando. Las hojas atacadas pronto se vuelven hinchadas, rizadas y deformes. Las plantas infectadas generalmente están achaparradas. Estas plantas pueden sobrevivir en esta condición de debilidad hasta mediados de la época de cultivo o hasta la cosecha, pero generalmente mueren al mes de haber nacido la planta.

DISTRIBUTION: A., Aust., Eur., N.A., C.A., S.A.

Ananas comosus (Saccharinum officinarum, Musa sapientum) + Ceratocystis paradoxa (Dade) Moreau (Ceratostomella paradoxa (De S.) Dade) =

ENGLISH	Black Rot, Shoot Rot
FRENCH	Pourriture Molle du Fruit, Maladie des Boutures
SPANISH	Podredumbre negra, Podredumbre del Brote
PORTUGUESE	Podridão negra
ITALIAN	
RUSSIAN	гниль плодов мягкая и белая пятнистость оснований листьев
SCANDINAVIAN	
GERMAN	Weisse Blattfleckenkrankheit, Fruchtnassfäule
DUTCH	
HUNGARIAN	
BULGARIAN	
TURKISH	
ARABIC	
PERSIAN	
BURMESE	
THAI	
NEPALI	
HINDI	
VIETNAMESE	
CHINESE	Fèng lí hei fŭ bìng 鳳梨黑腐病, Fèng lí míao fŭ bìng 鳳梨苗腐病
JAPANESE	

Infection usually occurs in the field or in the packing house as a result of rough handling. Soft spots develop on the surface of the fruit, and may spread and coalesce until the entire fruit is involved. Early internal symptoms are blackened areas under the skin which may later spread until the flesh of the fruit is a black rotted mass of tissue. The disease appears on the leaves as small water-soaked spots.

Le infection occurre usualmente in le campo o in le pacchetteria de conservas, como resultato de maltractamento rude. Punctos molle se developpa super le superfacie del fructo e illos pote extender se e coalescer usque a involver tote le fructo. Symptomas precoce interne es areas nigrate sub le pelle, le quales pote, plus tarde, extender se de maniera que le pulpa del fructo es un massa nigre de texito putrefacte. Le morbo se manifesta in le folios como minute maculas saturate de aqua.

L'infection survient généralement dans le champ ou au poste d'emballage suite à des manipulations peu délicates. Le fruit porte des marques ramollies qui peuvent s'étendre et se fusionner jusqu'à couvrir toute sa surface. Les premiers symptômes internes prennent la forme de plages noircies sous l'écorce. Ces plages peuvent s'agrandir et faire de tout le fruit une masse noire de tissus décomposés. Sur les feuilles, la maladie se présente sous forme de petites taches délavées.

La infección generalmente ocurre en el campo o al tiempo de empacar como resultado de un manejo inapropiado. En la superficie de la fruta se desarrollan manchas suaves, las cuáles pueden diseminarse y juntarse hasta que todo el fruto está atacado. Los primeros síntomas internos son algunas áreas ennegrecidas bajo de la cáscara, las cuáles más tarde crecen hasta que la carne del fruto se convierte en una masa negra podrida de tejido. La enfermedad aparece sobre las hojas como pequeñas manchas empapadas de agua.

DISTRIBUTION: Af., A., Aust., Eur., N.A., C.A., S.A.

10

Ananas spp. (<u>Cinchona</u> spp., <u>Cinnamomum</u> spp., <u>Castanea</u> spp., <u>Persea</u> spp., <u>Pinus</u> spp., <u>Rhododendron</u> and others) + <u>Phytophthora cinnamomi</u> Rands =

ENGLISH	Heart Rot of Stems and Buds, Root Rot
FRENCH	Pourriture du coeur de l'ananas
SPANISH	Podredumbre del corazón de tallos y capullos, Podredumbre de la raíz
PORTUGUESE	
ITALIAN	Seccume della frutescenza dell'ananas
RUSSIAN	гниль сердцевидная фитофторозная
SCANDINAVIAN	
GERMAN	Phytophthora-Herzfäule
DUTCH	
HUNGARIAN	
BULGARIAN	
TURKISH	
ARABIC	
PERSIAN	
BURMESE	
THAI	
NEPALI	
HINDI	
VIETNAMESE	
CHINESE	Fèng lǐ shin fǔ bìng 鳳梨心腐病
JAPANESE	

The fungus responsible for this disease also is associated with over 100 other hosts, some of which are important crops. The symptoms vary for the different crops, but generally they can be characterized by heart rot of stems and buds and root rots. The heart rot of stems and buds occurs internally and the death of these tissues results in stunting of the plants. The fungus can be readily cultured on artificial media.

Le fungo que causa iste morbo se associa con plus que 100 altere hospites, alicunes del quales es de importantia alimentari. Le symptomas se varia in le plantas differente, ma generalmente illos es characterisate per un putrefaction del corde de stirpes e de buttones o per putrefaction de radices. Le putrefaction del corde de stirpes e de buttones occurre internemente e le morte de iste texito resulta in maldeveloppamento del plantas. Le fungo se cultiva facilemente super medios artificial.

Le champignon responsable de cette maladie se retrouve chez plus de 100 autres hôtes, dont certains sont des plantes cultivées importantes. Les symptômes varient avec les différentes cultures, mais ils peuvent en général se caractériser par une pourriture du coeur et des bourgeons et par la pourriture des racines. La pourriture du coeur, des tiges et des bourgeons se produit a l'intérieur, mais la mort de ces tissus se traduit par le rabougrissement de la plante. Le champignon est facile à cultiver en milieu artificiel.

El hongo causante de esta enfermedad también está asociado con más de 100 otras hospederas, algunas de las cuáles son cultivos importantes. Los síntomas varían según los diferentes cultivos, pero generalmente pueden caracterizarse por una podredumbre del corazón de los tallos y capullos y podredumbre de las raíces. La podredumbre del corazón de tallos y capullos ocurre internamente y la muerte de estos tejidos da como resultado un achaparramiento de las plantas. El hongo puede ser facilmente cultivado en media artificial.

DISTRIBUTION: Af., A., Aust., Eur., C.A., N.A., S.A.

<u>Apium</u> spp. + <u>Septoria apiicola</u> Speg. (<u>Septoria apii</u> (Briosi & Cav.) Chester) =

ENGLISH	Leaf Spot, Late Blight
FRENCH	Brûlure Septorienne, Septoriose, Taches Foliaires
SPANISH	Mancha foliar, Tizón tardío, Septoriosis, Viruela, Tizón foliar
PORTUGUESE	
ITALIAN	Ticchiolatura delle foglie, Septoriosi delle foglie del sedano
RUSSIAN	септориоз
SCANDINAVIAN	Bladpletsyge, Selleribladpletsyge, Bladfläcksjuka på selleri, Selleribladflekk
GERMAN	Septoria-Blattfleckenkrankheit des Sellerie, Blattfleckenkrankheit
DUTCH	Bladvlekkenziekte
HUNGARIAN	Szeptóriás levélfoltosság
BULGARIAN	листни петна, септориоза
TURKISH	
ARABIC	Lafha Motaakhera لفحة متأخــرة
PERSIAN	Septõrēyãyē káráfs سپتوریای کرفس
BURMESE	
THAI	
NEPALI	
HINDI	
VIETNAMESE	
CHINESE	Chíen tsày yèh ku bìng 芹菜斑點病
JAPANESE	

The lesions begin on the lower outer leaves and gradually spread upward until the entire foliage is affected. The size of the lesions depends upon the amount of moisture present. A leaf with two or three spots may turn yellow and die. After the spot is formed, minute black fruiting bodies, just large enough to be seen by the unaided eye, are dotted near the center of the spot.

Le lesiones comencia super le exterior folios inferior e per stadios se extende in alto usque a affliger tote le foliage. Le dimension del lesiones depende del quantitate del humiditate presente. Un folio que ha 2 o 3 maculas pote devenir jalne e morir. Depost que le macula se manifesta, structuras sporal, nigre e tanto minute que illos es a pena observate per le oculo sin instrumento, es punctillate presso le centro del macula.

L'infection commence sur les feuilles dégagées du bas pour se propager graduellement vers le haut jusqu'à ce que tout le feuillage soit envahi. La grandeur des lésions varie avec l'humidité. Une feuille qui contient deux ou trois taches peut jaunir et mourir. Une fois la tache formée, de petites fructifications noires, à peine visibles à l'oeil nu, forment de petits points vers le centre de la tache.

Las lesiones comienzan en las hojas de afuera y gradualmente se diseminan hasta que todo el follaje está atacado. El tamaño de las lesiones depende de la cantidad de humedad que haya. Una hoja con dos o tres manchas puede volverse amarilla y morir. Después de que la mancha se haya formado, aparecen cerca del centro de la mancha cuerpos fructíferos del hongo, tan minúsculos que apenas pueden verse a simple vista.

DISTRIBUTION: Af., A., Aust., Eur.

Arachis hypogaea + Cercospora arachidicola Hori (Mycosphaerella arachidicola W. A. Jenkins)=

ENGLISH	Brown Leaf Spot, Halo Spot
FRENCH	Mycosphaerellose, Taches Foliaires
SPANISH	Mancha cercospora, Mancha café o viruela , Mancha de halo
PORTUGUESE	Mancha castanha, Cercosporiose
ITALIAN	
RUSSIAN	пятнистость микосфереллезная, микосфереллез
SCANDINAVIAN	
GERMAN	
DUTCH	
HUNGARIAN	
BULGARIAN	
TURKISH	
ARABIC	
PERSIAN	
BURMESE	Ywet-mee-pyauk roga သာ ကို ● ပို း ရ ၁ း သ ၁ ရ ၆ ၁် ၁် ကို လ
THAI	
NEPALI	
HINDI	Tikka Rog टिक्का रोग
VIETNAMESE	Đốm lá xuất hiện sớm hơn bệnh trên
CHINESE	Luòh hwa sheng hor ban bìng 落 花 生 褐 斑 病
JAPANESE	

This and the following disease look alike initially, but differ later. The first symptoms are small pale or blanched spots on the lower leaflets. These soon turn yellow on the upper surface and later necrotic in the center of the lesion. From this point, the symptoms are different for the two diseases. This one immediately forms a yellow halo at the margin and produces spores on the upper surface. The mature spots are reddish-brown to black above and brown to tan on the lower surface.

Iste morbo e le sequente maladia sembla esser similar al initio, ma se differentia plus tarde. Le prime symptomas es minute maculas pallide o blanchite super le foliettos inferior. Illos rapidemente deveni jalne al superfacie superior e, plus tarde, necrotic in le centro del lesion. Postea, le symptomas es differente in le duo morbos. Iste maladia immediatemente forma un halo jalne al margine e produce sporas al superfacie superior. Le maculas maturate es brun-rubiette o nigre supra, e brun o bronzate infra.

Cette maladie et la suivante sont semblables au début et diffèrent par la suite. Les premiers symptômes sont de petites taches pâles ou décolorées sur les folioles du bas. Ces dernières ne tardent pas à jaunir à la face supérieure, après quoi il y a nécrose au centre de la lésion. A partir de là, les symptômes des deux maladies diffèrent. La maladie qui nous occupe présentement forme aussitôt un halo jaune à la marge et produit des spores à la face supérieure. Les taches mûres sont brun rougeâtre à noires au-dessus et brunes à couleur du tan à la face inférieure.

Esta enfermedad y la siguiente parecen iguales inicialmente, pero más tarde difieren. Los primeros síntomas son pequeñas mánchas pálidas o blanquecinas sobre las hojuelas inferiores. Estas pronto se tornan amarillas en el haz y luego necróticas en el centro de la lesión. Desde este punto, los síntomas difieren en estas dos enfermedades. En ésta, immediatamente forma un halo amarillo en el margen y produce esporas en la superficie. Las manchas maduras son de color café-rojizo a negro encima y café o crema en la parte inferior.

DISTRIBUTION: Af., A., Aust., N.A., C.A., S.A.

Arachis hypogaea + Cercospora personata (Berk. & Curt.) Ell. & Ev. (Mycosphaerella berkeleyi Jenkins) =

ENGLISH	Leaf Spot
FRENCH	Cercosporiose, Maladie des Taches Noires
SPANISH	Mancha de la hoja, Mancha café, Mancha del tallo, Mancha foliar
PORTUGUESE	Mancha escura, Pinta preta, Mancha preta, Cercosporiose
ITALIAN	
RUSSIAN	церкоспороз
SCANDINAVIAN	
GERMAN	Cercospora-Blattfleckenkrankheit
DUTCH	
HUNGARIAN	
BULGARIAN	
TURKISH	
ARABIC	Tabbakoo Awrak تبقع اوراق
PERSIAN	
BURMESE	Ywet-mee-pyauk roga ဃၢတ္ၱိ၀ၛၡၟ၃ဃ၆ပၟၡ၎၄ၜာ၆
THAI	Bi-Jut ใบจุด
NEPALI	Badam Ko'kalothople rog बदामको कालोथोप्ले रोग
HINDI	
VIETNAMESE	Đóm lá
CHINESE	Luðh hwa sheng hei sèḥ bìng 落花生黑澀病
JAPANESE	Kurosibu-byð 黒渋病

This disease appears similar to the previous one in the early stages, but later the lesions are dark brown to black on both surfaces and produce the yellow halo only as they mature; spores are formed on the lower surface. This disease usually appears about three weeks later than the previous one and is more sporadic in its occurence.

Iste morbo sembla esser similar al maladia precedente in le stadios initial, ma, plus tarde, le lesiones es brun obscur o nigre a tote le duo superfacies. Illos produce le halo jalne solmente quando illos se matura; sporas se forma al superfacie inferior. Iste morbo usualmente se manifesta circa 3 septimanas plus tarde que le precedente e es plus sporadic in su occurrentia.

Cette maladie ressemble à la précédente aux premières phases, mais par la suite les lésions sont brun foncé a noires sur les deux faces et ne produisent le halo jaune qu'en arrivant à maturité; les spores se forment sur la face inférieure. La maladie apparaît généralement environ trois semaines plus tard que la précédente et se présente de façon plus sporadique.

Esta enfermedad se asemeja a la anterior en las primeras etapas, pero más tarde las lesiones son café obscuro a negras en ambas superficies y producen el halo amarillo sólo cuando maduran; las esporas se forman en el envés. Esta enfermedad generalmente aparece unas tres semanas más tarde que la anterior y es más esporádica.

DISTRIBUTION: Af., A., Aust., Eur., N.A., C.A., S.A.

Arachis hypogaea (and many other hosts) + _Corticium rolfsii_ Curzi (_Sclerotium rolfsii_ Sacc.), (_Pellicularia rolfsii_ (Curzi) West) =

ENGLISH Stem Rot, Southern Blight, Black Rot of Coffee, Leaf Rot of Coffee, Thread Blight of Cacao
FRENCH Pourriture du Pied et des Noix
SPANISH Pudrición basal, Podredumbre del tallo, Podredumbre foliar del café, Mal de hilacha del cacao
PORTUGUESE Murcha de Sclerotium, Murcha de esclerócio
ITALIAN
RUSSIAN гниль южная склероциальная
SCANDINAVIAN
GERMAN Stengel-und Hülsenfäule
DUTCH
HUNGARIAN
BULGARIAN
TURKISH
ARABIC
PERSIAN
BURMESE Myit-pok, Myit-swe roga ၊ အ၁လိုၵ်ၵ်ၵ၇ ၂ၵ်ၵ် စ်ယိၗင်
THAI Rak- Nao รากเน่า
NEPALI
HINDI Collar angmaari कालर अंगमारी
VIETNAMESE Hạch nấm làm thối gốc
CHINESE Luòh hwa sheng bór juan bìng 落花生白絹病
JAPANESE Sirakinu-byô 白絹病

This pathogen causes root and crown rot of the plants and rot of the "pegs", the underground roots upon which the nuts are formed. Also produces a blue-black stain on certain varieties. The roots and crown are usually attacked near the ground line where a light brown decay develops. If only a portion of the plant is wilted or killed, the remainder may remain healthy and vigorous. The leaves usually wilt slowly and appear to recover at night. Eventually they turn brown and die.

Iste pathogeno causa putrefaction del radices e del pedes del stripes in le plantas afflicte e anque putrefaction de ille radices subterranee in le quales se developpa le nuces. Illo produce anque un discoloration blau-nigre in alicun varietates. Le radices e le stirpe es usualmente attaccate presso le terreno, in qual loco un corruption brun clar se developpa. Si solmente un portion del planta marcesce o mori, le resto pote remaner salubre e vigorose. Le folios generalmente marcesce lentemente e sembla recuperar se de nocte. Al fin, illos deveni brun e mori.

Ce pathogène cause une pourriture de la racine et de la couronne des plantes ainsi qu'une pourriture des racines souterrines sur lesquelles se forment les arachides. Il produit aussi sur certaines variétés un bleuissement foncé. D'ordinaire, les racines et la couronne subissent l'attaque à la ligne de contact avec le sol où se produit une pourriture brun pâle. Lorsqu'une partie seulement de la plante se flétrit ou meurt, les reste de la plante peut demeurer sain et rigoureux. D'ordinaire, les feuilles se flétrissent lentement et semblent se rétablir la nuit. Elles finissent par brunir et mourir.

Este patogéno causa podredumbre de las raíces y corona de las plantas y de las raíces subterráneas sobre las cuáles se forman los granos. También produce un mancha azul-negra sobre ciertas variedades. Las raíces y corona generalmente son atacados cerca de la superficie del suelo, desarrollándose un podredumbre de color café claro. Si sólo una parte de la planta se marchita y muere, el resto puede permanecer sano y vigoroso. Las hojas generalmente se marchitan despacio y parecen recuperarse durante la noche. Eventualmente se tornan café y mueren.

DISTRIBUTION: Af., A., Aust., Eur., N.A., C.A., S.A.

Arachis hypogaea + Puccinia arachidis Speg. =

ENGLISH	Leaf Rust
FRENCH	Rouille
SPANISH	Roya
PORTUGUESE	
ITALIAN	
RUSSIAN	ржавчина
SCANDINAVIAN	
GERMAN	Rost
DUTCH	
HUNGARIAN	
BULGARIAN	
TURKISH	
ARABIC	
PERSIAN	
BURMESE	Than-chi roga ပတ်စိုးနှယားသာရက်စိုဒစိ
THAI	
NEPALI	
HINDI	
VIETNAMESE	
CHINESE	Luòh hwa sheng shìou bìng 落花生銹病
JAPANESE	Sabi -byô さび病

The first evident symptoms of this disease are whitish flecks on the lower surface of the leaves. About two days later, yellowish-green flecks become visible on the upper surface. A narrow zone of chlorotic tissue frequently surrounds each lesion. Pustules on the upper surface of a leaflet tend to be smaller than the lower. Disease appears on all above-ground plant parts except for the flower. Heavy infection of leaves result in small leaf size and premature shedding.

Le prime symptomas evidente de iste morbo es maculas blanchette al superfacie inferior del folios. Post circa 2 dies maculas verde a jalnette es visibile anque al superfacie superior. Cata lesion es frequentemente circumcincte per un zona stricte de texito chlorotic. Pustulas al super-facie superior de un folietto es generalmente plus minute que illos que se trova al superfacie inferior. Le morbo se manifesta in tote le partes del planta le quales es super le terreno, salvo le flor. Infection sever de folios resulta in dimension diminute e defoliation prematur.

Les premiers symptômes évidents de cette maladie sont des mouchetures blanchâtres à la face inférieure des feuilles. Une couple de jours après, des mouchetures vert jaunâtre apparaissent à la face supérieure. Une bande étroite de tissus chlorotiques entoure fréquemment chaque lésion. Les pustules de la face supérieure ont tendance à être plus petites que celles de la face inférieure. La maladie apparaît sur toutes les parties aériennes de la plante, sauf les fleurs. Une infection grave des feuilles se traduit par une réduction de leur taille et par leur chute prématurée.

Los primeros síntomas evidentes de esta enfermedad son puntos blanquecinos que aparecen en el envés de las hojas. Unos dos días más tarde se pueden ver algunos puntos verde-amaril-losos en el haz de las hojas. Frecuentemente cada lesión está rodeada de una zona de tejido clorótico. Las pústulas en el haz de la hojita tienden a ser más pequeñas que en el envés. La enfermedad aparece en todas las partes aéreas de la planta, exceptuando las flores. La infec-ción fuerte de las hojas da como resultado hojas muy pequeñas y caída prematura de las mismas.

DISTRIBUTION: A., Eur., N.A., C.A., S.A.

16

Avena sativa (Triticum aestivum, Zea mays and other Gramineae) + Gibberella zeae (Schw.)
Petch (Gibberella saubinetti (Mont.) Sacc.) =

ENGLISH   Head Mold of Oats, Scab, Seedling Blight, Head Blight
FRENCH   Flétrissement des Plantules des Céréales, Fusariose, Maladie d'enivrement
SPANISH   Moho de la espiga; roña, tizónde las plántulas, tizón de la espiga
PORTUGUESE
ITALIAN
RUSSIAN   гиббереллез, фузариоз, "пьяный хлеб"
SCANDINAVIAN
GERMAN   Fusariose, Taumelkrankheit
DUTCH   Kiemschimmel
HUNGARIAN
BULGARIAN
TURKISH
ARABIC
PERSIAN
BURMESE
THAI
NEPALI
HINDI
VIETNAMESE
CHINESE
JAPANESE   Akakabi-byô   赤かび病

The earliest and most conspicuous symptom of this disease occurs soon after flowering. Diseased
spikelets are a bleached, light straw color and are prematurely ripened, while healthy spikelets
are still a normal green. Later, diseased spikelets are ash-gray color. One or more spikelets
may be infected or the entire head may be diseased. Infected kernels are shrunken, wrinkled
and light in weight. The fungus may also cause seedling blight and stem blight.

Le prime e le plus remarcabile symptoma de iste morbo occurre brevemente post floration.
Spiculas morbide es blanchite, de color de palea pallide, e maturate ante le tempore ordinari;
spiculas salubre, al contrario, es verde e normal. Plus tarde, spiculas morbide es gris e
cinerose. Un o plures del spiculas pote esser afflicte o tote le partes capital pote esser morbide. Granos infectate es diminute, corrugate, e legier de peso. Le fungo pote causar angue
plaga de plantas juvene e plaga de stirpes.

Le symptôme le plus précoce et le plus frappant de cette maladie se manifeste peu après la
floraison. Les épillets malades prennent une teinte paille claire, comme blanchie, et mûris sent
prématurément, alors que les épillets sains sont encore d'un vert normal. Par la suite, les
épillets malades deviennent gris cendré. Un ou plusieurs épillets peuvent être infectés ou bien
toute la tête peut être malade. Les grains infectés se rapetissent, se rident et sont plus légers.
Le champignon peut aussi causer la brûlure des semis et des tiges.

Los más conspicuous y primeros síntomas de este enfermedad aparecen poco después de la floración. Las espículas enfermas tienen un color pajizo claro, desteñido y maduran prematuramente,
mientras que las sanas tienen aún un color verde normal. Más tarde, las espículas enfermas toman un color gris ceniza. Puede ser que sólo una espícula o más estén infectadas o toda la espiga
puede estar enferma. Los granos infectados están encogidos, arrugados y pesan poco. El hongo
también puede causar tizón en las plántulas y tizón del tallo.

DISTRIBUTION: Af., A., Eur., N.A., C.A., S.A.

Avena sativa (and other Gramineae) + Helminthosporium nodulosum (Berk. & Curt.) Sacc.
(Cochliobolus nodulosus Luttrell) =

ENGLISH   Leaf Spot
FRENCH
SPANISH   Mancha foliar
PORTUGUESE
ITALIAN
RUSSIAN
SCANDINAVIAN
GERMAN
DUTCH
HUNGARIAN
BULGARIAN
TURKISH
ARABIC
PERSIAN
BURMESE
THAI
NEPALI
HINDI
VIETNAMESE
CHINESE
JAPANESE

Long rectangular spots form on the leaves, becoming coalescent, straw colored with narrow
red-brown margins. Spikes are infected and may collapse; grain does not form or is underdeveloped. Many infected plants die in the seedling stage, but others show only stunting.

Longe maculas rectangular se forma al folios, coalesce, e deveni de color paleate con margines
stricte de color rubie e brun. Spiculas es infectate e pote suffrer collapso; le grano non se
forma o se subdeveloppa. Un grande numero de plantas mori ancora multo juvene, ma alteres
suffre solmente obstruction de crescimento.

De longues taches rectangulaires se forment sur les feuilles, se fusionnent et prennent une
couleur de paille avec une bordure étroite rouge brunâtre. Les épis sont infectés et peuvent
s'affaisser, les grains ne se forment pas ou se développent peu. Plusieurs plantes infectées
meurent au stade de semis, alors que d'autres sont simplement rabougries.

Sobre las hojas se forman manchas largas, rectangulares que luego se unen, de color paja con
márgenes angostos de color café-rojo. Las espículas se infectan y pueden sufrir un colapso;
el grano no se forma o no se desarrolla bien. Muchas plantas infectadas mueren cuando están
pequeñas, mientras que otras sólo muestran achaparramiento.

DISTRIBUTION: Af., A., Aust., N.A.

18

Avena sativa (Hordeum vulgare, Triticum spp.) + Helminthosporium sativum Pam., King & Bakke (Cochliobolus sativus (Ito & Kurib.) Drechsl. ex Dastur) =

ENGLISH   Seedling Blight, Root Rot, Foot Rot
FRENCH   Pourriture des Racines
SPANISH   Tizón de las plántulas; podredumbre de la raíz, podredumbre del pié
PORTUGUESE
ITALIAN
RUSSIAN   гниль корней гельминтоспориозная
SCANDINAVIAN
GERMAN   Helminthosporium-Wurzelfäule
DUTCH
HUNGARIAN
BULGARIAN
TURKISH
ARABIC
PERSIAN
BURMESE
THAI
NEPALI
HINDI
VIETNAMESE
CHINESE
JAPANESE

Chocolate-brown to black spots appear near the soil line on the sheaths that cover the seedling leaves. Infections may progress until the seedlings turn yellow and die either before or, more frequently, after emergence and thus reduce stands. Affected seedlings may be dwarfed and produce dark green leaves. Seedlings commonly have weakened, dark brown rotted crowns and roots. When seedling infections are severe, plants may be dwarfed, heads may not emerge and kernels are poorly filled.

Maculas de color de chocolate o nigre se manifesta presso le terreno super le vaginas que coperi le folios del plantas juvene. Infectiones pote avantiar se de maniera que le plantulas deveni jalne e mori o ante o, plus frequentemente, post emergentia, causante un diminution del plantation. Plantas afflicte pote esser obstructe in lor crescimento e pote producer folios verde obscur. Generalmente, plantas juvene monstra pedes del stirpes e radices infirme, putreface, e brun obscur. Quando le plantulas es afflicte per infectiones sever, le plantas matur es a vices nanos, partes capital frequentemente non emerge, e granos es minus replete que normalmente.

Des taches brun chocolat à noires apparaissent au ras du sol sur les gaines qui recouvrent les feuilles des plantules. L'infection peut se propager jusqu'à ce que les plantules jaunissent et meurent, soit avant, ou plus fréquement, après l'émergence, et ainsi réduire les peuplements. Les plantules infectées peuvent rester naines et former des feuilles vert foncé. Les plantules ont communément des couronnes et des racines affaiblies, atteintes d'une pourriture brun foncé. Lorsque l'infection des plantules est grave, les plantes peuvent rester naines, les têtes ne pas se montrer, et les grains sont pauvrement pourvus.

Manchas de color café-chocolate a negras aparecen cerca de la superficie del suelo sobre la envoltura que cubre las hojas de la plántula. Las infecciones pueden progresar hasta que las plantitas se vuelven amarillas y mueren, ya sea antes o más frecuentemente después de la emergecia, reduciendo así la población. Las plantitas afectadas pueden achappararse y producir hojas de color verde obscuro. Las plantitas generalmente tienen coronas y raíces debilitadas, de color café obscuro y podridas. Cuando la infección de las plántulas es severa éstas pueden achapparrarse, las espigas pueden no emerger y los granos están pobremente llenos.

DISTRIBUTION: Af., A., Aust., Eur., N.A., C.A., S.A.

Avena sativa + Leptosphaeria avenaria G. F. Weber =

ENGLISH	Speckled Blotch
FRENCH	Septoriose
SPANISH	Mancha moteada
PORTUGUESE	
ITALIAN	
RUSSIAN	
SCANDINAVIAN	Havre-septoria
GERMAN	
DUTCH	
HUNGARIAN	
BULGARIAN	
TURKISH	
ARABIC	
PERSIAN	
BURMESE	
THAI	
NEPALI	
HINDI	
VIETNAMESE	
CHINESE	
JAPANESE	

Round or diamond-shaped spots or blotches develop on the leaves. The lesions are yellowish to light brown or chocolate brown, usually surrounded by a band of dull brown that changes to yellow as it blends into the green leaf. Grayish-brown to black lesions develop on the stems. Infection at the joints appear as dark brown blotches and may girdle the nodes. Heads and kernels may be infected. These infections resemble those occuring on the leaves.

Maculas e pustulas rotunde o de forma como un diamante se developpa super le folios. Le lesiones es jalnette a brunette o de color de chocolate, usualmente circumcincte per un banda de color brun mat que se cambia in jalne quando illos se misce con le verde del folio. Lesiones brun o gris a nigre se developpa al stirpes. Infectiones in le juncturas se manifesta como pustulas brun obscur e pote circumcinger le nodes. Partes capital es frequentemente afflicte. Iste infectiones sembla similar a los que occurre al folios.

Des taches rondes ou en losange ou des éclaboussures se forment sur les feuilles. Les lésions sont jaunâtres à brun pâle ou brun chocolat, généralement entourées d'une bande brun terne qui tourne au jaune en se fondant avec le vert du feuillage. Des lésions brun grísâtre à noires se forment sur les tiges. Sur les noeuds, l'infection apparaît sous forme d'éclaboussures brun foncé qui peuvent les encercler. Les têtes peuvent être infectées. Ces infections ressemblent à celles des feuilles.

Sobre las hojas se desarrollan manchas redondas o en forma de diamante. Las lesines son de color amarilloso a café claro o café chocolate, generalmente rodeadas por una banda de café opaco que cambia a amarillo cuando se mezcla dentro de la hoja verde. En los tallos se desarrollan lesiones grisáceas-café a negras. Las infecciones en las uniones aparacen como manchas café obscuro y pueden ceñir los nudos. Las espigas pueden infectarse. Estas infecciones semejan a las que ocurren en las hojas.

DISTRIBUTION: Af., Aust., Eur., N.A.

20

Avena sativa + Pseudomonas coronafaciens (Elliott) Stevens =

ENGLISH   Halo Blight
FRENCH   Tache Aréolée, Bactériose
SPANISH   Tizón de halo
PORTUGUESE
ITALIAN
RUSSIAN   пятнистость бурая бактериальная, бактериоз бурый (красный) пятнистый
SCANDINAVIAN
GERMAN   Braune(rote) Fleckenbakteriose, Bakterielle Blattdürre
DUTCH
HUNGARIAN
BULGARIAN   бактериоза
TURKISH
ARABIC
PERSIAN
BURMESE
THAI
NEPALI
HINDI
VIETNAMESE
CHINESE
JAPANESE   Kasagare-byô   かさ枯病

The lesions are more common on leaves, but do occur on floral parts in severe late infections. The initial lesions are oval to oblong water-soaked small spots. The tissue surrounding the spots lose their green color and become slightly water-soaked and light yellow color. The light yellow zone forms a halo area around the brown lesions. As infection increases the lesions coalesce, forming an irregular halo area. The tissues dry out and fade to light brown and straw-colored mottling. No exudate is present on the lesions.

Le lesiones se trova plus frequentemente in folios, ma occurre anque in partes floral in infectiones tarde e sever. Le lesiones initial es minute maculas saturate de aqua, de forma oval a oblonge. Le texito que circumcinge le maculas perde su color verde e deveni un pauc aquose e de color jalne clar. Le zona jalne clar forma un area de halo circum le lesiones brun. Quando le infection se avantia, le lesiones coalesce, formante un area irregular de halo. Le texitos se desicca e lor color se attenua, deveniente un mixtura brunette e paleate. Nulle exsudato es presente al lesiones.

Les lésions sont plus communes sur les feuilles, mais se produisent aussi sur les parties florales lors des infections tardives graves. Les lésions initiales sont de petites taches délavées, ovales à oblongues. Les tissus qui entourent les taches perdent leur couleur verte et deviennent légèrement aqueux et de couleur jaunâtre. La zone jaune pâle forme un halo autour des lésions brunes. A mesure que l'infection avance, les lésions se fusionnent de façon à former une aréole irrégulière. Les tissus se dessèchent et s'affadissent pour former une marbrure de couleur brun pâle à paille. Il n'y a pas d'exsudat sur ces lésions.

Las lesiones son más comunes en las hojas, pero también atacan las partes florales en infecciones tardías severas. Las lesiones iniciales son pequeñas manchas ovaladas u oblongas, empapadas de auga. El tejido que rodea las manchas pierde su color verde y se vuelve algo empapado en agua y de color amarillo pálido. La zona amarillo claro forma un área de halo alrededor de las lesiones café. A medida que aumenta la infección las lesiones se van uniendo, formando un área irregular de halo. Los tejidos se secan y se destiñen hasta adquirir un color café claro o paja moteado. No hay exudaciones en estas lesiones.

DISTRIBUTION: Aust., Eur., N.A., S.A.

<u>Avena</u> <u>sativa</u> (<u>Hordeum</u> <u>vulgare</u>) + <u>Pseudomonas</u> <u>striafaciens</u> (Elliott) Starr & Burkholder =

ENGLISH	Bacterial Stripe
FRENCH	Strie Bactérienne, Bactériose
SPANISH	Listas bacteriales, Raya bacterial
PORTUGUESE	
ITALIAN	
RUSSIAN	бактериоз бурый (красный) полосатый, ожог полосатый
SCANDINAVIAN	
GERMAN	Braunstreifenbakteriose, Bakterielle Blattstreifenkrankeit
DUTCH	
HUNGARIAN	
BULGARIAN	
TURKISH	
ARABIC	
PERSIAN	
BURMESE	
THAI	
NEPALI	
HINDI	
VIETNAMESE	
CHINESE	
JAPANESE	Suzigare-saikin-byô　すじ枯細菌病

Lesions first appear as sunken water-soaked dots on the leaves. The dots enlarge and may coalesce to form long water-soaked streaks or blotches that may extend the entire length of the leaf. The stripes or streaks often have narrow yellowish margins. Later they become rusty-brown. Halo-like borders do not form around these lesions. During moist weather, droplets of bacterial ooze that later dry to thin white scales appear on the leaf along the lesions. Disease may kill the entire top of the plant.

Lesiones se manifesta primarimente como punctos, submerse e saturate de aqua, al folios. Illos cresce e pote coalescer de maniera que illos forma o elongate venas colorate e saturate de aqua o pustulas, le quales pote extender se per tote le longitude del folio. Le bandas o venas sovente ha stricte margines jalnette. Plus tarde, illos deveni de color ferruginose. Bordos similar a un halo non se forma circum iste lesiones. In tempore humide, guttas de exsudation bacterial se manifesta al folio in connexion con le lesiones e, plus tarde, se desicca, deveniente blanc scalias tenue. Le morbo pote causar le morte de tote le partes superior del planta.

Les lésions apparaissent d'abord comme des points déprimés et détrempés sur les feuilles. Les points s'agrandissent et peuvent se fusionner pour former de longues stries ou éclaboussures détrempées qui peuvent s'étendre sur toute la longueur de la feuille. Les stries ou rayures ont souvent des bords jaunâtres étroits. Par la suite, elles deviennent brun rouille. Il ne se forme pas d'aréola autour de ces lésions. Par temps humide apparaissent sur la feuille, le long des lésions, des gouttelettes d'exsudat bactérien qui sèchent ensuite pour former des pellicules blanches. La maladie peut détruire tout le sommet de la plante.

Las lesiones aparecen primero como puntos hundidos y empapados de agua sobre las hojas. Los puntos se agrandan y pueden unirse para formar vetas empapadas de agua o manchas que pueden extenderse a todo lo largo de la hoja. Las rayas o vetas frecuentemente tienen márgenes amarillentos y angostos. Más tarde se vuelven café-herrumbroso. Alrededor de estas lesiones no se forman bordes como halo. Durante el tiempo húmedo, gotitas de la exudación bacterial que luego se tornan delgadas escamas blancas, aparecen en la hoja a lo largo de las lesiones. La enfermedad puede matar toda la parte superior de la planta.

DISTRIBUTION: Aust., Eur., N.A., S.A.

Avena sativa + Pyrenophora avenae Ito & Kuribay =

ENGLISH	Leaf Stripe, Leaf Blotch
FRENCH	Helminthosporiose de l'avoine
SPANISH	Helmintosporiosis, Roya de la hoja, Mancha de la hoja
PORTUGUESE	
ITALIAN	Seccume delle foglie di avena
RUSSIAN	
SCANDINAVIAN	Havrebrunflekk, Bladpletsyge
GERMAN	
DUTCH	Strepenziekte, Vlekkenziekte
HUNGARIAN	
BULGARIAN	
TURKISH	
ARABIC	
PERSIAN	
BURMESE	
THAI	
NEPALI	
HINDI	Shlath Kand श्लाथ कंद
VIETNAMESE	
CHINESE	
JAPANESE	Hagare-byô 葉枯病

The blotches are oblong to linear with an irregular margin. They are light reddish-brown, frequently with a sunken center. The lesions are generally on the leaves which turn yellow and dry out as the infection advances. When the disease is severe the spikelets drop and the stem may break.

Le pustulas es oblonge usque a linear con un margine irregular; illos es clar, burn e rubiette; e frequentemente illos ha un centro submerse. Le lesiones occurre usualmente al folios le quales deveni jalne e se desicca quando le infection se avantia. Quando le morbo es sever, le spiculas declina e le stirpe pote rumper se.

Les éclaboussures sont oblongues à linéaires, à contours irréguliers. Elles sont brun rougeâtre et souvent deprimees au centre. Les lésions se forment généralement sur les feuilles, qui jaunissent et sèchent à mesure que l'infection progresse. Quand la maladie est grave, les épillets tombent et la tige peut se casser.

Las manchas son oblongas o lineales con un margen irregular. Son de color café rojizo claro, frecuentemente con el centro hundido. Las lesiones están generalmente sobre las hojas, las cuáles se vuelven amarillas y se secan a medida que la infección avanza. Cuando la enfermedad es severa las espículas se caen y el tallo se puede quebrar.

DISTRIBUTION: Af., A., Aust., Eur., N.A.

Avena sativa + Ustilago avenae (Pers.) Rostr.

ENGLISH	Loose Smut
FRENCH	Charbon nu de l'avoine
SPANISH	Carbón desnudo, Carbón volador, Carbón descubierto
PORTUGUESE	Carvão
ITALIAN	Carbone volante, Carbone dell'avena
RUSSIAN	головня пыльная
SCANDINAVIAN	Nøgen havrebrand, Havreflygsot
GERMAN	Flugbrand Hafer, Haferflugbrand, Staubbrand, Nackter Haferbrand
DUTCH	Haverstuifbrand, Stuifbrand
HUNGARIAN	Zab-porüszög
BULGARIAN	праховита главня
TURKISH	Açık rastık
ARABIC	Taphahom تفحم
PERSIAN	
BURMESE	
THAI	
NEPALI	
HINDI	Shlath Kand श्लाथ कंद
VIETNAMESE	
CHINESE	
JAPANESE	Hadakakuroho-byŏ 裸黒穂病

The most noticeable symptom of this disease is the appearance of black spore masses replacing the floral parts of the plant. This loose spore mass is held together by a thin gray membrane which breaks and disintegrates shortly after the floral parts emerge. Usually all of the spikelets and heads of an infected plant are diseased. Occasionally one portion may escape and appear healthy. Diseased plants are shorter than normal plants and are often overlooked at harvesttime, since they have then lost the black spore mass.

Le symptoma le plus remarcabile de iste morbo es le apparition de massas de sporas nigre, reimplaciante le partes floral del planta. Iste massa disligate de sporas resta unite per un membrana tenue gris que se frange e se disintegra brevemente depost que le partes floral emerge. Usualmente, tote le spiculas e tote le partes capital de un planta infectate es morbide. Occasionalmente, un portion pote salvar se e pote semblar esser salubre. Plantas morbide es plus curte que le plantas normal e es sovente non remarcate durante le rendimento perque illos ha alora, post un periodo de tempore, perdite le massa nigre sporal.

Le symptôme le plus apparent de cette maladie est l'apparition de masses de spores noires qui remplacent les parties florales de la plante. Cette masses de spores nue est maintenue en place par une mince menbrane grise qui se brise et se désintègre peu après la sortie des parties florales. D'ordinaire, tous les épillets et toutes les têtes d'une plante infectée sont malades. Occasionnellement, une partie peut échapper à la maladie et paraître saine. Les plantes malades sont plus petites que les plantes saines et passent souvent inaperçues à la récolte, du fait qu'elles ont perdu leur masses de spores noires.

El síntoma más notorio de esta enfermedad es la aparición de masas negras de esporas en reposición de las partes florales de la planta. Esta masa suelta de esporas se mantiene unida gracias a una delgada membrana gris que se rompe y desintegra poco después de que las partes florales emergen. Generalmente todos las espículas y espigas de una planta infectada están enfermos. Ocasionalmente puede escapar una porción y parecer sana. Las plantas enfermas son de menor tamaño que las normales y frecuentemente no se ven al tiempo de la cosecha, ya que para entonces han botado la masa negra de esporas.

DISTRIBUTION: Af., A., Aust., Eur., N.A., S.A., C.A.

Beta vulgaris + Alternaria tenuis Nees =

ENGLISH	Leaf Blight
FRENCH	Alternariose
SPANISH	Alternariosis, Mancha circular, Tizón de la hoja
PORTUGUESE	
ITALIAN	
RUSSIAN	альтернариоз
SCANDINAVIAN	
GERMAN	Alternaria-Blattbraune, Alternaria-Fleckenkrankheit
DUTCH	
HUNGARIAN	
BULGARIAN	
TURKISH	
ARABIC	
PERSIAN	Ältĕrnārēyā chŏghŏndàr ghånd آلترناریای چغندرقند
BURMESE	
THAI	
NEPALI	
HINDI	
VIETNAMESE	
CHINESE	
JAPANESE	Hagare-byŏ 葉枯病

Spots on the leaves at first are small, irregular, and reddish-brown. As they grow larger they become circular and have concentric rings; the center turns from tan to gray and the margin becomes dark in color. The centers of the oldest spots may fall out, leaving a shot-hole effect. Only in severe infections are whole leaves killed.

Maculas super le folios es al initio minute, irregular, e brun e rubiette. Quando illos cresce, illos deveni circular con anellos concentric; le centro se varia ab bronzate a gris; e le margine deveni de color obscur. Le centros pote separar se ab le maculas le plus vetule e cader, causante le aspecto de un explosion. Le folios integre mori solmente in infectiones sever.

Les taches sur les feuilles sont d'abord petites, irrégulières et brun rougeâtre. A mesure qu'elles grossissent, elles deviennent circulaires et ont des cercles concentriques; le centre va du beige au gris et les bords prennent une teinte foncée. Le centre des plus vieilles taches peut tomber, laissant comme un trou de balle. Ce n'est que dans les infections graves que des feuilles entières sont détruites.

Las manchas sobre las hojas son primero pequeñas, irregulares y de color café-rojizo. A medida que crecen se vuelven circulares y tienen anillos concéntricos; el centro cambia de color crema a gris y los márgenes se vuelven de color más obscuro. Los centros de las manchas más viejas pueden caerse, dejando un agujero como de bala. Solamente en infecceiones severas mueren las hojas enteras.

DISTRIBUTION: Af., A., Eur., N.A., S.A.

<u>Beta vulgaris</u> + <u>Cercospora beticola</u> Sacc. =

ENGLISH	Leaf Spot
FRENCH	Tache Cercosporéene, Cercosporiose, Taches Foliaires
SPANISH	Cercosporiosis, Mancha cercospora, Mancha de la hoja, Mancha foliar
PORTUGUESE	
ITALIAN	Cercospora della bietola, Nebbia, Vaiolatura delle foglie
RUSSIAN	церкоспороз, пятнистость церкоспорозная
SCANDINAVIAN	Bladpletsyge hos bederoe, Bladfläcksjuka på betor
GERMAN	Blattfleckenkrankheit
DUTCH	Bladvlekken, Gewone Bladvlekkenziekte
HUNGARIAN	Cerkospórás levélfoltosság
BULGARIAN	церкоспороза
TURKISH	Pancar yaprak lekesi
ARABIC	Tabbakoo Awrak تبغع أوراق
PERSIAN	Làkêh bárgêyê chôghôndár لكة برگى جغندر
BURMESE	
THAI	
NEPALI	Sukandar Ko Khairo thople rog सुकन्दरको खैरो थोप्ले रोग
HINDI	Parn dhabba पर्ण धब्वा
VIETNAMESE	
CHINESE	Tíen tsày hor ban bìng 甜菜褐斑病
JAPANESE	Kappan-byô 褐斑病

This disease appears usually on the older leaves. It is characterized by small, brownish spots with reddish-purple borders which give the leaf a speckled appearance. As the disease advances the spots enlarge and turn gray. The center tissue of old lesions drop out, leaving ragged holes and eventually the leaf dies. The root from an infected plant will usually be long and have a rough crown. The fungus does not attack the root itself.

Iste morbo se manifesta usualmente al folios plus vetule. Illo se characterisa per minute maculas brunette con bordos de color rubiette e purpuree, le quales face le folio semblar punctillate. Quando le morbo se avantia, le maculas cresce in dimension e deveni gris. Le texito central del lesiones plus vetule se separa e cade, lassante cavitates con bordos zigzag, e eventualmente le folio mori. Le radice de un planta infectate es usualmente elongate e monstra un collar asperate. Le fungo non attacca le radice ipse.

Cette maladie apparaît généralement sur les vieilles feuilles. Elle se caractérise par de petites taches brunâtres à contours rouge pourpré qui donnent à la feuille un aspect moucheté. A mesure que la maladie pregresse, les taches grossissent et deviennent grises. Le tissu central des vieilles lésions tombe, laissant des trous à bords déchiquetés, et la feuille finit par mourir. Habituellement, la racine d'une plante infectée est longue et sa couronne est rugueuse. Le champignon n'attaque pas la racine elle-même.

Esta enfermedad generalmente aparece en las hojas más viejas. Se caracteriza por manchas pequeñas, cafesosas con bordes rojizo-purpúreos que le dan a la hoja una apariencia moteada. A medida que la enfermedad avanza, las manchas se agrandan y se vuelven de color gris. El tejido del centro de las lesiones viejas se cae, dejando agujeros de bordes desiguales y eventualmente la hoja muere. La raíz de una planta infectada generalmente es larga y tiene una corona áspera. El hongo no attaca la raíz en sí.

DISTRIBUTION: Af., A., Aust., Eur., N.A., C.A., S.A.

Beta spp. + Erysiphe betae (Vañha) Weltzien =

ENGLISH	Powdery Mildew
FRENCH	
SPANISH	Mildiú polvoriento
PORTUGUESE	
ITALIAN	
RUSSIAN	
SCANDINAVIAN	
GERMAN	Echter Mehltau
DUTCH	
HUNGARIAN	
BULGARIAN	
TURKISH	
ARABIC	
PERSIAN	
BURMESE	
THAI	
NEPALI	
HINDI	
VIETNAMESE	
CHINESE	
JAPANESE	Udonko-byð うどんこ病

Talcum-like growth of the fungus appears first on the upper surface of the older, but still bright green, leaves. These whitened areas enlarge, often until they cover the entire surface of the leaf. The affected foliage gradually loses its luster and tends to curl, so that finally some of the leaves turn yellow, then brown,and finally die. Only rarely do black, minute fruiting bodies appear near the veins amid the white mold.

Un apparition del fungo, similar a talco, se manifesta primarimente al superfacie superior del folios plus vetule, le quales es, nonobstante, ancora brillante e verde. Iste areas blanchite cresce in dimension, frequentemente usque a coperir tote le superfacie del folio. Le foliage afflicte pauc a pauc perde su lustro e propende inrolar se, de maniera que al fin alicun folios deveni jalne, plus tarde brun, e alora illos mori. Multo rarmente se manifesta minute structuras sporal nigre presso le venas inter le mucor blanc.

Au début, le champignon a l'aspect d'une poudre de talc à la face supérieure des vieilles feuilles encore bien vertes. Ces plages blanches s'agrandissent, souvent jusqu'à recouvrir la surface entière de la feuille. Le feuillage atteint perd graduellement son lustre et tend à s'incurver, de sorte qu'à la fin, certaines feuilles jaunessent, puis brunissent et meurent. Ce n'est que rarement que de minuscules fructifications noires apparaissent près des nervures parmi la "moisissure blanche".

El crecimiento polvoriento del hongo aparece primero en el haz de las hojas viegas, pero que aún tienen un color verde brillante. Estas áreas blanqueadas se agrandan, frecuentemente hasta que cubren toda la superficie de la hoja. El follaje afectado gradualmente pierde su lustre y tiende a enroscarse, hasta que finalmente algunas de las hojas se vuelven amarillas, luego café y finalmente mueren. Solamente en raras ocasiones aparecen cuerpos fructíferos diminutos cerca de las venas enmedio del moho blanco.

DISTRIBUTION: Af., A., Eur.

Beta vulgaris (Beta spp.) + Peronospora schachtii Fuckel (Peronospora farinosa Kiessl.) =

ENGLISH	Downy Mildew
FRENCH	Mildiou
SPANISH	Mildiú lanoso
PORTUGUESE	
ITALIAN	Peronospora della barbabietola, Mal secco
RUSSIAN	пероноспороз, ложная мучнистая роса
SCANDINAVIAN	Bedeskimmel, Betbladmögel
GERMAN	Falscher Mehltau
DUTCH	Valse Meeldauw
HUNGARIAN	
BULGARIAN	
TURKISH	
ARABIC	Bayad Dakiky بياض دقيق
PERSIAN	
BURMESE	
THAI	
NEPALI	
HINDI	
VIETNAMESE	
CHINESE	
JAPANESE	Beto-byō べと病

This disease appears on all above-ground parts of the plant. When seedlings are infected they turn pale in color, become covered with the fungus on both surfaces and curl sharply downward. Usually such plants die on developing leaves. Irregular, pale green areas form on the upper surface while the fungus fruits in a white to gray layer on the corresponding lower surface. Similar lesions form on flowers and seed stalks and cause distortion on these parts.

Iste morbo se manifesta in tote le portiones del planta que es super le terreno. Quando plantulas es infectate, illos deveni de color pallide, es coperte con le fungo in tote le duo superfacies, e se inrola acutemente al basso. Usualmente tal plantas mori quando le folios emerge. Verde e pallide areas irregular evolve al superfacie superior durante que le fungo produce sporas in un strato blanc o gris al correspondente superfacie inferior. Lesiones similar evolve al flores e al pedunculos, e illos causa distortion de iste partes.

Cette maladie apparait sur toutes les parties aériennes de la plante. Les plantules infectées pâlissent, se couvrent du champignon des deux côtés et s'enroulent abruptement vers le bas. Généralement, ces plantules meurent en produisant leurs feuilles. Des plages irrégulières vert pâle se forment a la face supérieure, pendant que le champignon fructifie sur une couche blanche à grise à la face inférieure correspondante. Des lésions semblables se forment sur les fleurs, les pédoncules, et causent la distorsion de ces parties.

Esta enfermedad aparece en todas las partes aéreas de la planta. Cuando las plántulas están infectadas se vuelven de color pálido, se cubren con el hongo en ambas superficies y se enroscan bastante hacia abajo. Generalmente tales plantas mueren cuando están desarrollando las primeras hojas. En el haz se forman áreas irregulares, color verde pálido mientras el hongo fructifica en una capa blanca a gris en el envés correspondiente. Lesiones similares se forman en las flores, tallos de semilla y causa deformación en estas partes.

DISTRIBUTION: Af., A., Aust., N.A., S.A.

Beta spp. + <u>Phoma betae</u> Frank (Pleospora bjoerlingii Byford, <u>Mycosphaerella tabifaca</u> (Prill. & Del.) Lind.) =

ENGLISH   Seedling Root Rot, Leaf Spot, Black Rot of Growing Roots, Heart Rot of Mature Roots
FRENCH   Pied Noir de la Betterave, Maladie du Coeur
SPANISH   Podredumbre de la raíz de las plántulas, mancha foliar
PORTUGUESE
ITALIAN   Mal del Cuore, Moria della bietola, Mal del piede
RUSSIAN   фомоз, пятнистость зональная
SCANDINAVIAN   Rodbrand, Groddbrand, Rotbrand, Hjärtröta
GERMAN   Wurzelbrand, Herzfäule, Trockenfäule, Bormangelkrankheit,
           Zonale Blattfleckenkrankheit
DUTCH   Zwartrot, Bietenwortelbrand
HUNGARIAN
BULGARIAN   фомоза
TURKISH
ARABIC
PERSIAN
BURMESE
THAI
NEPALI
HINDI
VIETNAMESE
CHINESE   Tíen tsày shéor yǐan bìng   甜菜蛇眼病
JAPANESE   Zyanome-byô   じゃのめ病

This disease attacks all parts of the plant. It produces a typical damping-off of seedlings. The bases of the young stems and roots turn black and then the plants fall over and die. It causes a distinct leaf spot, brown in color and sometimes with concentric rings. On weakened roots the inner younger leaves are infected first, followed by a gradual progress to the older leaves as they become weakened and yellowed. When the complete top is destroyed, rosettes of small foliage may form.

Iste morbo attacca tote le partes del planta. Typicamente illos produce un putrefaction de plantulas per fungos. Le pede del stirpe juvene e le radices deveni nigre e alora le planta suffre collapso e mori. Le morbo causa un macula distinctive al folio, de color brun e alicun vices con anellos concentric. Con radices infirme, le interior folios plus juvene es infectate al initio e un progression gradual del infection se continua al folios plus vetule, le quales deveni infirme e jalnette. Quando tote le partes apical es destructe, rosettas de foliage minute pote evolver.

Cette maladie attaque toutes les parties de la plante. Elle produit une fonte des semis typique. La base de la jeune tige et les racines noircissent, après quoi la plante tombe et meurt. Elle cause une tache des feuilles bien définie, de couleur brune, avec parfois des cercles concentriques. Sur des racines affaiblies, les jeunes feuilles du centre sont les premières infectées, suivies progressivement par les plus vieilles, qui s'affaiblissent et jaunissent. Lorsque la tête entière est détruite, il peut se former des rosettes de petites feuilles.

Esta enfermedad ataca todas las partes de la planta. Produce un mal de talluelo típico de las plántulas. La base del tallo tierno y de las raíces se vuelve negro y luego la planta se dobla y muere. Causa una mancha foliar distinta, de color café y algunas veces con anillos concéntricos. En los tubérculos debilitados las hojitas tiernas internas son infectadas primero, sequido de un progreso gradual hacia las hojas maduras mientras se van debilitando y amarillando. Cuando toda la parte aérea se destruye, pueden formarse rosetas de follaje pequeño.

DISTRIBUTION: Af., A., Aust., Eur., N.A.

Beta vulgaris + Uromyces betae (Pers.) Lév. =

ENGLISH	Rust
FRENCH	Rouille
SPANISH	Roya
PORTUGUESE	
ITALIAN	Ruggine della barbabietola
RUSSIAN	
SCANDINAVIAN	
GERMAN	Rost, Rübenrost
DUTCH	Roest der Biet
HUNGARIAN	
BULGARIAN	
TURKISH	Pancar pası
ARABIC	
PERSIAN	Zàngê chôghôndàr زنگ چغندر
BURMESE	
THAI	
NEPALI	
HINDI	
VIETNAMESE	
CHINESE	
JAPANESE	Sabi-byô さび病

Small cinnamon-brown pustules are the first symptom observed, and in susceptible plants the disease quickly spreads over the entire foliage causing the older leaves to wilt, wither and die prematurely. The younger leaves remain erect but their blades become crumpled, drooping and yellowish. Badly rusted plants with blisters on leaf blades and petioles finally collapse.

Parve pustulas de colore de cannella es le prime symptoma observate, e in plantas susceptible le morbo rapidemente extende se super tote le foliage, facente le folios plus vetule languer, marcescer, e morir prematuremente. Le folios plus juvene resta erecte, ma lor laminas deveni corrugate, clinante, e satis jalne. Plantas multo oxydate, que ha ampullas super le laminas del folios e super le petiolos, finalmente suffre collapso.

De petites pustules brun cannelle sont les premiers symptômes observés, puis, chez les plantes sensibles, la maladie gagne rapidement tout le feuillage, causant le flétrissement, la décomposition et la mort prématurée des vieilles feuilles. Les jeunes feuilles demeurent dressées, mais leur limbe se recroqueville, penche et jaunit. Les plantes fortement rouillées et portant des vésicules sur les limbes et les pétioles finissent par s'affaisser.

Los primeros síntomas que se observan son pequeñas pústulas de color café-canela y en plantas susceptibles la enfermedad se propaga pronto sobre todo el follaje causando que las hojas viejas se marchiten, se ajen y mueran prematuramente. Las hojas más jóvenes permanecen erectas pero sus láminas se vuelven arrugados, caídos y amarillos. Las plantas severamente atacadas por la roya con pústulas en las hojas y pecíolos, sufren finalmente un colapso.

DISTRIBUTION: Af., A., Aust., Eur., N.A., S.A.

Brassica spp. (and other Cruciferae)   + Alternaria brassicae (Berk.) Sacc. =

ENGLISH   Gray Leaf Spot
FRENCH   Tache Grise, Alternaria des Crucifères, Taches Noires
SPANISH   Mancha de la hoja, Tizón de la hoja, Tizón temprano, Alternariosis, Mancha
               zonal, Lancha, Mancha gris de la hoja
PORTUGUESE   Podridão do colo, Podridão parda, Mancha de Alternaria
ITALIAN   Alternariosi delle crocifere, Macchie flogliari, Marciume nero del cavolfiore
RUSSIAN   пятистость черная, плесень черная, альтернариоз
SCANDINAVIAN   Skulpesvamp
GERMAN   Rapsverderber, Dürrfleckenkrankheit, Kohlhernie, Schwarzbeinigkeit, Kohl-
               schotenschwärze, Schwarzpilz an Blumen
DUTCH   Spikkelziekte
HUNGARIAN   Alternáriás szárazfoltosság
BULGARIAN   чернилка
TURKISH   Lahana yaprak lekesi
ARABIC   Tabbakoo Awrak   تبقع أوراق
PERSIAN   Älternärëyä kàlàm   آلترناریای کلم
BURMESE   Ah-net-kwet, Ah-net-pyauk roga   အာ့နက်ကွက် အာ့နက်ပြောက်ရောဂါး
THAI   Bi-Jut ใบจุด
NEPALI   Banda Ko kalothople rog वन्दाको कालो थोप्ले रोग
HINDI   Alternaria Angmaari   आल्टरनेरिया अंगमारी
VIETNAMESE   Đốm thâm trên cái hoa
CHINESE   Bái tsày hei dǐan bìng 白菜黑點病 ，Gan lán hei ban bìng 甘藍黑斑病
JAPANESE   ·Kokuhan-byð 黒斑病

The first symptom is a minute dark spot on the seedling stem immediately after germination, causing a damping-off effect or stunting of the young plant. Such plants, when set into the field, never grow as large as normal nor do they yield well. It not only affects seed pods, but also shrivels the seeds, kills the pod stalks before seed forms, and with severe infection kills the whole seed plant.

Le prime symptoma es un minute macula obscur al pedunculo del plantula immediatemente post germination, causante putrefaction per fungos o nanismo del plantula. Tal plantas, quando on los situa in le campo, non cresce a esser tanto grande como plantas normal, ni rende multo. Le morbo non solmente affice le siliquas, ma anque face le semines corrugar se, face le pedunculos morir ante que le semines se forma, e, in casos de infection sever, face tote le planta morir.

Le premier symptôme est une minuscule tache sombre sur la tige de la plantule, immédiate-ment après la germination, qui produit un effet semblable à la fonte ou un rabougrissement de la jeune plante. Transplantées dans de champ, ces plantes n'atteignent jamais une taille normale et ne donnent pas un bon rendement. Cette maladie, non seulement s'attaque aux siliques, mais aussi fait ratatiner les graines, mourir les pédoncules avant que les graines ne se forment, et même, dans les cas graves, tue tout le porte-graine.

El primer síntoma es una manchita obscura en el tallo de la plantita inmediatamente después de la germinación, causando un efecto de "mal del talluelo" o achaparramiento en la misma. Estas plantas, al llevarlas al campo, nunca crecen tanto como las normeles, ni producen bien. En las plantas para producir semilla no sólo afecta las vainas, sino que también se ajan las semillas, mueren los pedúnculos de las vainas antes de que se forme la semilla y cuando la infección es severa mata toda la planta.

DISTRIBUTION: Af., A., Aust., Eur., N.A., C.A., S.A.

Brassica spp. (Cruciferae and others) + Alternaria brassicicola (Schw.) Wiltsh. =

ENGLISH	Black Spot of Leaves and Pods, Brown Rot, Head Browning
FRENCH	Tache Noire
SPANISH	Alternariosis, Mancha negra de las hojas y vainas, tostado de la cabeza
PORTUGUESE	
ITALIAN	
RUSSIAN	
SCANDINAVIAN	
GERMAN	
DUTCH	
HUNGARIAN	
BULGARIAN	
TURKISH	
ARABIC	
PERSIAN	
BURMESE	Ah-net-kwet, Ah-net-pyauk roga · အာ‌ာ်လ‌က ၄ ဇ်ဃ‌‌ာ‌‌ရ‌ဝ ‌စ ‌‌ိ‌က‌ိ‌လ
THAI	Bi-Jut ใบจุก
NEPALI	
HINDI	Parn Dhabba परण घव्वा
VIETNAMESE	
CHINESE	
JAPANESE	Kurosusu-byð 黒すす病

Symptoms vary depending upon crops attacked. They may have pale yellow, white, tan, gray, brown, dark green, violet, black or water-soaked spots on leaves, stalks and seed pods. Certain leaf spots are concentrically zoned, or drop out leaving shot-holes. Leaves may wilt or shrivel and die early. Seedlings may be killed and heads discolored. This disease also causes considerable discoloration to heads in transit or in storage.

Le symptomas se varia secundo le plantas attaccate. Maculas jalne clar, blanc, bronzate, gris, brun, verde obscur, violette, nigre, o saturate de aqua pote evolver al folios, pedunculos, o siliquas. Alicun maculas al folios ha zonation concentric; alteres pote separar se e cader, causante le apparition de un explosion. Folios pote marcescer, crispar se, o morir prematur-mente. Plantulas pote morir e lor partes apical pote esser discolorate. Iste morbo, in plus, causa discoloration extensive al partes apical o durante le transportation o in magasinage.

Les symptômes varient selon les cultures attaquées. Ils peuvent consister en de petites taches jaune pâle, blanches, beiges, grises, brunes, vert foncé, violettes, noires, ou d'aspect dé-trempé, sur les feuilles, les pédoncules et les siliques. Certaines taches des feuilles ont des cercles concentriques ou tombent en laissant des trous de cible. Les feuilles peuvent se flétrir ou se retatiner et mourir prématurément. Les semis peuvent être tués et les pommes brunir. Cette maladie fait aussi brunir considérablement les pommes de chou durant le transport e l'entreposage.

Los síntomas varían dependiendo del cultivo atacado. Pueden tener manchas de color amarillo pálido, blanco, crema, gris, café, verde obscuro, violeta o manchas empapadas de agua, sobre las hojas, tallos y vainas. Ciertas manchas foliares están localizadas concéntricamente o pueden caerse dejando agujeros como de bala. Las hojas pueden marchitarse o arrugarse y mueren tem-prano. Las plántulas pueden morir y las cabezas se decoloran. Esta enfermedad también causa considerable decoloración en las cabezas que están en tránsito o almacenadas.

DISTRIBUTION: Af., A., Aust., Eur., N.A., C.A., S.A.

32

Brassica oleracea (Brassica spp.) + Cercosporella brassicae (Fautr. & Roum.) Höhnel =

ENGLISH   White Spot
FRENCH   Tache Blanche, Taches Foliaires Blanches, Cercosporellose
SPANISH   Mancha blanca
PORTUGUESE
ITALIAN
RUSSIAN   церкоспореллез, пятнистость белая
SCANDINAVIAN
GERMAN   Cercosporella-Blattfleckenkrankheit
DUTCH
HUNGARIAN
BULGARIAN
TURKISH
ARABIC
PERSIAN
BURMESE
THAI
NEPALI
HINDI
VIETNAMESE
CHINESE
JAPANESE

The pathogen causes lesions on leaves, stems and seed pods. The spots are gray or brown with almost paper-white centers and with slightly darkened margins. When the spots are numerous, the affected foliage may turn yellow and finally drop. Seedlings are killed when the disease appears early in its most severe form. Later infection reduces the size of edible parts.

Le pathogeno causa lesiones al folios, pedunculos, e siliquas. Le maculas es gris o brun e ha centros tanto blanc como nive con margines un pauc obscurate. Quando existe maculas numerose, le foliage afflicte pote devenir jalne e al fin se separar e cader. Plantulas mori si le morbo se manifesta precocemente in su stato le plus sever; infection plus tarde diminue le dimension del partes comestibile.

Le pathogène produit des lésions sur les feuilles, les tiges et les siliques. Les taches sont grises ou brunes; leur centre est d'une blancheur approchant celle du papier et leurs marges sont légèrement assombries. Lorsque les taches sont nombreuses, les feuilles atteintes peuvent jaunir et finalement tomber. Lorsque la maladie sévit tôt, elle fait mourir les plantules. Les infections subséquentes diminuent le volume des parties comestibles.

El patógeno causa lesiones en las hojas, pedúnculos y vainas. Las manchas son de color gris o café, con centros casi blancos y orillas un poco más obscuras. Cuando las manchas son numerosas, el follaje afectado puede volverse de color amarillo y finalmente se cae. Las plantitas mueren cuando la enfermedad aparece temprano en su forma más severa. Las infecciones tardías reducen el tamaño de las partes comestibles.

DISTRIBUTION: A., Aust., Eur., N.A.

Brassica spp. (Phaseolus vulgaris) + Erysiphe polygoni de Candolle =

ENGLISH   Powdery Mildew
FRENCH   Blanc
SPANISH   Oidium, Oidio, Mildiú polvoriento
PORTUGUESE
ITALIAN
RUSSIAN
SCANDINAVIAN
GERMAN   Echter Mehltau
DUTCH   Meeldauw
HUNGARIAN
BULGARIAN
TURKISH
ARABIC
PERSIAN   Sêfêdàkê hàghēghēyê kàlàm   سفیدک حقیقی چغندر
BURMESE
THAI
NEPALI
HINDI
VIETNAMESE
CHINESE
JAPANESE   Udonko-byô   うどんこ病

This disease produces a white fungal growth over parts of the foliage, especially on the upper surface. After the fungus has developed for some time, the affected leaf areas become pale green to yellow or tan. In a severe attack the whole leaf curls and dies and many drop off. Black fruiting bodies frequently form on stalks but almost never on the leaf itself.

Iste morbo produce un developpamento del fungo blanc super partes del foliage, specialmente al superfacie superior. Depost que le fungo ha developpate se per qualque tempore, le areas afflicte del folios deveni verde pallide o jalne o bronzate. In un attacco sever, tote le folio se inrola e mori; e multe folios se separa e cade. Nigre structuras sporal frequentemente evolve al pedunculos, ma quasi nunquam al folios ipse.

Cette maladie se manifeste par la croissance d'un mycélium blanc sur les feuilles, particulière-ment à la face supérieure. Au bout de quelque temps, les parties atteintes de la feuille devien-nent vert pâle à jaune ou couleur du tan. Lors d'une attaque grave, la feuille entière se recroque-ville et meurt. Plusieurs feuilles tombent. Des fructifications noires du champignon apparaissent fréquemment sur les tiges, mais presque jamais sur la feuille elle-même.

Esta enfermedad produce un crecimiento fungoso de color blanco sobre partes del follaje, es-pecialmente en la superficie superior. Después de que el hongo se has desarrollado por algún tiempo, las áreas foliares afectadas se vuelven de color verde pálido, amarillas or crema. En ataques severos la hoja entera se enrolla y muere y muchas se caen. En los tallos frecuente-mente se forman cuerpos fructíferos negros pero casi nunca en la propia hoja.

DISTRIBUTION: Af., A., Aust., Eur., N.A., C.A., S.A.

34

Brassica oleracea (Brassica spp.) + Fusarium conglutinans Wollenweber =

ENGLISH   Yellows
FRENCH   Fusariose, Flétrissement Fusarien de Signe
SPANISH   Marchitez, Amarillamiento
PORTUGUESE   Podridão do colo, Fusariose, Murcha de Fusarium
ITALIAN
RUSSIAN   увядание фузариозное, фузариоз, желтизна
SCANDINAVIAN
GERMAN   Welkekrankheit, Fusarium Welke
DUTCH
HUNGARIAN
BULGARIAN
TURKISH
ARABIC
PERSIAN
BURMESE
THAI   Hiaw โรคเหี่ยว
NEPALI
HINDI
VIETNAMESE   Vàng bắp-Héo cây
CHINESE
JAPANESE   Iō-byō 萎黄病

The first and most apparent symptom is a dull green to yellowish-green color of the leaves of plants in the seedbed or in the field within a month after transplanting. Young plants may turn yellow and die rapidly. The lower leaves develop a one-sided warping or curling, and the yellow is more intense on one side of the leaf midrib than on the other. The stem may be twisted toward one side. Yellowing, browning and dying of the leaves progresses up the leaf. Affected leaves drop prematurely and may die within a few weeks.

Le prime symptoma (e le plus remarcabile) es le coloration verde mat o jalnette del folios del plantas o in le seminario o in le campo intra un mense post transplantation. Plantas jevene pote devenir jalne e morir rapidemente. Le folios inferior monstra un torsion o crispation unilateral, e le color jalne es plus intense a un latere del nervatura central del folio que al altere latere. Le pedunculo es frequentement distorte a un latere. Le necrotic discoloration jalne o brun se avantia in le folios. Folios afflicte se separa e cade ante le tempore ordinari e frequentemente le planta mori intra alicun septimanas.

Le symptôme initial le plus apparent est l'aspect terne ou jaunâtre des feuilles des plantes dans les couches ou dans le champ au cours du mois qui suit la transplantation. Les jeunes plantes peuvent jaunir et mourir rapidement. Les feuilles inférieures ondulent ou s'enroulent sur un côté et le jaunissement est plus intense sur une moitié de la nervure principale que sur l'autre. La tige peut se tordre vers un côté. Le jaunissement, le brunissement et la mort des feuilles progressent vers le sommet. Les fueilles atteintes tombent prématurément et peuvent mourir en deça de quelques semaines.

El primer y más aparente síntoma es un color verde opaco a verde-amarilloso en las hojas de las plantas en el almácigo o en el campo después de un mes de haberlas transplantado. Las plantitas jóvenes pueden volverse amarillas y morir rápidamente. Las hojas inferiores desarrollan un enrollamiento en un lado y el amarillo es más intenso en una mitad de la hoja que en la otra, partiendo de la vena central. El tallo puede estar retorcido hacia un lado. El amarillamiento, obscurecimiento y muerte de las hojas progresa hacia arriba de la hoja. Las hojas afectadas se caen prematuramente y pueden morir en unas pocas semanas.

DISTRIBUTION: Af., A., Eur., N.A., C.A.

<u>Brassica</u> <u>oleracea</u> (Brassica spp.) + <u>Mycosphaerella</u> <u>brassicicola</u> (Duby) Johanson ex Oudem. =

ENGLISH	Ring Spot
FRENCH	Tache Annulaire
SPANISH	Mancha de anillo, Mancha de la hoja
PORTUGUESE	Padridão do colo, Mancha com pontos
ITALIAN	Tacche fogliari del cavolfiore
RUSSIAN	**пятнистость кольцевая микосфереллезная**
SCANDINAVIAN	Kaalens bladpletsyge
GERMAN	Blattfleckenkrankheit, Ringfleckenkrankheit
DUTCH	Bladvlekkenziekte
HUNGARIAN	
BULGARIAN	
TURKISH	Lahana yaprak halka lekesi
ARABIC	
PERSIAN	
BURMESE	
THAI	
NEPALI	
HINDI	
VIETNAMESE	
CHINESE	
JAPANESE	Rinmon-byô 輪紋病

The leaves and seed pods of the plants are attacked. The leaf veins as well as the parts between may show spotting. The outer lower leaves are more seriously affected than are the inner ones. The brown to tan areas are bordered by a green zone that retains its color even after the remainder of the leaf turns yellow. The spots vary greatly in size and are marked by numerous fruiting bodies arranged in concentric rings.

Le folios e le siliquas del plantas es attaccate. Le venas del folios e le portiones intra le venas frequentemente es maculate. Le inferior folios exterior es afflicte plus severmente que le folio del interior. Le areas brun o bronzate es delimitate per un zona verde le qual retine su coloration mesmo quando le resto del folio ha devenite jalne. Le maculas se varia multo in dimension e es marcate per numerose structuras sporal situate in anellos concentric.

Les feuilles et les siliques des plantes sont atteintes. Les nervures de la feuille aussi bien que les tissus adjacents peuvent porter des taches. Les feuilles inférieures du centre sont plus gravement atteintes que ne le sont celles de l'intérieur. Les plages brunes à beige sont bordées d'une zone verte qui conserve sa couleur même après que le reste de la feuille a jauni. La grosseur des taches varie beaucoup. De nombreuses fructifications disposées en anneaux concentriques caractérisent ces taches.

Las hojas y vainas de las plantas son atacadas. Las venas de la hoja, así como las partes entre ellas pueden mostrar manchas. Las hojas inferiores del centro son más seriamente afectadas que las de adentro. Las áreas café a canela están rodeadas por una zona verde que retiene su color aún después de que el resto de la hoja se vuelva amarilla. Las manchas varían grandemente en tamaño y están marcadas por numerosos cuerpos fructíferos dispuestos en anillos concéntricos.

DISTRIBUTION: Af., A., Aust., Eur., N.A., C.A., S.A.

36

Brassica oleracea (Lactuca sativa and other hosts) + Olpidium brassicae (Wor.) Dang. =

ENGLISH   Seedling Disease, Big Vein of Lettuce
FRENCH   Maladie des Plantules, Fonte des Semis, "Toile", Pied Noir
SPANISH   Enfermedad de la plántula
PORTUGUESE
ITALIAN   Marciume delle piantine di cavolo
RUSSIAN   "нерная ножка", ольпидиозная
SCANDINAVIAN
GERMAN   Schwarzbeiningkeit des Kohls, Umfallen, Olpidium-Fusskrankheit, Halsbrand
DUTCH
HUNGARIAN
BULGARIAN
TURKISH   Lahana fide yanıklığı
ARABIC
PERSIAN
BURMESE
THAI
NEPALI
HINDI
VIETNAMESE
CHINESE
JAPANESE   Konae-tatigare-byô   子苗立枯病

This pathogen causes a damping-off of seedlings by attacking the stem at or near the surface of the soil. The seeds may rot and the seedlings may decay before the plants emerge. Plants that survive the damping-off turn pale, then curl, wilt and collapse. The stems of the plants that damp-off are generally water-soaked at first, then turn gray to brown or black. Some plants are girdled by brown or black sunken cankers. Transplants make slow growth or die.

Iste pathogeno causa un putrefaction del plantulas, attaccante le pedunculo in o presso le super-facie del solo. Le semines pote putrer e le plantulas pote decader ante que le plantas emerge. Le plantas que supervive le putrefaction deveni pallide, e alora se crispa, marcesce, e suffre collapso. Le pedunculos de plantas putrefacte es generalmente saturate de aqua al principio e plus tarde de color gris usque a brun o nigre. Alicun plantas es circumcincte con canceres submerse de color brun o nigre. Si on los transplanta, lor crescimento es retardate o illos pote morir.

Ce pathogène cause une fonte des semis en attaquant la tige près de la surface du sol. Les graines peuvent pourrir et les plantules se décomposer avant la levée. Les plants qui survivent à la fonte deviennent pâles, puis s'enroulent, se flétrissent et s'affaissent. Les tiges des plants atteintes de la fonte sont imbibées d'eau au départ, puis tournent du gris au brun ou noir. Quelques plants sont ceinturées par des chancres profonds, bruns ou noirs. Les plants croissent lentement ou meurent.

Este patógeno causa mal de talluelo en las plántulas al atacar el tallo sobre o cerca de la superficie del suelo. Las semillas pueden podrirse y las plántulas también, antes de que las plantas emer-jan. Las plantas que sobreviven al mal de talluelo se palidecen, luego se enrollan, se marchitan y sufren un colapso. Los tallos de las plantas con mal de talluelo están generalmente empapa-das de agua primero, luego se vuelven gris a café o negras. Algunas plantas están rodeadas por chancros hundidos de color café o negro. Las plantas transplantadas crecen despacio o mueren.

DISTRIBUTION: Af., A., Aust., Eur., N.A., S.A.

Brassica spp. + Peronospora brassicae Gäumann (Peronspora parasitica Pers. ex Fr.) =

ENGLISH	Downy Mildew
FRENCH	Mildiou, Oïdium, Blanc
SPANISH	Mildiú, Mildiú felpudo, Mildiú velloso, Mildiú lanoso
PORTUGUESE	Mildio
ITALIAN	Peronospora delle crocifere, Mal secco
RUSSIAN	ложная мучнистая роса, пероноспороз
SCANDINAVIAN	Kålskimmel, Kålbladmögel, Kålbladskimmel
GERMAN	Falscher Mehltau
DUTCH	Valse Meeldauw
HUNGARIAN	Káposztaperonoszóra
BULGARIAN	мана
TURKISH	Lahana mildiyösü
ARABIC	Bayad Zaghaby بياض زغبى
PERSIAN	Sĕfĕdákĕ dŏrōoghēyĕ kálăm سفیدک دروغی کلم
BURMESE	Ywet-chauk roga ပ ဆ ၌ ၎ ၀ ၄ ၄ ၁ ၁ ၁ ၆ ၆ ၊ း
THAI	Ra-Num-Karng รานำคาง
NEPALI	Tori Ko dunshe rog तोरीको धुसे रोग
HINDI	Mriduromil Phaphundi मृदुरोमिल फाफुन्दी
VIETNAMESE	Ðốm phấn-Úa cây
CHINESE	Bái tsày lùh jìunn bìng 白菜露菌病, Gan lán lùh jìunn bìng 甘藍露菌病
JAPANESE	Beto-byð べと病

This disease is readily distinguished by the white fuzzy fungus growing in the lesions. The infections appear first as small dark spots which enlarge to indefinite, yellow areas. All aerial parts of the plant may be attacked. Infections in the lower leaves may result in internal discoloration and breakdown of the tissue. This type of symptom also may develop in the heads in storage.

Iste morbo ha un multo distinctive fungo lanuginose blanc que cresce in le lesiones. Le infectiones se manifesta al principio como minute maculas obscur le quales cresce a indefinite areas jalne. Tote le partes aeree del planta pote suffrer attacco. Infectiones in le folios inferior pote resultar in discoloration interne e collapso del texito. Tal symptomas pote evolver anque in le testas in magasinage.

Cette maladie se distingue facilement par le champignon blanc floconneux qui croît sur les lésions. Les infections apparaissent d'abord comme de petites taches foncées qui s'agrandissent pour former des plages jaunes indéfinies. Toutes les parties aériennes de la plante peuvent être attaquées. Les infections sur les feuilles inférieures peuvent causer la décoloration et la désintégration des tissus internes. Pareil symptôme peut aussi se présénter sur les pommes (de chou) en entrepôt.

Esta enfermedad se distingue fácilmente por el hongo blanco peludo que crece en las lesiones. Las infección aparece primero como pequeñas manchas obscuras que crecen y llegan a ser áreas amarillas indefinidas. Todas las partes aéreas de la planta pueden ser atacadas. Las infecciones de las hojas inferiores pueden causar una decoloración interna y descomposición del tejido. Este tipo de síntomas también puede desarrollarse en las cabezas que están almacenadas.

DISTRIBUTION: Af., A., Aust., Eur., N.A., C.A., S.A.

38

Brassica oleracea (Brassica rapa) + Phoma lingam (Fr.) Desm. =

ENGLISH	Black Leg, Leaf Spot
FRENCH	Jambe Noire, Pourriture Sèche et des Pieds, Phomose, Chou Moellier
SPANISH	Mancha de la hoja, Pié negro, Mancha foliar
PORTUGUESE	
ITALIAN	Cancro del fusto, Annerimento del gambo del cavelo, Marciume secco del navone
RUSSIAN	фомоз, гниль стеблей сухая
SCANDINAVIAN	
GERMAN	Umfallkrankheit, Fallsucht, Umfallen, Krebsstrünke, Trockenfäule
DUTCH	Vallers, Kankerstronken, Bladvlekken
HUNGARIAN	
BULGARIAN	сухо стъблено гниене
TURKISH	
ARABIC	
PERSIAN	Fōōmāye kàlàm فوُماى كلم
BURMESE	
THAI	Khang-Dum แฃงกำ
NEPALI	
HINDI	
VIETNAMESE	
CHINESE	Gan lán gen fǔ bìng 甘藍根腐病
JAPANESE	Nekuti-byō 根朽病

The plants may become infected in the seed bed or any time later. Usually the first symptom is an oval, depressed, light brown canker near the base of the stem. The canker enlarges until the stem is girdled. Circular, light brown spots appear on the leaves. Similar lesions elliptical in shape occur on the seed stalks and pods of seed plants. When the plants are severely diseased, they wilt or the edges of the leaves turn bluish-red in color. Older plants may lean or fall over.

Le plantas pote esser infectate o in le seminario o plus tarde. Usualmente le prime symptoma es un cancer oval, submerse, e brunette presso le pedunculo del stirpe. Iste cancer cresce usque a circumcinger le pedunculo. Maculas circular brunette se manifesta al folios. Lesiones similar de forma elliptic evolve al pedunculos e al siliquas de plantas. Quando le plantas es severmente infectate, illos marcesce o le margines del folios deveni de color blau e rubie. Le plantas plus vetule se inclina o cade a basso.

Les plantes peuvent subir l'infection en couche ou n'importe quand plus tard. D'ordinaire, le premier symptôme apparaît comme un chancre ovale, brun pâle, en dépressions près de la base de la tige. Le chancre s'agrandit jusqu'à ceinturer la tige. De petites taches circulaires brun pâle se forment sur les feuilles. Des lésions semblables de forme elliptique apparaissent sur les pédoncules et les siliques des porte-graine. Lorsque les plantes sont gravement atteintes, elles se flétrissent, ou la marge des feuilles devient rouge bleuâtre. Les vieilles plantes peuvent s'incliner et tomber.

Las plantas pueden infectarse en el semillero o en cualquier época después. Generalmente el primer síntoma es un cáncer ovalado, hundido, de color café claro cerca de la base del tallo. El cáncer crece hasta que rodea el tallo. Manchas circulares, de color café aparecen sobre las hojas. Lesiones elípticas similares aparecen en los tallos de la semilla y en las vainas de las plantas para semilla. Cuando las plantas están severamente enfermas, se marchitan o las orillas de las hojas se tornan de color rojo-azulado. Las plantas más viejas pueden inclinarse o caerse.

DISTRIBUTION: Af., A., Aust., Eur., N.A., C.A.

Brassica spp. + Plasmodiophora brassicae Woronin =

ENGLISH	Club Root
FRENCH	Hernie, Gros Pied, Maladie Digitoire
SPANISH	Hernia de la col, Hernia de la raíz
PORTUGUESE	Hérnia das cruciferas
ITALIAN	Ernia dei cavoli, Tubercolosi, Mal del gozzo
RUSSIAN	кила
SCANDINAVIAN	Kålbrok, Klumprotsjuka, Klumprot
GERMAN	Kohlhernie, Fingerkrankheit, Kropfkrankheit, Knotensucht, Hernie der Kohlgewachse
DUTCH	Knolvoet
HUNGARIAN	
BULGARIAN	гуша, кила
TURKISH	Boğaz uru
ARABIC	
PERSIAN	
BURMESE	
THAI	
NEPALI	
HINDI	Mudgar Mool Rog मदगर मूल रोग
VIETNAMESE	
CHINESE	Bái tsày gen lío bìng 白菜根瘤病
JAPANESE	Nekobu-byô 根こぶ病

Plants affected by this disease have yellowish or green leaves that wilt on hot days. Both the main and lateral roots are completely clubbed, resembling large nodules like those diseased by nematodes. The aerial parts of the plant usually wilt or are poorly developed. The infected roots disintegrate in the soil. The decaying roots give off a very characteristic odor which can be detected for a long distance.

Plantas afflicte con iste morbo ha folios jalnette or verde, le quales marcesce durante dies calorose. E le principal e le lateral radices es totalmente deformate, e appare similar a grande nodulos infectate per nematodos. Le partes aeree del planta usualmente marcesce o es mal developpate. Le radices infectate se disintegra in le terra. Le radices putrescente emitte un odor multo distinctive le qual es facilemente perceptibile de longe.

Les plantes atteintes de cette maladie ont des feuilles jaunâtres ou vertes qui se flétrissent durant les jours chauds. La racine principale et les racines secondaires sont complètement couvertes de hernies qui ressemblent à de grosses nodosités semblables à celles causées par les nématodes. Ordinairement, les parties aériennes de la plante se flétrissent ou sont peu développées. Les racines infectées se désintègrent dans le sol. Les racines en décomposition émettent une odeur très particulière qu'on peut reconnaître à distance.

Las plantas infectadas por esta enfermedad tienen hojas amarillentas o verdes que se marchitan en los días cálidos. Tanto la raíz principal como la lateral están llenas de hernias, semejando grandes nódulos, igual que las atacadas por nemátodos. Las partes aéreas de la planta generalmente se marchitan o se desarrollan pobremente. Las raíces infectadas se desintegran en el suelo. Las raíces podridas tienen un olor muy característico que puede sentirse a una gran distancia.

DISTRIBUTION: Af., A., Aust., Eur., N.A., S.A.

40

Brassica spp. + Pseudomonas maculicola (McCull.) Stev. =

ENGLISH	Leaf Spot
FRENCH	Tache Bactérienne, Taches des Feuilles
SPANISH	Mancha foliar
PORTUGUESE	
ITALIAN	Maculatura del cavolfiore
RUSSIAN	пятнистость бактериальная (цветной капусты)
SCANDINAVIAN	
GERMAN	Blattfleckenkrankheit, Baterielle Blattfleckenkrankheit
DUTCH	
HUNGARIAN	
BULGARIAN	бактериоза
TURKISH	Karnibahar yaprak leke hastalığı
ARABIC	
PERSIAN	
BURMESE	
THAI	
NEPALI	
HINDI	
VIETNAMESE	
CHINESE	
JAPANESE	Kokuhan-saikin-byō　黒斑細菌病

The diseased leaves are covered with numerous small, brown to purplish spots ranging from mere points to large lesions having a yellow halo around them. The dead areas unite and produce irregular blotches. The spots appear first on the lower side of the leaf. The infection frequently occurs in the vein of the leaf, retards growth and causes puckering. Severe infection causes the leaves to drop off.

Le folios morbide es coperte de numerose maculas minute, brun o purpuree, le quales se varia in dimension ab punctos minime usque a grande lesiones circumcincte per un halo jalne. Le areas morte se conjunge e produce pustulas irregular. Le maculas se manifesta al initio in le latere inferior del folio. Frequentemente le infection occurre in le vena del folio, impedi le crescimento, e causa corrugation. Infection sever causa que le folios cade.

Les feuilles malades sont couvertes de nombreuses petites taches brunes à pourpres formant de simples points jusqu'à de grandes lésions entourées d'un halo jaune. Les plages de tissus morts s'unissent pour former des éclaboussures irrégulières. Les taches apparaissent d'abord à la face inférieure de la feuille. L'infection se produit fréquemment sur la nervure de la feuille, retarde la croissance et cause la gauffrure. Une infection grave provoque la chute des feuilles.

Las hojas enfermas están cubiertas con numerosas manchas pequeñas, de color café a púrpura, que varían desde apenas unos puntos pequeñas hasta grandes lesiones que tienen un halo amarillo alrededor. Las áreas muertas se unen y producen parches irregulares. Las manchas aparecen primero en el envés de la hoja. La infección frecuentemente ocurre en la vena de la hoja, retarda el crecimiento y causa arrugas. La infección severa causa que las hojas se caigan.

DISTRIBUTION: Af., A., Aust., Eur., N.A., C.A.

Brassica oleracea (Brassica spp.) + Xanthomonas campestris (Pammel) Dowson (Pseudomonas campestris (Pammel) E. F. Smith) =

ENGLISH   Black Rot
FRENCH   Nervation Noire, Taches Foliaires Bactériennes
SPANISH   Podredumbre negra, Bacteriosis, Podredumbre bacteriana, Pudrición negra
PORTUGUESE   Podridão negra
ITALIAN   Batteriosi delle crocfere, Marciume nero
RUSSIAN   бактериоз сосудистый, гниль черная
SCANDINAVIAN   Brunbakteriose
GERMAN   Adernschwärze des Kohls, Brunfäule, Schwarzaderigkeit, Gefässbakteriose
DUTCH   Zwart Rot, Bacterieziekte, Zwartnervigheid
HUNGARIAN
BULGARIAN   черно гниене
TURKISH   Lahana siyah damar çürüklüğü
ARABIC   Affan Asswad غن اسود
PERSIAN   Bäktĕrĕuzĕ kálám باکتریوز کلم
BURMESE
THAI   Nao-Dum เน่าดำ
NEPALI   Banda Ko sadne rog बन्दाको सडने रोग , Tori Ko kuhine rog तोरीको कुहिने रोग
HINDI   Kala Viglan काला विगलान
VIETNAMESE   Thoi thâm-Héo lá
CHINESE   Gan lán hei fŭ bìng 甘藍黑腐病
JAPANESE   Kurogusare-byō 黒腐病

The first sign of the disease appears as a blackening of the veins upon the edges of the leaves. The affected region rapidly enlarges, the blackening extending toward the stalk and then throughout the stem tissue; thus gaining entrance to other leaves. Usually many leaves are infected simultaneously. Affected leaves soon yellow and wilt. The disease continues to develop while plants are in storage, resulting in complete loss of economic value.

Le prime symptoma de iste morbo es un nigration del venas al margines del folios. Le region afflicte rapidemente cresce, le color nigre extendente se usque al pedunculo e alora per tote le texito del pedunculo de maniera que illo penetra in altere folios additional. Usualmente un grande numero de folios es simultaneemente infectate. Folios afflicte rapidemente deveni jalne e marcesce. Le maladia continua su progresso in le plantas conservate in magasinage, causante perdition total de valor economic.

La maladie se manifeste d'abord par un noircissement des nervures à la marge des feuilles. La région affectée s'agrandit rapidement, le noircissement gagne la tige et il en envahit tous les tissus pour ensuite atteindre ainsi d'autres feuilles. D'ordinaire, plusieurs feuilles sont infectées simultanément. Les feuilles atteintes jaunissent tôt et se flétrissent. La maladie continue d'evoluer sur les plantes entreposées, ce qui entraîne une perte complète.

El primer síntoma de esta enfermedad aparece como un ennegrecimiento de las venas sobre las orillas de las hojas. La región afectada crece rápidamente, lo negro se extiende hacia el tronco y luego a través del tejido del tallo, logrando así entrar en las otras hojas. Generalmente bastantes hojas se ven infectadas simultáneamente. Las hojas afectadas se amarillan pronto y se marchitan. La enfermedad continúa desarrollándose mientras las plantas están almacenadas, dando como resultado una pérdida completa de valor económico.

DISTRIBUTION: Af., A., Aust., Eur., N.A., S.A.

42

Capsicum spp., (Cucurbitaceae, Lycopersicon esculentum) + Phytophthora capsici Leonian =

ENGLISH   Phytophthora Blight, Fruit Rot
FRENCH   Pourriture des Fruits, Mildiou du Piment
SPANISH   Podredumbre del fruto, Tizón, Mildiú, Pudrición basal, Marchitez, Tizón tardío
PORTUGUESE   Requeima
ITALIAN   Cancrena pedale del peperóne
RUSSIAN   гниль плодов фитлфторзная, фитофтороз
SCANDINAVIAN
GERMAN   Phytophthora-Fäule
DUTCH
HUNGARIAN
BULGARIAN
TURKISH   Biber mildiyösü
ARABIC
PERSIAN
BURMESE
THAI
NEPALI
HINDI
VIETNAMESE
CHINESE
JAPANESE   Eki-byð   疫病

All parts of the plant are affected. The invaded seedlings are damped-off and the older plants
have a root rot, stem canker, leaf blight, and fruit rot. The stem canker may be anywhere on
the plant from the small twigs to the base of the main stem. If the canker advances far enough,
the parts of the plant above the lesion wither and die. The leaf becomes water-soaked in a
rapidly enlarging spot which is covered by the white fruiting layer of the pathogen. The same
white mold marks the lesion on the fruit. Later the fruit lesion is depressed and bleached and
might be confused with sun scald.

Tote le partes del planta es afflicte. Le plantas juvene es invase per putrefaction fungal al pede,
e le plantas plus vetule monstra putrefaction del radices, cancere del stirpes, plaga del folios,
e putrefaction del fructos. Le cancere del stirpes pote situar se in qualcunque locos, ab le
rametttos le plus minute usque al pede del pedunculo principal. Si le cancere es assatis avantiate, le partes del planta que se trova super le lesiones marcesce e mori. Le folio appare
saturate de aqua in un macula le qual rapidemente se extende e le qual es coperte del blanc
strato sporal continente le pathogeno. Le mesme mucor blanc es observate in le lesion del
fructo. Plus tarde, le lesion del fructo es depresse e blanchite e facilemente se confunde
con escaldatura del sol.

Toutes les parties de la plante sont atteintes. Les plantules envahies subissent la fonte, alors que
les plantes adultes contractent la pourriture des racines, le chancre de la tige, la brûlure des
feuilles et la pourriture du fruit. Le chancre de la tige peut se présenter n'importe où, depuis les
petits rameaux jusqu'à la base de la tige principale. Si le chancre progresse assez loin, les parties
situées au-dessus de la lésion dépérissent et meurent. Sur les feuilles, des plages délavées
s'agrandissent rapidement et se couvrent du feutrage blanc fructifère du pathogène. Le même
moisissure blanche couvre la lésion sur le fruit. Plus tard la lésion du fruit se déprime et
blanchit, et risque de passer pour de l'échaudage.

Todas las partes de la planta son afectadas. Las plántulas afectadas tienen mal de talluelo y las
plantas más viejas tienen podredumbre de la raíz, chancro del tallo, tizón de la hoja y podredumbre del fruto. El chancro del tallo puede aparecer en cualquier parte de la planta desde
las pequeñas ramillas hasta la base del tallo principal. Si el chancro avanza lo suficiente,
las partes de la planta arriba de la lésion se ajan y mueren. La hoja se empapa en agua en
una mancha que crece rápidamente, la cuál está cubierta por la capa fructífera blanca del
patógeno. El mismo moho blanco marca la lésion en la fruta. Más tarde la lesión de la
fruta se hunde y decolora, pudiendo confundirse con una quemadura de sol.

DISTRIBUTION: A., Aust., Eur., N.A., C.A., S.A.

Carthamus tinctorius + Alternaria carthami Chowdhury =

ENGLISH   Leaf Spot
FRENCH   Tache Alternarienne, Alternariose, Taches Foliaires
SPANISH   Mancha foliar
PORTUGUESE
ITALIAN
RUSSIAN   альтернариоз
SCANDINAVIAN
GERMAN   Alternaria-Blattfleckenkrankheit
DUTCH
HUNGARIAN
BULGARIAN
TURKISH
ARABIC
PERSIAN
BURMESE
THAI
NEPALI
HINDI   Alternaria Rog   आल्टरनारिया रोग
VIETNAMESE
CHINESE
JAPANESE

Symptoms appear before flowering on all parts, but especially on the leaves. The brown to dark brown spots expand to one centimeter in diameter, the center, lighter in color, being surrounded by dark rings. Shot-holes may develop and unite if the spots lead to large irregular lesions. Spotting is less severe on the stems. Infected flower buds do not open, and they become shriveled.

Symptomas se manifesta ante floration in tote le partes, ma specialmente in le folios. Le maculas brun o obscur se extende usque a 1 centimetro de diametro. Le centro, de color plus clar, es circumcincte de anellos obscur. Perforationes pote occurrer e unir se, se le maculas cresce in grande lesiones irregular. Le pedunculos es minus severmente maculate. Buttones infectate non se aperi e es crispate.

Les symptômes apparaissent avant la floraison sur tous les organes, surtout sur les feuilles. Des taches brunes à brun foncé se forment, qui peuvent atteindre un centimètre de diamètre. Leur centre, plus pâle, est entouré d'anneaux foncés. Des trous peuvent apparaître et s'unir si les taches se transforment en de grandes lésions irrégulières. Les taches sont moins graves sur les tiges. Les boutons floraux infectés ne s'ouvrent pas et se ratatinent.

Los síntomas aparecen en todas partes antes de la floración, pero especialmente sobre las hojas. Las manchas de color café a café obscuro crecen hasta alcanzar un centímetro de diámetro; el centro es de color más claro y está rodeado de anillos obscuros. Pueden desarrollarse hoyos como de bala y unirse si las manchas llegar a ser lesiones irregulares grandes. Las manchas son menos severas en los tallos. Los botones florales infectados no se abren y se ajan.

DISTRIBUTION: Af., A., Eur., N.A.

44

Carthamus tinctorius (Carthamus spp.) + Puccinia carthami Corda =

ENGLISH	Rust
FRENCH	Rouille
SPANISH	Roya
PORTUGUESE	
ITALIAN	
RUSSIAN	ржавчина
SCANDINAVIAN	
GERMAN	Rost
DUTCH	
HUNGARIAN	
BULGARIAN	
TURKISH	
ARABIC	Sadaa صدا
PERSIAN	Zånge gôlèràng زنگ گلرنگ
BURMESE	
THAI	
NEPALI	
HINDI	Ratua रतुवा
VIETNAMESE	
CHINESE	
JAPANESE	Sabi-byô さび病

This disease occurs chiefly on leaves, but is also prevalent on roots and tender stems. Seedlings and older plants are equally susceptible. Affected plants initially show discoloration of leaves and this is later followed by drooping or wilting. In certain cases seedlings die without showing above-ground symptoms. Infection causes the older plants frequently to become girdled. Severe infection leads to reduction in stands and serious damage to production.

Iste morbo occurre principalmente al folios, ma se trova frequentemente anque in radices e in le pedunculos tenere. Le plantulas e le plantas plus vetule es equalmente susceptibile. Plantas afflicte monstra al initio un discoloration de folios e, plus tarde, le planta langue o marcesce. In alicun casos, plantulas mori sin monstrar symptomas in le partes supra le terreno. Le plantas plus vetule es frequentement circumcincte de lesiones. Infection sever diminue le plantation e causa un grande reduction de production.

Cette maladie apparaît surtout sur les feuilles, mais elle sévit aussi sur les racines et les tiges tendres. Les plantules et les plantes plus agées sont également susceptibles. Au début, les plantes atteintes accusent une décoloration des feuilles et par la suite, s'affaissent ou se flétreissent. Dans certains cas, les plantules meurent sans présenter de symptômes au-dessus du sol. Souvent, les plantes adultes deviennent complètement recouvertes de rouille. Une infection grave diminue les peuplements et déprécie beaucoup la récolte.

Esta enfermedad ocurre principalmente en las hojas, pero también puede atacar las raíces y tallos tiernos. Las plántulas y las plantas grandes son igualmente susceptibles. Las plantas afectadas inicialmente muestran decoloración de las hojas y ésto más tarde es seguido por decaimiento o marchitez. En ciertos casos las plántulas mueren sin mostrar síntomas en la parte aérea. Las plantas mayores frecuentemente se ven rodeadas con una infección severa. La infección severa causa una reducción de la población y daños en la producción.

DISTRIBUTION: Af., A., Eur., N.A.

Castanea dentata + Endothia parasitica (Murrill) Anderson & Anderson =

ENGLISH	Blight, Canker
FRENCH	Brûlure du Châtaigner, Maladie Chancreuse, Chancre
SPANISH	Tizón, Cancro
PORTUGUESE	
ITALIAN	Seccume del castágno, Cancro americano, Mal della corteccia
RUSSIAN	ожог коры, рак коры эндотиевый
SCANDINAVIAN	
GERMAN	Kastaniensterben
DUTCH	
HUNGARIAN	Endotiás rákosodás
BULGARIAN	
TURKISH	Kestane kanseri
ARABIC	
PERSIAN	
BURMESE	
THAI	
NEPALI	
HINDI	
VIETNAMESE	
CHINESE	
JAPANESE	Dōgare-byō 胴枯病

The attack occurs upon the bark through wounds, but twigs and leaves are not directly affected. From the point of attack the fungus spreads in all directions until the diseased parts meet on the opposite side of the branch, thus girdling the twig. Dead, discolored, sunken patches with numerous yellow, orange, or reddish-brown pustules are produced. Cankers enlarge rapidly and death soon occurs. If small branches are first diseased, the tree may continue to live for a few years.

Le attacco occurre in le cortice per vulneres, ma le ramettos e le folios non es directmente afflicte. Ab le loco initial de attacco, le fungo se extende in omne directiones de maniera que le partes morbide coalesce al latere opposite del ramo, circumcingente lo. Areas morte, discolorate, e depresse evolve con numerose pustulas jalne, orange, o brun e rubiette. Canceres cresce rapidemente e le arbore mori in breve tempore. Si le attacco comencia in le ramettos plus minute, le arbore pote superviver per alicun annos.

Le pathogène envahit l'écorce par des blessures, mais les rameaux et les feuilles ne sont pas atteints de façon directe. Du point d'entrée, le champignon croît dans tous les sens jusqu'à ce que les parties malades se rejoignent sur le côté opposé de la branche, encerclant ainsi le rameau. Il se produit alors des plages nécrotiques, décolorés et creuse, portant de nombreuses pustules jaunes, orange ou braun rougeâtre se produisent. Les chancres s'agrandissent rapidement et la mort ne tarde pas. Si les petites branches sont attaquées les premières, l'arbre peut continuer à vivre encore quelques années.

El ataqué ocurre sobre la corteza a través de heridas, pero las ramitas y hojas no son directamente afectadas. Desde el punto de ataque el hongo se disemina en todas las direcciones hasta que las partes enfermas se encuentran en la parte opuesta de la rama, rodeando así la ramita. Se producen parches muertos, decolorados, hundidos, con numerosas pústulas amarillas, anaranjadas o rojizas. Los cancros crecen rápidamente y pronto ocurre la muerte. Si las ramas pequeñas son atacadas primero, el árbol enfermo puede continuar viviendo por algunos años.

DISTRIBUTION: A., Eur., N.A.

46

Castanea spp. + Phytophthora cambivora (Petri) Buisman =

ENGLISH	Ink Disease
FRENCH	Maladie de l'encre, Pourriture
SPANISH	Enfermedad de tinta
PORTUGUESE	
ITALIAN	Mal dell'inchiostro del castágno
RUSSIAN	**гниль корней фитофторозная**
SCANDINAVIAN	
GERMAN	Tintenkrankheit, Phytophthora-Wurzelfäule
DUTCH	
HUNGARIAN	
BULGARIAN	
TURKISH	Kestane mürekkep hastalığı
ARABIC	
PERSIAN	
BURMESE	
THAI	
NEPALI	
HINDI	
VIETNAMESE	
CHINESE	
JAPANESE	

The fungus advances along roots and root collars in wedge-shaped streaks, turning the invaded bark tissues brown or green and girdling roots and stems. Root lesions exude an inky-blue substance that stains the soil close to the roots. Probably the most noticeable symptom of this disease is the dying of large numbers of trees growing in certain types of soil at low elevations and where there is no evidence of a specific cause.

Le fungo se avantia per le radices e le collares del radices como cuneate venas de color, tintante le texitos invase del cortice brun o verde e circumcingente radices e truncos. Lesiones del radices exsuda un substantia blau o nigre le qual colora le solo presso le radices. Lo que es probabilemente le symptoma le plus remarcabile de iste morbo es le mortification, sin evidente causa specific, de un grande numero de arbores le quales es situate in un type particular de solo in terrenos basse.

Le champignon progresse le long des racines et du collet en stries cunéiformes. Il colore en brun ou vert les tissus infectés de l'écorce, et encercle les racines et les tiges. Les lésions sur les racines produisent une substance bleu noir qui colore le sol près des racines. Le symptôme le plus remarquable de cette maladie est probablement la mortalité d'un grand nombre d'arbres sur certains types de sols en terrains bas, où l'on ne peut soupçonner aucune cause spécifique.

El hongo avanza a lo largo de las raíces y cuello de las raíces en vetas de forma de cuña; el tejido invadido de la corteza se vuelve café o verde y rodea las raíces y tallos. Las lesiones radiculares exudan una substancia de color azul negro que mancha el suelo cercano a las raíces. Probablemente el síntoma más notorio de esta enfermedad es la muerte de muchos árboles que crecen en ciertos tipos de suelo a baja elevación y donde no hay evidencias de una causa específica.

DISTRIBUTION: Af., A., Aust., Eur., N.A.

Citrus spp. + Alternaria citri (Ellis & Pierce) Pierce (Alternaria mali Roberts) =

ENGLISH	Black Rot of Oranges, Fruit Rot of Lemons and Tangerines, Leaf Spot of Rough Lemon and Emperor Mandarin
FRENCH	Pourriture Noire des Fruits, Alternariose
SPANISH	Mancha de ojo, Podredumbre interna del fruto, Pudrición negra, Mancha foliar
PORTUGUESE	
ITALIAN	
RUSSIAN	гниль плодов черная, альтернариоз
SCANDINAVIAN	
GERMAN	Schwarzfäule, Alternaria-Fäule
DUTCH	
HUNGARIAN	
BULGARIAN	
TURKISH	
ARABIC	
PERSIAN	
BURMESE	
THAI	
NEPALI	
HINDI	
VIETNAMESE	
CHINESE	
JAPANESE	kurogusare-byô 黒腐病

On fruit in storage, the fungus usually proceeds from the stem end down the core producing rot in the center in which the fruit may appear sound on the surface. These fruits may be easily crushed and made to exude juice and disintegrated tissue on slight pressure. Some fruits show a soft to firm stem-end rot. In the orchard the fruit may drop prematurely and this condition seems to be associated with certain unfavorable weather.

In fructo conservate in magasinage, le fungo usualmente progrede ab le pedunculo e penetra in le pulpa, causante putrefaction in le centro ben que le fructo pote semblar esser salubre al superfacie. Iste fructos es facilemente contundibile, e on facilemente face los exsudar succo e texito disintegrate per un pression legier. Alicun fructos monstra un putrefaction o molle o dur presso le pedunculo. In le verdiero, le fructo pote separar se prematurmente e cader-- un condition que se associa probabilemente con tempore infavorabile.

Sur le fruit en entrêpot, le champignon se développe à partir du pédoncule jusqu'au coeur a l' intérieur duquel il produit une pourriture, alors que le fruit reste apparemment sain à l'ex- térieur. Les fruits peuvent facilement s'écraser et, sous une faible pression, libérer leur jus avec des tissus désintégrés. Quelques fruits exhibent une pourriture du calice molle à ferme. Dans le verger, le fruit peut tomber prématurément, ce qui semble lié à des conditions clima- tiques défavorables.

En las frutas almacenadas, el hongo generalmente progresa desde el final del tallo abajo hacia el corazón, produciendo podredumbre en el centro en el cuál la fruta puede aparecer sana en la superficie. Estás frutas pueden apretarse fácilmente y hacer que exuden jugo y tejido desinte- grado con poca presión. Algunas frutas muestran una podredumbre de suave a firme al final del tallo. En el huerto la fruta puede caerse prematuramente y esta condición parece estar asociada con tiempo no favorable.

DISTRIBUTION: Af., Aust., N.A.

Citrus spp., (Hevea brasiliensis, Musa sapientum, Theobroma cacao) + Botryodiplodia theobromae Pat. (Diplodia natalensis P. Evans) =

ENGLISH	Die-back, Stem-end Rot, Botryodiplodia Rot of Fruit, Charcoal Rot
FRENCH	Pourriture des Fruits, Desséchement des Branches
SPANISH	Muerto de brote terminales, Muerte progresive, Pudrición basal de la fruta
PORTUGUESE	Podridão preta, Podridão de Diplodia
ITALIAN	
RUSSIAN	диплоиоз, гниль плодов диплодиозная и усыхание ветвей диплодиозное
SCANDINAVIAN	
GERMAN	Diplodia-Fruchtfäule, Diplodia-Zweigsterben
DUTCH	
HUNGARIAN	
BULGARIAN	
TURKISH	
ARABIC	
PERSIAN	
BURMESE	
THAI	Yang-Lai ยางไหล
NEPALI	
HINDI	
VIETNAMESE	Thối thân - Rỉ nhựa Khô cành
CHINESE	Gan jýu tỉ fủ bìng 柑橘蒂腐病, Gan jýu lio jaw bìng 柑橘流膠病
JAPANESE	Hetagusare-byō へた腐病

This disease occurs as small patches of dead bark accompanied by gum, followed by partial healing, leaving a wound or scar. In severe cases large patches of bark are killed, and the wood underneath is blackened and killed. The fruit exhibit a slight off-color and increased pliability at the stem which is soon followed by a leathery condition with a brown water-soaked discoloration. These symptoms often progress over the sides in wide dark bunches corresponding to the division of internal segments of the fruit.

Iste morbo se manifesta como minute areas de cortice morte in conjunction con gumma, le quales partialmente se cura, lassante un vulnere o un cicatrice. In casos sever, extensive areas de cortice mori e le ligno que se trova intra es anque nigrate e morte. Le fructos monstra un pauc de discoloration e un augmentation de flexibilitate al pedunculo. In breve tempore, lor pelles semblar esser coriacee, con un brun discoloration saturate de aqua. Iste symptomas frequentemente se avantia super le pelle in gruppos large e obscur, le quales corresponde al divisiones interne del segmentos del fructo.

Cette maladie se manifeste sous forme de petits morceaux d'écorce morte portant de la gomme. Suit une guérison partielle qui laisse une plaie ou une cicatrice. Dans les cas graves, de grandes pièces d'écorce sont détruites, et le bois sous l'écorce noircit et meurt. La couleur du fruit s'affadit un peu, et le pédoncule devient plus flexible. La pelure devient coriace, comme délavée et brune. Souvent, les symptômes évoluent sur la pelure de façon à former des secteurs sombres qui correspondent aux segments internes du fruit.

Esta enfermedad aparece con pequeñas porciones de corteza muerta acompañada de goma, seguida de una cicatrización parcial que deja una herida o cicatriz. En casos severos, mueren porciones grandes de corteza y la madera debajo se ennegrece y muere. La fruta muestra un color un poco desteñido y aumento de flexibilidad en el pedúnculo y pronto es seguida por una condición coriácea con una decoloración color café y acuosa. Estos síntomas progresan frecuentemente hacia los lados en manojos anchos y obscuros correspondiendo a la división de segmentos internos del fruto.

DISTRIBUTION: Af., N.A., S.A.

Citrus spp. + Diaporthe citri (Fawcett) Wolf (Phomopsis citri Fawc.) =

ENGLISH	Melanose of Fruit and Foliage, Phomopsis Rot, Gummosis
FRENCH	Melanose, Taches Foliaires, Pourriture des Fruits
SPANISH	Manchas de las hojas, Melanosis, Muerte descendente, Podredumbre de Phomopsis
PORTUGUESE	Melanose, Podridão peduncular
ITALIAN	Melanosi degli agrumi
RUSSIAN	меланоз листьев и побегов, гниль плодов фомопсисная
SCANDINAVIAN	
GERMAN	Melanose, Fleckenbildungen, Gummose, Phomopsis-Fruchtfäule
DUTCH	
HUNGARIAN	
BULGARIAN	
TURKISH	
ARABIC	
PERSIAN	
BURMESE	
THAI	Melanose เมลาโนส
NEPALI	
HINDI	
VIETNAMESE	
CHINESE	Gan jýu sha péa bǐng 柑橘沙皮病
JAPANESE	Kokuten-byô 黒点病

This disease occurs on fruit, leaves and small twigs. The symptoms consist of small, raised, superficial dots or pustules made up of gum, often arranged in lines, curves, rings; the irregularly shaped spots around the margin lines of breakage give an appearance, in miniature, of dry, caked mud. Lesions are wax-like and amber-brown, to dark brown to nearly black. To the touch they suggest coarse sandpaper. The pustules often form a tear-streaking pattern.

Iste morbo occurre in fructos, folios, e ramettos minute. Le symptomas es minime punctillos o pustulas elevate, composite de gumma e frequentemente situate in lineas, curvas, o anellos, e maculas irregular circum le margines. Lineas interrupte causa le apparition, in miniatura, de fango disiccate e incrustate. Lesiones es ceree e del color brun de ambra, o de color brun obscur usque a quasi nigre. Al tacto illos evoca le sensation de papiro abrasive. Le pustulas frequentemente pinctura un designo de tracias de lacrimas.

Cette maladie apparaît sur le fruit, les feuilles et les petits rameaux. Les symptômes consistent en de petits points superficiels ou en des pustules soulevées, formées de gomme et souvent disposées en rangées, en courbes ou en cercles; des taches de forme irrégulière, à la marge de ces groupes de points ou pustules, rappellent des îlots de boue sèche. Les lésions sont cireuses, ambre brun à brun foncé ou presque noires. Au toucher, elles donnent l'impressions de papier sablé grossier. Souvent, la disposition des pustules évoque une traînée de larmes.

La enfermedad ataca los frutos, hojas y brotes pequeños. Los síntomas consisten en pequeños montículos o pústulas pequeñas llenos de goma. Estas son levantadas, superficiales y frecuentemente dispuestas en líneas, curvas y anillos. Las manchas irregulares alrededor de las líneas les dan una apariencia, en miniatura, de un pastel de lodo seco. Las lesiones parecen ser de cera y color café ambar hasta café obscuro o casi negro. Al tocarlas dan la impresión de estar hechas de lija. Las pústulas frecuentemente forman un patrón como de líneas de lágrimas.

DISTRIBUTION: Af., A., Aust., Eur., N.A., C.A., S.A.

Citrus spp. + Elsinoe fawcettii Bitancourt & Jenkins (Sphaceloma fawcettii Jenkins) =

ENGLISH	Scab
FRENCH	Gale des Agrumes, Maladie des Verrues
SPANISH	Roña de los cítricos, Sarna
PORTUGUESE	Verrugose, Sarna dos citro
ITALIAN	Scabbia del limone e dell'arancio
RUSSIAN	бородавчатость, парша, антракпоз пятнистый
SCANDINAVIAN	
GERMAN	Schorf, "Verrucosis", Flecken-Anthraknose
DUTCH	
HUNGARIAN	
BULGARIAN	
TURKISH	
ARABIC	
PERSIAN	
BURMESE	
THAI	Scab สะแคป
NEPALI	Kagati Ko dade rog कागतिको दढे रोग
HINDI	
VIETNAMESE	Ghe nhám trên cành lá và quả
CHINESE	Gan jyu chwong já bìng 柑橘瘡痂病
JAPANESE	Sôka-byô そうか病

On fruit, the lesions consist either of corky projections with distortions or of slightly raised scabs without distortion depending upon the particular type of fruit involved. The surface of lesions of considerable size often breaks up into small scabs. On leaves the lesions consist, when young, of small dots that become sharply defined, flat or somewhat depressed at the center. Leaves often become distorted, wrinkled, stunted, and misshapen. The surface of the lesions becomes a dusky color with age.

In fructos, le lesiones es composite o de projectiones corcose con distortiones o de crustas un pauc elevate sin distortiones, secundo le genere particular del fructo. Le superfacie del plus grande lesiones frequentemente se disintegra in crustas minute. In folios, le lesiones juvene es composite de minime punctillos le quales deveni acutemente definite, platte, o un pauc depresse al centro. Folios frequentement deveni distorte, corrugate, e deformate e monstra obstruction de crescimento. Le superfacie del lesiones deveni de color obscur post qualque tempore.

Sur le fruit, les lésions consistent soit en projections liégeuses avec distorsion, soit en galles légèrement soulevées sans distorsion, selon le type particulier de fruit impliqué. La surface des grandes lésions se sépare souvent en petites galles. Sur les feuilles, les lésions, lorsque jeunes, consistent en petits points qui deviennent très définis, plats et quelque peu déprimés au centre. Souvent les feuilles deviennent distorses, plissées, rabougries et malformées. La surface des lésions s'assombrit avec l'âge.

En los frutos las lesiones consisten ya sea de proyecciones corchosas con distorsiones o de roñas un poco protuberantes sin distorsión, dependiendo del tipo particular de fruta involucrado. La superficie de las lesiones de tamaño considerable frecuentemente se parte en pequeñas roñas. En las hojas las lesiones consisten, cuando son tiernas, en pequeños puntos que se vuelven bien definidos, planos o algo hundidos en el centro. Las hojas frecuentemente se vuelven torcidas, arrugadas, achaparradas y mal formadas. La superficie de las lesiones se vuelve de color pardo al envejecer.

DISTRIBUTION: Af., A., Aust., Eur., N.A., C.A., S.A.

Citrus aurantifolia + Gloeosporium limetticola Clausen =

ENGLISH	Anthracnose, Withertip
FRENCH	Anthracnose
SPANISH	Muerte de brotes terminales, Antracnosis, Punta mustia, Marchita
PORTUGUESE	Antracnose do limoeiro Galego.
ITALIAN	
RUSSIAN	антракноз
SCANDINAVIAN	
GERMAN	Anthraknose, Zweigspitzendürre
DUTCH	
HUNGARIAN	
BULGARIAN	
TURKISH	
ARABIC	
PERSIAN	
BURMESE	
THAI	Anthracnose แอนแทรคโนส
NEPALI	
HINDI	
VIETNAMESE	
CHINESE	
JAPANESE	Syôko-byô しょう枯病

The fungus attacks young shoots, leaves and fruits while they are still tender. The infected twigs wither and shrivel at the tips. They sometimes are girdled by an attack farther back, fall over and hang lifelessly. Young leaves which are sometimes infected but not entirely killed back may have dead areas on the margins or tips. Unopened buds, when attacked, may fail to develop. The petals turn brown and the buds fall off. Fruits that are attacked but not shed show lesions varying from scab-like spots to cankers.

Le fungo attacca sectiones crescente, folios, e fructos dum illos es ancora tenere. Le ramettos infectate marcesce e se crispa al extremitates. A vices illos es circumcincte per un attacco que se trova plus basse, in qual casos illos se inclina e pende sin vigor. Folios juvene que es a vices infectate, ma non totalmente mortefacte, pote monstrar areas morte al margines o al extremitates. Buttones inaperte, si attaccate, pote non developpar se. Le petalos deveni brun e le buttones se separa e cade. Le fructos que esseva attaccate ma non ha essite jectate monstra lesiones le quales se varia ab maculas similar a crustas a canceres.

Le champignon attaque les jeunes tiges, les feuilles et les fruits, lorsqu'ils sont encore tendres. Les bouts des rameaux infectés se flétrissent et se dessèchent. Parfois, ils sont encerclés à la suite d'une infection venant de plus bas; ils tombent alors et pendent sans vie. Les jeunes feuilles qui sont parfois infectées mais pas complètement mortes peuvent présenter des plages mortes sur les bords ou aux extrémités. S'ils sont attaqués avant l'éclosion les boutons floraux peuvent ne pas s'ouvrir, les pétales deviennent bruns et les bourgeons tombent. Les fruits malades mais non tombés montrent des lésions variant depuis des points de tavelure jusqu'aux chancres.

El hongo ataca los brotes jóvenes, las hojas y frutos cuando aún están tiernos. Las ramitas infectadas se marchitan y ajan en las puntas. Algunas veces están rodeadas por un ataque hacia abajo, se caen y cuelgan sin vida. Las hojas jóvenes que algunas veces están infectadas pero no del todo muertas pueden tener áreas muertas en los márgenes o puntas. Los capullos sin abrir, cuando son atacados, puede que no lleguen a abrirse, los pétalos se vuelven de color café y los capullos se caen. Las frutas que están atacadas pero no desprendidas muestran lesiones que varían desde manchas parecidas a las de la roña hasta cancros.

DISTRIBUTION: Af., A., Aust., Eur., N.A., C.A., S.A.

52

Citrus spp. + Guignardia citricarpa Kiely (Phoma citricarpa McAlp.) =

ENGLISH	Black Spot
FRENCH	Taches sur les Fruits, Phomose
SPANISH	Mancha negra
PORTUGUESE	
ITALIAN	
RUSSIAN	**пятнистость плодов черная фомозная**
SCANDINAVIAN	
GERMAN	Phoma, Fruchtfleckigkeit
DUTCH	
HUNGARIAN	
BULGARIAN	
TURKISH	
ARABIC	
PERSIAN	
BURMESE	
THAI	
NEPALI	
HINDI	
VIETNAMESE	
CHINESE	Gan jýu hei shing bǐng 柑橘黑星病
JAPANESE	Kurohosi-byô 黒星病

This disease is characterized by three different types of symptoms on the fruit. At first there are numerous circular brown spots with slight depressions in the center which turn gray-white with black margins and are surrounded by a ring of green tissue. Another type of spot develops after the hard spot phase with abundant lesions, small deep orange to brick red, finally brown, lacking a green ring. The third type of spot is irregular and spreads rapidly, black in the center, brown near the edge, finally red forming the margin of the sunken lesions. It also occurs on leaves, twigs and flowers with varied symptoms.

Iste morbo se characterisa per tres generes differente de symptomas in le fructos. Al initio, il ha numerose maculas circular brun, un pauc depresse al centro, le quales deveni gris o blanc con margines nigre e circumcincte per anellos de texito verde. Post le stadio del maculas dur, un secunde genere de maculas se developpa, le quales monstra un grande numero de minute lesiones orange, rubie, e al fin brun, sin le anello verde. Le tertie genere de maculas es irregular e se extende rapidemente, nigre al centro, brun presso le bordo, e al fin rubie al margines del lesion depresse. Iste morbo occurre anque in folios, ramettos, e flores con symptomas diverse.

Cette maladie est caractérisée par trois types différents de symptômes sur le fruit. Il se forme d'abord de nombreuses taches circulaires brunes, légèrement déprimées au centre, qui deviennent gris blanc à marges noires et sont entourées d'un anneau de tissu vert. Après la dure phase de ces taches, un autre type de taches se dessine. Il consiste en des lésions abondantes, petites, de couleur orange foncé à rouge brique, finalement brunes et dépourvues d'anneau vert. Les taches du troisième type sont irrégulières et s'étendent rapidement. Elles sont noires au centre, brunes près de la marge et finalement rouges pour former la marge de la lésion déprimée. La maladie apparait aussi sur les feuilles, les rameaux et les fleurs avec des symptômes variés.

Esta enfermedad se caracteriza por tres tipos diferentes de síntomas en el fruto. Primero aparecen numerosas manchas circulares de color café, con depresiones leves en el centro, las cuáles se vuelven de color gris-blanco, con márgenes negros y están rodeadas por un anillo de tejido verde. Después de la fase de la mancha dura se desarrolla otro tipo de mancha, con lesiones abundantes, pequeñas de color anaranjado intenso a rojo ladrillo, finalmente café, sin un anillo verde. El tercer tipo de mancha es irregular y crece rápidamente; es negra en el centro, café cerca de la orilla y finalmente roja formando el margen de la lesión hundida. También aparece sobre las hojas, ramitas y flores con síntomas variados.

DISTRIBUTION: Af., A., Aust., Eur., N.A., C.A., S.A.

Citrus spp. + Oospora citri-aurantii (Ferraris) Sacc. & Syd. =

ENGLISH	Sour Rot
FRENCH	Pourriture Amère des Fruits, Pourriture Acide
SPANISH	Podredumbre amarga del fruto, Podredumbre ácida
PORTUGUESE	Podridão amarga
ITALIAN	Marciume acido degli agrumi, Marciume amaro
RUSSIAN	гниль плодов ооспорозная
SCANDINAVIAN	
GERMAN	Berührungsfäule, Oospora-Fruchtfäule
DUTCH	
HUNGARIAN	
BULGARIAN	
TURKISH	
ARABIC	
PERSIAN	
BURMESE	
THAI	
NEPALI	
HINDI	
VIETNAMESE	
CHINESE	Gan jýu swuan fú bìng 柑橘酸腐病
JAPANESE	Sirakabi-byô 白かび病

The decaying areas on mature normally colored fruit may at first become creamy yellow, then dark yellow and finally light buff. The fruit surface then becomes wrinkled and water-soaked. The disease readily spreads to other fruit by contact. Diseased fruit usually attracts insects, especially fruitflies. The fungus probably enters the fruit through injuries.

Le areas putrescente in le maturate fructos de color normal pote devenir al initio jalne como crema, alora jalne obscur, e al fin castanie clar. Le superfacie del fructo alora es corrugate e saturate de aqua. Le morbo facilemente se extende a altere fructos per contacto. Le fructos morbide es attractive a insectos, specialmente muscas mediterranee de fructos. Le fungo probabilemente entra in le fructo per lesiones.

Les plages en décomposition sur un fruit mûr de couleur normale peuvent d'abord devenir jaune crème, puis jaune foncé et finalement chamois pâle. La surface du fruit se plisse alors et devient aqueuse. Par contage, la maladie se propage rapidement à d'autres fruits. En général, les fruits malades attirent les insectes, particulièrement les mouches à fruits. Il est probable que le champignon pénétre dans le fruit par des blessures.

El área podrida en frutos maduros normalmente coloreados puede primero volverse de color amarillo cremoso, luego amarillo obscuro y finalmente ante claro. La superficie de la fruta se vuelve ajada y empapada de agua. La enfermedad se disemina fácilmente hacia otras frutas en contacto. La fruta enferma generalmente atrae insectos, especialmente moscas de la fruta. El hongo probablemente entra en la fruta a través de lesiones.

DISTRIBUTION: Af., A., Aust., Eur., N.A., C.A., S.A.

Citrus spp. + *Penicillium italicum* Wehmer =

ENGLISH	Blue Mold
FRENCH	Moisissure Bleue des Agrumes et Fruits
SPANISH	Podredumbre azul, Pudrición del fruto, Moho azul
PORTUGUESE	Podridão azul
ITALIAN	Muffa azzurra
RUSSIAN	гниль плодов плесневидная годубая
SCANDINAVIAN	
GERMAN	Blau Fruchtfäule
DUTCH	
HUNGARIAN	
BULGARIAN	
TURKISH	
ARABIC	Affan غــن
PERSIAN	
BURMESE	
THAI	Rok-Pol-Noa โรคผลเน่า
NEPALI	
HINDI	
VIETNAMESE	
CHINESE	
JAPANESE	aokabi-byð 青かび病

At first a water-soaked, soft area appears on the fruit, which is easily punctured by pressure of the finger. Usually the initial spots covered by the blue fungus growth enlarge slowly and develop into a mixture of white and blue color which eventually covers large areas surrounded by a narrow white border. It is not uncommon for this disease to be associated on the same fruit with another related fungus which produces a green mold.

Al initio se manifesta in le fructo un area molle e saturate de aqua, le qual es facilemente puncturate per presion de un digito. Usualmente le maculas initial, le quales es coperte del fungo blau, cresce lentemente e deveni un mixtura del colores blanc e blau que, plus tarde, coperi grande areas circumcincte per un stricte bordo blanc. Frequentemente, iste morbo occurre al mesme fructo con un altere fungo associate le qual produce un mucor verde.

Au départ, une plage molle imbibée d'eau apparaît sur le fruit qui peut être facilement percé sous la pression du doigt. Habituellement, les taches initiales couvertes par la croissance bleue du champignon s'agrandissent lentement et se développent en un mélange de couleurs blanche et bleue. Eventuellement, elles couvrent de larges étendues entourées par une bordure étroite blanche. Il n'est pas rare que cette maladie soit associée, sur le même fruit, avec un champignon apprenté qui produit une moisissure verte.

Primero aparecen en el fruto áreas suaves, empapadas de agua, que se perforan fácilmente con la presión del dedo. Generalmente las manchas iniciales cubiertas con el crecimiento azul del hongo se agrandan despacio y se desarrollan en una mezcla de color blanco y azul, que eventualmente cubre grandes áreas rodeadas por un angosto borde blanco. Es común que esta enfermedad esté asociada, en el mismo fruto, con otro hongo parecido el cuál produce un moho verde.

DISTRIBUTION: Af., A., Aust., Eur., N.A., C.A., S.A.

Citrus spp. (and related genera) + Phytophthora citrophthora (R. E. Smith & E. H. Smith) Leonian (Phytophthora citricola Sawada) =

ENGLISH   Brown Rot of Fruit, Brown Rot Gummosis
FRENCH   Gommose, Pourriture Brun des Fruits, Taches Foliaires, Pourriture du Collet
SPANISH   Podredumbre morena del fruto, Gomosis, Podredumbre café del fruto
PORTUGUESE   Gomose de Phytophthora, Podridão do pé das laranjeiras
ITALIAN   Marciume radicale degli agrumi, Gommosi dei limoni, Marciume nero
RUSSIAN фитофтороз, гоммоз, гниль корневой шейки фитофторозная
SCANDINAVIAN
GERMAN   Bräunfaule der Früchte, Phytophthora-Blattfleckenkrankheit, Gummosis
DUTCH
HUNGARIAN
BULGARIAN
TURKISH
ARABIC
PERSIAN
BURMESE
THAI
NEPALI
HINDI
VIETNAMESE
CHINESE   Gan jýu jýu fǔ bǐng   柑橘裾腐病
JAPANESE   Kassyokuhuhai-byð   褐色腐敗病

The symptoms caused by this disease are varied, depending on the part of the plant affected. The first visable sign of infection is gum on the bark of the trees which appears near the groundline. If the bark is scraped off, brown infected areas reaching to the wood will be seen. The spots eventually dry and crack. A distinguishing feature is that one side of the tree may die and the other remain sound. Infection appears on the fruit as small dark spots which turn various shades of brown and the tissue becomes leathery. A characteristic odor is present.

Le symptomas de iste morbo es diverse, secundo le parte afflicte del planta. Le prime symptoma evidente de infection es gumma que se manifesta in le cortice del arbores presso le terreno. Si on remove le cortice, on vide brun areas infectate que se extende in le ligno. Plus tarde, le maculas e desicca e se finde. Un characteristica distinctive es le mortification de un latere del arbore durante que le altere resta salubre. Le infection se manifesta in le fructo como minute maculas obscur. Iste maculas se cambia in diverse tintas del color brun e le texito deveni coriacee. Es presente un odor characteristic.

Les symptômes causés par cette maladie varient selon la partie atteinte de la plante. Le premier signe visible d'infection est l'apparaition de gomme sur l'écorce de l'arbre au niveau du sol. Si l'on enlève l'écorce, on aperçoit des plages brunes infectées qui se rendent jusqu'au bois. Les taches finissent par sécher et craquer. Une caractéristique particulière est qu'un côte de l' arbre peut mourir tandis que l'autre reste sain. L'infection apparaît sur le fruit comme de petites taches foncées qui peuvent prendre plusieurs teintes de brun et les tissus deviennent coriaces. Une odeur caractéristique se dégage.

Los síntomas causados por esta enfermedad son variados, dependiendo de la parte de la planta que sea afectada. El primer síntoma visible de la infección es la goma que aparece en la corteza de los árboles, cerca de la superficie del suelo. Si la corteza se raspa se pueden ver áreas infectadas de color café que llegan hasta la madera. Las manchas eventualmente se secan y se rajan. Un rasgo distinctivo es que un lado del árbol puede morir y el otro permanece sano. La infección aparece en el fruto como pequeñas manchas café que adquieren diversos tonos de color café y el tejido se vuelve cuerudo. Hay un olor característico.

DISTRIBUTION: Af., A., Aust., Eur., N.A., C.A., S.A.

<u>Citrus</u> spp. (and many other hosts) + <u>Pseudomonas</u> <u>syringae</u> van Hall =

ENGLISH	Citrus blast, Black Pit, Bacterial Rot, Leaf Blotch, Bacterial Rot, Bacterial Canker
FRENCH	Nécrose Bactérienne
SPANISH	Quemadura de los brote terminales, Requemo , Corazón negro, Mancha foliar
PORTUGUESE	
ITALIAN	Batteriosi dei rami e frutti degli agrumi
RUSSIAN	
SCANDINAVIAN	
GERMAN	
DUTCH	
HUNGARIAN	
BULGARIAN	
TURKISH	Turunçgil yaprak, Sürgün yanikliği, Bakteriyel sürgün çürüklüğü
ARABIC	
PERSIAN	Poosedeguëyë toghëyë morakkëbat پوسیدگی طوقۀ مرکبات
BURMESE	
THAI	
NEPALI	
HINDI	
VIETNAMESE	
CHINESE	
JAPANESE	Netu-byô 熱病

This bacterial organism produces symptoms on foliage and fruit. The lesions on foliage are at first brown to black, most often starting at the base of the leaf and then to the twig at the place of attachment. Later the infected tissue becomes reddish-brown to chestnut-colored dry scab. Leaves often wither rapidly while still attached or fall off. Severe attacks kill entire branches. The fruit lesions appear as sunken, black spots on the fruit surface or as small pits or specks. This type of lesion becomes larger and produces loss of fruit in storage.

Iste organismo bacterial produce symptomas in le foliage e in le fructos. In foliage, le lesiones es, al initio, brun o nigre, e le plus frequentemente illos comencia al base del folio e se extende al rametto al puncto de junctura. Plus tarde, le texito infectate deveni un crusta desiccate, brun-rubiette or castanie. Folios attachate sovente marcesce rapidemente o se separa e cade. Attaccos sever causa le morte de ramos integre. Le lesiones in le fructos se manifesta como depresse maculas nigre al superfacie del fructo o como minute punctos. Ille lesiones cresce e causa perdition del fructo conservate in immagasinage.

Cette bactérie cause des symptômes sur le feuillage et le fruit. Les lésions sur le feuillage sont d'abord de brunes à noires; elles débutent le plus souvent à la base de la feuille et vont ensuite sur le rameau au point d'attache. Plus tard, les tissus infectés deviennent des galles sèches d'un brun rougâtre à châtains. Souvent, les feuilles se déssèchent rapidement lorsque encore attachées ou tombent. Les attaques graves font mourir des branches entières. Les lésions du fruit apparaissent comme des taches deprimées à la surface du fruit ou comme de petites cavités ou mouchetures. Ce type de lésion s'agrandit et entraîne la perte du fruit en entrepôt.

Este organismo bacterial produce síntomas en el follaje y fruto. Las lesiones en el follaje son primero de color café a negras, frecuentemente comenzando en la base de la hoja y luego avanza hacia la ramita en el lugar donde se juntan. Luego, el tejido infectado se vuelve de color café-rojizo a castaño como escama seca. Las hojas frecuentemente se marchitan rápidamente mientras están aún pegadas al árbol o se caen. Los ataques severos matan ramas enteras. Las lesiones del fruto aparecen como manchas hundidas, negras sobre la superficie o como pequeños hoyos o motas. Este tipo de lesion   se agranda y causa pérdida de la fruta almacenada.

DISTRIBUTION: Af., A., Aust., Eur., N.A., C.A., S.A.

Citrus aurantifolia + Sphaeropsis tumefaciens Hedges =

ENGLISH	Branch Knot
FRENCH	
SPANISH	Nudo de la rama
PORTUGUESE	
ITALIAN	
RUSSIAN	
SCANDINAVIAN	
GERMAN	
DUTCH	
HUNGARIAN	
BULGARIAN	
TURKISH	
ARABIC	
PERSIAN	
BURMESE	
THAI	
NEPALI	
HINDI	
VIETNAMESE	
CHINESE	
JAPANESE	Tengusu-byô てんぐ巣病

This disease produces woody hard enlargements on the branches. These knots are usually rounded, abnormal growths, at first covered with normal bark, but later they become deeply furrowed. Sometimes these areas become erupted and extend along the stems. When cut into, the surface of the knot is found to have a soft crumbling character as compared with the normally firm, greenish bark of healthy branches. Sometimes the knots cause girdling of the stems.

Iste morbo produce lignose excrescentias dur in le ramos. Iste nodos es usualmente rotunde tumores anormal, coperte al initio de cortice normal, ma plus tarde illos es profundemente cannellate. A vices iste areas se erumpe e se extende al truncos. Si on los sectiona, le superfacie del nodo monstra se esser characteristicamente molle e friabile in contradistinction al normal cortice verde e dur de ramos salubre. A vices le truncos es cincte per le nodos.

Cette maladie produit des protubérances ligneuses sur les branches. Ces noeuds sont habituellement des excroissances anormales, arrondies, d'abord couvertes par l'ecorce, mais qui par la suite deviennent profondément encastrées. Parfois ces plages font éruption et s'étendent le long des tiges. Lorsque sectionnée, la surface du noeud montre une texture friable comparée à l' écorce normale ferme et verdâtre des branches saines. Parfois, les nodules encerclent les tiges.

Esta enfermedad produce nudos de madera dura en las ramas. Estos nudo generalmente son crescimientos anormales, redondeados, cubiertos primero con corteza normal, pero luego se vuelven bastante agrietados. Algunas veces estas áreas se vuelven eruptas y se extienden a lo largo de los tallos. Cuando se les hace un corte, se encuentra que la superficie del nudo tiene una característica blanda y desmenuzable si se compara con la corteza sana de las ramas que es firme, verdosa. Algunas veces los nudos ciñen los tallos.

DISTRIBUTION: Af., A., N.A., C.A., S.A., W.I.

58

Citrus spp. + Xanthomonas citri (Hasse) Dowson (Phytomonas citri (Hasse) Dowson) =

ENGLISH   Citrus Canker, Bacterial Canker
FRENCH   Chancre Bactérien des Citrus
SPANISH   Cancrosis, Cancrosis bacteriana, Cancro bacterial, Cancro de los cítricos
PORTUGUESE   Cancro cítrico
ITALIAN   Cancro degli agrumi
RUSSIAN   рак бактериальный
SCANDINAVIAN
GERMAN   Bakterienkrebs
DUTCH
HUNGARIAN
BULGARIAN
TURKISH
ARABIC
PERSIAN
BURMESE   Gyi-but-nar, Nou-nar roga, ကငြုဖ့ၐ်ၕြၐ၆ကာၕၜြၚၚ
THAI   Canker   แคงเกอร
NEPALI   Kagati Ko dhabbe rog   कागतिको धब्बे रोग
HINDI
VIETNAMESE   Ghẻ nhám trên cành lá và quả
CHINESE   Gan jýu kwèi yáng bìng   柑橘潰瘍病
JAPANESE   Kaiyo-byð   かいよう病

Canker lesions on fruit consist at first of spongy eruptions of tissue in the rind which usually show oily-appearing margins and lack of yellow halos of those on affected leaves. They often show a crater-like appearance. On leaves the lesions are small spongy eruptions, at first white, later becoming tan. A margin appears which becomes watery and greasy and is yellowish-brown or green in color.

Le lesiones del canceres in le fructo es al initio composite de eruptiones de texito spongiose in le pelle, le quales usualmente monstra un margine que sembla esser oleose e le quales monstra un carentia del halo jalne que se trova in folios afflicte. Sovente illos appare similar a crateres. In folios le lesiones es minute eruptiones spongiose, al initio blanc, plus tarde bronzate. Un margine evolve le qual deveni aquose e oleose, de color brun-jalnette o verde.

Les lésions chancreuses du fruit consistent d'abord en éruptions de tissus spongieux dans l'écorce (pelure). Elles ont une marge huileuse et n'ont pas le halo jaune qu'on voit sur les feuilles atteintes. Elles ont souvent l'apparence d'un cratère. Sur les feuilles, les lésions sont de petites éruptions spongieuses, d'abord blanches, puis couleur du tan. Une marge se forme qui devient aqueuse et graisseuse et de couleur brune ou verte.

Las lesiones de cancro en el fruto consisten primero de erupciones esponjosas de tejido en la cáscara, que generalmente muestran márgenes aceitosos y falta del halo amarillo de las hojas afectadas. Frecuentemente muestran una apariencia de cráter. En las hojas las lesiones son erupciones pequenas y esponjosas, primero blancas, que luego se vuelven de color crema. Aparece un margen que se vuelve acuoso y grasoso y   es de color amarillo-café o verde.

DISTRIBUTION: Af., A., Aust., S.A.

Coffea spp. + Cercospora coffeicola Berk. & Cooke =

ENGLISH	Brown Spot, Eyespot of Leaves and Berries
FRENCH	Maladie des Yeux Bruns du Caféier, Cercosporose
SPANISH	Mancha de hierro, Mancha del fruto, Mancha parda de la hoja, Chasparria
PORTUGUESE	Moléstia do olho pardo, Mancha do olho pardo, Olho de pomba, Cercosporiose
ITALIAN	Cercosporiosi del cafè
RUSSIAN	церкоспороз, пятнистость бурая глазковая
SCANDINAVIAN	
GERMAN	Braune Blattflekkenkrankheit
DUTCH	
HUNGARIAN	
BULGARIAN	
TURKISH	
ARABIC	
PERSIAN	
BURMESE	
THAI	Bi-Jut ใบจุก
NEPALI	
HINDI	Bhura Drik-Bindu भूरा द्रिक विन्दु
VIETNAMESE	Đốm lá
CHINESE	Ka fei hor yǐan bìng 咖啡褐眼病
JAPANESE	

This disease attacks the leaves and fruits, causing large blotches which at first are visible only on the upper surface. They are dark brown at first, becoming grayish above and clear below. The centers of these blotches become dead and covered with the fungus. It causes the leaves to fall, thus reducing the vitality of the plant and preventing the maturity of the berries. It attacks the twigs and also the berries which fall before ripening.

Iste morbo attacca le folios e le fructos, producente grande pustulas le quales es, al initio, evidente solmente al superfacie superior. Illos es, al initio, brun obscur, ma deveni gris supra e clar infra. Le centros de iste pustulas deveni morte e coperte del fungo. Illo face le folios cader, de iste modo diminuente le vitalitate del planta e impediente le maturation del baccas. Illo attacca le ramettos e anque le baccas, le quales se separa e cade prematurmente.

Cette maladie attaque les feuilles et les fruits et produit de grandes taches qui ne sont d'abord visibles qu'à la face supérieure. Elles sont brun foncé au début, devenant grisâtres en surface et plus pâles en dessous. Le centre de ces éclaboussures meurt et se couvre de mycélium. Les feuilles attaquées se détachent de la plante, ce qui en diminue la vitalité et prévient la maturation des graines. Cette maladie s'attaque aux rameaux et aussi aux graines, qui tombent avant de mûrir.

Esta enfermedad ataca las hojas y frutos, causando grandes manchas que primero son visibles sólo en la superficie de arriba. Primero son de color café obscuro, volviéndose grisosas encima y claras abajo. Los centros de estas manchas se mueren y se cubren con el hongo. Causa la caída de las hojas, reduciendo así la vitalidad de la planta y previniendo la maduración de los frutos. Ataca las ramillas y también los frutos que se caen antes de madurar.

DISTRIBUTION: Af., A., Aust., N.A., C.A., S.A.

Coffea spp. + Hemileia vastatrix Berk. & Br. =

ENGLISH Rust, Leaf Disease
FRENCH Rouille Vraie du Caféier
SPANISH Enfermedad negra de la raíz, Maya, Pudrición de raíz principal, Llaga negra
PORTUGUESE Ferrugem
ITALIAN Ruggine del caffè
RUSSIAN ржавчина
SCANDINAVIAN
GERMAN Kaffeerost
DUTCH
HUNGARIAN
BULGARIAN
TURKISH
ARABIC
PERSIAN
BURMESE Na-nwin, Sa-nwin roga ဟယ်မီလီယ ဝဘတ်စ်တတြစ်ကစ်
THAI Ra-Snim-Leck ราสนิมเหล็ก
NEPALI
HINDI Ratua रतुवा
VIETNAMESE Rỉ lá
CHINESE Ka fei shìou bìng 咖啡銹病
JAPANESE

This disease produces yellowish-orange, powdery, rounded blotches on the lower surface of the leaves. The spots may coalesce with others to form an irregularly shaped lesion which is accompanied by the loss of color on the upper surfaces of older leaves. The center of the leaf turns dark brown and dies and premature defoliation and die-back of branches follows. These symptoms sometimes occur on the berries and young shoots.

Iste morbo produce rotunde pustulas pulverose, de color orange-jalnette, al inferior superfacie del folios. Le maculas pote coalescer con alteres usque a formar un lesion de forma irregular le qual se associa con discoloration al superior superfacies de folios plus vetule. Le centro del folio deveni brun obscur e mori. Defoliation prematur e degeneration del ramos occurre. Iste symptomas a vices se manifesta in le baccas e in le juvene ramos germinante.

Cette maladie produit des éclaboussures jaune orange rondes et poudreuses sur la face inférieure des feuilles. Ces taches peuvent se fusionner pour former une lésion de forme irrégulière, ce qui entraîne la décoloration de la surface des vieilles feuilles. Le centre de la feuille tourne au brun foncé et meurt, après quoi les rameaux perdent prématurément leurs feuilles et dépérissent. Ces symptômes se retrouvent parfois sur les graines et les jeunes pousses.

Esta enfermedad produce manchas redondeadas, polvorientas, de color anaranjado-amarillento en el envés de las hojas. Las manchas pueden unirse con otras para formar una lesión de forma irregular la cuál es acompanada por la pérdida de color en el haz de las hojas maduras. El centro de la hoja se vuelve color café obscuro y muere, seguida de defoliación prematura y muerte progresiva de las ramas. Estos síntomas algunas veces aparecen en los frutos y en brotes jóvenes.

DISTRIBUTION: Af., A., Aust., S.A.

Coffea arabica (Thea sinensis, Citrus spp., Hevea brasiliensis, Theobroma cacao) +
Rosellinia bunodes (Berk. & Br.) Sacc. =

ENGLISH   Secondary Root Rot, Black Root of Citrus
FRENCH   Pourridié du Tronc dû à Rosellinia
SPANISH   Pudrición secundaria de la raíz
PORTUGUESE   Roseliniose, Podridão das raízes, Mal de Araraquara, Podridão do pião
ITALIAN
RUSSIAN   гниль стволов розоллиниозная
SCANDINAVIAN
GERMAN   Rosellinia-Stammfäule
DUTCH
HUNGARIAN
BULGARIAN
TURKISH
ARABIC
PERSIAN
BURMESE
THAI
NEPALI
HINDI   Kala Mool Viglan   काला मूल विगलान
VIETNAMESE
CHINESE
JAPANESE

Wilting and death of the entire plant or of single branches may be the first sign of attack. Under damp conditions a sheet of the fungus, cream-white to purplish-black, may extend from the groundline and collar of the main trunk to well above the soil surface. Firm, black branching strands with irregularly spaced thickened knots attach themselves to the root surface. In the outer layer of the wood, the strands have a black periphery and white core while the inner woody tissue appears thread-like and black.

Le prime symptoma del attacco frequentemente se compone de marcescentia e morte de tote le planta o de ramos individual. Sub conditiones humide, un strato del fungo, blanc como crema verso nigre o purpuree, pote extender se ab le terreno e le collar del trunco principal usque a un puncto satis alte. Firme cordas nigre con nodos inspissate, inequalmente spatiate, se ramifica e se attacha al superfacie del radices. In le strato exterior del ligno le cordas monstra un peripheria nigre e un centro blanc; ma le texito interior lignose es nigre e similar a filos.

Un flétrissement suivi de la mort de la plante entière ou de branches isolées peuvent marquer le début d'une infection. A l'humidité, un manchon de mycélium, blanc-crème à pourpre foncé, peut s'étendre de la surface du sol et du collet de la tige principale jusque beaucoup plus haut. Des cordons noirs, fermes et ramifiés, portant çà et là des renflements, s'attachent à la surface des racines. Dans les couches externes du bois, ces cordons sont noirs avec un centre blanc, alors que les tissus internes du bois paraissent filamenteux et noirs.

El primer síntoma de un ataque de la enfermedad puede ser la marchitez o muerte de toda la planta o de una sola rama. Bajo condiciones húmedas una lámina del hongo, de color crema-blanco a púrpura-negro, puede extenderse desde cerca de la superficie del suelo y cuello del tronco principal hasta bastante arriba de la superficie del suelo. Hebras firmes, negras, ramificadas con nudos espesos espaciadas irregularmente se adhieren sobre la superficie de las raíces. En la capa exterior de la madera, las hebras tienen una periferia negra y corazón blanco, mientras que el tejido leñose interno tiene apariencia hilachosa y negra.

DISTRIBUTION: Af., A., N.A., C.A., S.A.

62

Corchorus spp. + Colletotrichum corchorum Ikata et S. Tanaka =

ENGLISH   Anthracnose
FRENCH
SPANISH   Antracnosis
PORTUGUESE
ITALIAN
RUSSIAN
SCANDINAVIAN
GERMAN
DUTCH
HUNGARIAN
BULGARIAN
TURKISH
ARABIC
PERSIAN
BURMESE   Ywet-pyauk roga ✔ ꓲ   ‎ကြ�　လက္ဍြင္ခ္‌ခဲ ကြ ꩡꩾ‌ꩺ
THAI   Lum-Ton-Nao   ลำต้นเน่า
NEPALI
HINDI
VIETNAMESE
CHINESE   Huáng má tàn dzrǔ bìng   黄蔴炭疽病
JAPANESE   Tanso-byð   炭そ病

The disease causes small, moist, brownish-black spots on the internodes or nodes of the young stems of plants. A number of spots coalesce to form larger lesions consisting of dead tissue. The fungus may penetrate deep into the stem resulting in a break along the wound. In mature plants the spots are usually dry and cankerous, with numerous black dots produced by the fungus. Dark brown sunken spots appear on infected pods, causing the seeds to collapse and shrivel.

Iste morbo produce minute maculas humide, nigre o brun, in le internodos o in le nodos del pedunculos juvene de plantas. Alicun maculas coalesce, transformante se in plus grande lesiones composite de texito morte. Le fungo pote penetrar profundemente in le pedunculo, causante fissura del vulnere. In plantas maturate le maculas es usualmente desiccate e cancerose, con numeros punctos nigre del fungo. Depresse maculas brun obscur se manifesta in le siliquas infectate e face le granos putrescer e crispar se.

Cette maladie produit de petites taches délavées brun foncé sur les noeuds et entrenoeuds des jeunes tiges. Plusieurs taches se fusionnent pour former de grandes lésions de tissus morts. Le champignon peut pénétrer profondément dans la tige, ce qui rend la plaie irrégulière. Chez les plantes adultes, les taches sont généralement sèches et chancreuses, parsemées de nombreux petits points noirs qui sont les fructifications du champignon. Des taches profondes de couleur brun foncé apparaissent sur les siliques infecteés et font s'affaisser et ratatiner les graines.

La enfermedad causa pequeñas manchas húmedas, de color cafesoso a negro, en los internudos o nudos de los tallos jóvenes de las plantas. Varias manchas se unen para formar lesiones más grandes formadas por tejido muerto. El hongo puede penetrar bastante hondo dentro del tallo, resultando en una quebradura en la madera. En plantas maduras las manchas son generalmente secas y cancrosas, con numerosos puntos negros producidos por el hongo. Manchas de color café obscuro y hundidas aparecen en las vainas infectadas, causando que las semillas sufran un colapso y se ajen.

DISTRIBUTION: A., S.A.

Corchorus spp. + Helminthosporium corchorum Watanabe et Hara (Corynespora corchorum
(Watanabe et Hara) Goto) =

ENGLISH   Leaf Blight
FRENCH
SPANISH   Tizón de la hoja
PORTUGUESE
ITALIAN
RUSSIAN
SCANDINAVIAN
GERMAN
DUTCH
HUNGARIAN
BULGARIAN
TURKISH
ARABIC
PERSIAN
BURMESE
THAI
NEPALI
HINDI
VIETNAMESE
CHINESE   Huáng má yèh ku bìng   黃 蔴 葉 枯 病
JAPANESE   Hagare -byð   葉枯病

The disease appears as light reddish-brown blotches, frequently with a sunken center. The
blotches usually are on the leaf blade. The infected leaves turn yellow and dry out as the infection
advances. The disease also may cause seedling blight and root rot.

Le morbo se manifesta como pustulas brun-rubiette clar, frequentemente   con un centro
depresse, usualmente in le laminas del folios. Le folios infectate deveni jalne e desiccate
quando le infection se avantia. Le morbo pote causar anque plage de plantulas e putrefaction
de radices.

La maladie se présente sous forme d'éclaboussures brun rougeâtre pâle, dont le centre est
souvent déprimé. Les taches apparaissent le plus souvent sur le limbe de la feuille. A mesure
que l'infection progresse, les feuilles infectées jaunissent et sèchent. La maladie peut aussi
causer une brulure des semis et a carie des racines.

La enfermedad aparece como manchas de color café rojizo, frecuentemente con el centro hundido. Las manchas generalmente aparecen sobre la hoja. Las hojas infectadas se vuelven de
color amarillo y se secan a medida que la infección avanza. La enfermedad también puede
causar tizón de las plántulas y podredumbre de la raíz.

DISTRIBUTION: A.

64

Corchorus spp. + Microsphaera polygoni (de Candolle) Sawada (Sphaerotheca fuligena (Sch.) Sawada) =

ENGLISH   Powdery Mildew
FRENCH
SPANISH   Mildiú polvoriento
PORTUGUESE
ITALIAN
RUSSIAN
SCANDINAVIAN
GERMAN
DUTCH
HUNGARIAN
BULGARIAN
TURKISH   Külleme hastalığı
ARABIC
PERSIAN
BURMESE
THAI
NEPALI
HINDI
VIETNAMESE
CHINESE   Huáng má bór fěn bìng   黄 蔴 白 粉 病
JAPANESE   udonko-byô   うどんこ病

In the early stages the disease may be recognized by the presence of small groups of thin, light gray or white fungus growths spreading rapidly over the surface of the leaves. Reddening of the underlying leaf tissue is sometimes evident. Later the infected areas take on a white powdery appearance, as though coated with flour. Heavily coated leaves turn brown and drop off.

In le stadios initial le morbo se characterisa per le apparition de gruppamentos del tenue fungo gris o blanc, le quales se extende rapidemente al superfacie del folios. Rubification del texito infra es a vices evidente in le folios. Plus tarde, le areas afflicte es pulverose, quasi coperte de farina. Folios con un dense coperatura deveni brun, se separa, e cade.

A ses premières phases, la maladie peut se reconnaître à la présence de petites colonies minces gris pâle ou blanches du champignon, qui ne tarde pas à recouvrir les feuilles. On peut quelquefois observer le rougissement des tissus sous-jacents. Puis, les plages infectées semblent recouvertes d'une poudre blanche ou de farine. Les feuilles très infectées brunissent et tombent.

Al comienzo la enfermedad puede reconocerse por la presencia de pequeños grupos de crecimiento fungosos delgados, de color gris claro o blanco, que se diseminan rápidamente sobre la superficie de las hojas. A veces es evidente el enrojecimiento del tejido interno de la hoja. Más tarde las áreas infectadas adquieren una apariencia de polvo blanquecino, como si estuvieran cubiertas de harina. Las hojas muy atacadas se vuelven de color café y se caen.

DISTRIBUTION: A.

Corchorus spp. + <u>Xanthomonas nakatae</u> (Okabe) Dowson (<u>Bacterium nakatae</u> Takimoto) =

ENGLISH   Bacterial Leaf Spot
FRENCH
SPANISH   Mancha bacterial de la hoja
PORTUGUESE
ITALIAN
RUSSIAN
SCANDINAVIAN
GERMAN
DUTCH
HUNGARIAN
BULGARIAN
TURKISH
ARABIC
PERSIAN
BURMESE
THAI
NEPALI
HINDI
VIETNAMESE   Đốm thâm kim
CHINESE   huáng má jǐao ban bǐng 黃 蔴 角 斑 病 ,ban dǐan shǐng shǐh jǐunn bǐng斑 點 性 細 菌 病
JAPANESE   Hanten-saikin-byô   斑点細菌病

Minute water-soaked dots which enlarge to translucent blackish or brownish spots appear on the leaves. They are irregular in shape and frequently have light yellowish-green halos. When spots coalesce, they produce shot holes.

Minute punctos saturate de aqua, le quales crecse in nigre o brunette maculas translucide, se manifesta al folios. Illos es de forma irregular e frequentement ha halos verde-jalnette clar. Quando le maculas coalesce, perforationes resulta.

Des points minuscules et détrempés apparaissent sur les feuilles et grossissent pour devenir des taches translucides noirâtres ou brunâtres. Leurs contours sont irréguliers et l'on peut souvent observer des aréoles vert jaunâtre pâle. Lorsqu'il y a fusion des taches, il se forme des trous de cible dans les feuilles.

Sobre las hojas aparecen puntos minúsculos, empapados de agua, que se agrandan hasta formar manchas translúcidas cafesosas o negruzcas. Son de forma irregular y frecuentemente tienen halos claros amarillo-verdosos. Cuando las manchas se juntan, producen agujeros como de bala.

DISTRIBUTION: A.

66

Crotalaria juncea + Uromyces decoratus Syd. =

ENGLISH	Rust	
FRENCH		
SPANISH	Roya	
PORTUGUESE		
ITALIAN		
RUSSIAN		
SCANDINAVIAN		
GERMAN		
DUTCH		
HUNGARIAN		
BULGARIAN		
TURKISH		
ARABIC		
PERSIAN		
BURMESE		
THAI	Ra-Snim-Leck	ราสนิมเหล็ก
NEPALI		
HINDI	Ratua	रतुवा
VIETNAMESE		
CHINESE		
JAPANESE		

The disease appears on all aboveground parts of the plant, including pods, although the disease is generally more prevalent on leaves and stems. It appears first as small specks and later develops into large dark brown lesions. Quite often the lesions are surrounded by small, secondary, uniform rings which are composed of the fungus. As the disease advances the lesions become black in color.

Le morbo se manifesta in tote le partes epigee del planta, le siliquas incluse ben que le folios e le pedunculos es plus frequentemente infectate. Le morbo se manifesta al initio como minute punctos, e plus tarde illos se transforma in grande lesiones brun obscur. Frequentement, le lesiones es imbraciate per minime annelos secundari e uniforme, le quales se compone del fungo. Quando le morbo se avantia, le lesiones deveni de color nigre.

La maladie apparaît sur toutes les parties aériennes de la plante, y compris les gousses, bien qu'elle s'installe plus généralement sur les feuilles et les tiges. Elle apparaît d'abord comme de petits points qui se transforment par la suite en grandes lésions brun foncé. Assez souvent, les lésions sont entourées d'anneaux secondaires, petits et uniformes, formés par le champignon. A mesure que la maladie progresse, les lésions tournent au noir.

La enfermedad aparece en todas las partes aéreas de la planta, incluyendo vainas, aunque la enfermedad es generalmente más notoria sobre las hojas y tallos. Primero aparece como pequeñas manchitas y luego se desarrolla y llegan a formar grandes lesiones de color café obscuro. Frecuentemente las lesiones están rodeadas por anillos pequeños, secundarios, uniformes que están compuestos de hongos. A medida que la enfermedad avanza las lesiones se vuelven de color negro.

DISTRIBUTION: Af., A., C.A., S.A.

Cucurbitaceae + <u>Alternaria</u> <u>cucumerina</u> (Ellis & Everh.) Elliott =

ENGLISH   Leaf Spot
FRENCH   Alternariose, Taches Foliaires
SPANISH   Mancha de la hoja, Tizón de la hoja, Mancha foliar
PORTUGUESE
ITALIAN
RUSSIAN   альтернариоз
SCANDINAVIAN
GERMAN   Alternaria-Blattfleckenkrankheit
DUTCH
HUNGARIAN
BULGARIAN
TURKISH
ARABIC
PERSIAN
BURMESE
THAI   Bi-Jut   ใบจุด
NEPALI
HINDI
VIETNAMESE
CHINESE
JAPANESE   Kokuhan-byô   黒斑病

The first symptoms of this disease usually appear on the leaves nearest the center of the hill during the middle of the growing season. Small, circular, water-soaked areas appear on the leaves. These sometimes enlarge into definite concentric rings with margins. The rings usually appear only on the upper surfaces. Lesions vary in size from a mere point to 1/2 inch in diameter. Often the spots run together and cover 1/4 of the leaf surface. The disease causes damage by defoliating the vines, causing premature ripening of fruit and storage injury.

Le prime symptomas de iste morbo usualmente se manifesta in le folios que se trova presso le centro del collina al medie-saison. Minute areas circular e saturate de aqua se manifesta al folios. A vices illos cresce in definite anellos concentric con margines. Le anellos usualmente occurre solmente in le superfacie superior. Lesiones se varia in dimension ab un puncto minute usque a 1,25 cm in diametro. Sovente le maculas coalesce e coperi 1/4 del superfacie del folio. Le morbo face mal per defoliation del vites e causa maturation del fructo ante le tempore ordinari e perdition del fructo in immagasinage.

Les premiers symptômes de cette maladie apparaissent généralement au milieu de la période de croissance, sur les feuilles les plus rapprochées du centre de la butte. De petites zones circulaires et détrempées apparaissent sur les feuilles. Elles s'agrandissent parfois en anneaux concentriques définis avec marges. En general, les anneaux ne sont visibles que sur la face supérieure des feuilles. La grosseur des lésions varie d'un simple point à 1/2 pouce de diamètre. Souvent, les taches se rejoignent et couvrent 1/4 de la surface de la feuille. La maladie cause des pertes par la défoliation prématurée des tiges, la maturation précoce des fruits et les détériorations dans l'entrepôt.

Los primeros síntomas de esta enfermedad generalmente aparecen sobre las hojas más cercanas al centro del montículo durante la mitad de la época de crecimiento. Sobre las hojas aparecen áreas pequeñas, circulares, empapadas de agua. Estas algunas veces se agrandan hasta formar anillos concéntricos con márgenes. Los anillos generalmente aparecen solamente en los haces. Las lesiones varían en tamaño desde sólo un puntito hasta 1/2 pulgada de diámetro. Frecuentemente las manchas se juntan y cubren una cuarta parte de la superficie de la hoja. La enfermedad causa daño al defoliar las guías, cuasando una maduración prematura del fruto y daños durante el almacenamiento.

DISTRIBUTION: Af., A., Aust., Eur., N.A., C.A., S.A.

68

Cucurbitaceae + <u>Cladosporium</u> <u>cucumerinum</u> Ell. & Arth. =

ENGLISH	Scab
FRENCH	Gale, Cladosporiose du concombre, Nuile, Gommose
SPANISH	Sarna de las hojas, Roña
PORTUGUESE	Queima de cladosporio
ITALIAN	Gommosi, Cladosporiosi dei cetrioli
RUSSIAN	кладоспориоз, пятнистость плодов оливковая
SCANDINAVIAN	Gummiflod, Gurkfläcksjuka, Agurkgummiflod
GERMAN	Gummifluss, Krätze, Gurkenkrätze
DUTCH	Vruchtvuur der Komkommers
HUNGARIAN	
BULGARIAN	краста
TURKISH	Kellik
ARABIC	
PERSIAN	
BURMESE	
THAI	
NEPALI	
HINDI	
VIETNAMESE	Thối gốc hoặc lá
CHINESE	
JAPANESE	Kurohosi-byô 黒星病

Lesions appear on all parts of the vine which are above ground. Most of the injury occurs on the fruit which is especially susceptible when young. The spots, which exude a sticky substance, are gray to white and slightly sunken. The canker darkens with age and collapses even further until a distinct cavity, lined with a dark green velvety layer, is formed. On the foliage, water-soaked or pale green lesions appear between or on the veins. Later the leaves wilt and become a decaying mass.

Lesiones se manifesta in tote le partes epigee del vite. Le injuria afflige le plus severmente le fructo, le qual es specialmente susceptibile quando illo es juvene. Le maculas, le quales exsuda un materia glutinose, es gris o blanc e un pauc depresse. Le cancere se obscura con etate e putresce de plus in plus usque a formar un cavitate distincte, con un revestimento interior verde obscur, similar a villuto. In le foliage, lesiones saturate de aqua o verde clar se manifesta o inter o in le venas. Plus tarde le folios marcesce e tote le planta putresce.

Des lésions apparaissent sur toutes les parties aériennes des plantes. Les plus grands dégâts se produisent sur les fruits, qui sont particulièrement sensibles lorsqu'ils sont jeunes. Les taches, qui exsudent une substance gluante, sont grises à blanches et légèrement déprimées. Avec le temps, le chancre devient plus foncé et déprimé jusqu'à ce que se forme une cavité nette, bordée d'une couche veloutée vert foncé. Sur le feuillage, des lésions translucides ou vert pâle apparaissent entre ou sur les nervures. Plus tard, le feuillage se flétrit et se décompose.

Las lesiones aparecen en todas las partes aéreas de las planta. El daño mayor ocurre en el fruto el cuál es especialmente susceptible cuando está tierno. Las manchas, que exudan una substancia pegajosa, son de color gris a blanco y un poco hundidas. El cancro se obscurece con la edad y ataca aún más hasta formar una cavidad distinguible, forrada con una capa aterciopelada de color verde obscuro. En el follaje, entre o sobre las venas, aparecen lesiones empapadas de agua o de color verde pálido. Más tarde la hoja se marchita y se vuelve una masa podrida.

DISTRIBUTION: Af., A., Eur., N.A., C.A.

Cucurbitaceae + <u>Colletotrichum</u> lagenarium (Pass.) Ell. & Halst. =

ENGLISH	Anthracnose
FRENCH	Anthracnose, Nuile, Nuile Rouge
SPANISH	Antracnosis
PORTUGUESE	Antracnose
ITALIAN	Antracnosi, Nebbia del melone
RUSSIAN	антракноз
SCANDINAVIAN	Gurkröta, Skiveplet
GERMAN	Brennflecken, Anthraknose der Gurkengewächse
DUTCH	Vruchtvuur, Bladvuur, Brandvlekkenziekte
HUNGARIAN	Fenésedés
BULGARIAN	антракноза
TURKISH	Bostanlarda antraknoz
ARABIC	
PERSIAN	Änträknōōzĕ khårbŏzĕh انتراکنوز خربزه
BURMESE	Ywet-pyauk roga တော်ၚ်လက် တိုက်ပြုင်္ ၚ် ၃ ၆လက်ရိ ၄ ၄ ၄ ၊ ၊ ၆
THAI	
NEPALI	Pharsi Ko rate rog फरसिको राते रोग
HINDI	
VIETNAMESE	
CHINESE	Hú gwua tàn dzú bìng 胡瓜炭疽病
JAPANESE	Tanso-byŏ 炭そ病

This disease affects all aboveground parts of the plant with the most noticeable symptoms appearing on the fruit. On the foliage, spots begin as small yellowish or water-soaked areas which enlarge rapidly and turn brown or black. Sometimes the entire leaf dies and, when petioles are attacked, defoliation of the vine occurs. When these elongated lesions occur on the stem as well as the leaves, the entire vine dies. Circular, black, sunken cankers appear on the fruit. The black center is lined with a salmon-colored gelatinous mass when moisture is present.

Iste morbo afflige tote le partes epigee del planta, le symptomas le plus remarcabile manifestante se in le fructos. In le foliage, maculas minute, jalnette o saturate de aqua, cresce rapidemente e deveni brun o nigre. A vices, tote le folio mori, e, quando le petiolos es attaccate, defoliation del vite occurre. Se iste lesiones elongate occurre non solmente al folios ma anque in le pedunculo, tote le vite mori. Circular canceres nigre e depresse se manifesta in le fructos. Se humiditate es presente, le centro nigre monstra un revestimento gelatinose de color de salmon.

Cette maladie atteint toutes les parties aériennes de la plante, mais les symptômes les plus perceptibles apparaissent sur les fruits. Sur le feuillage, les taches commencent par de petites plages jaunâtres ou translucides qui s'agrandissent rapidement et tournent au brun ou au noir. Parfois, la feuille entière meurt et, lorsque les pétioles sont attaqués, il y a défoliation. Lorsque ces lésions allongées se présentent à la fois sur la tige et les feuilles, la tige entière meurt. Des chancres noirs, circulaires et creux apparaissent sur le fruit. A l'humidité, le centre noir est bordé d'une masse gélatineuse de couleur saumon.

Esta enfermedad afecta todas las partes aéreas de la planta, pero los síntomas más notorios aparecen sobre el fruto. En el follaje, las manchas comienzan como pequeñas áreas amarillosas o empapadas de agua, las cuáles se agrandan rapidámente y se vuelven de color café o negras. Algunas veces toda la hoja muere y cuando los pecíolos son atacados, hay defoliación. Cuando aparecen estas lesiones alargadas sobre el tallo y también sobre las hojas, la planta entera muere. En el fruto aparecen cancros circulares, negros y hundidos. El centro negro está forrado con una masa gelatinosa de color salmón cuando hay humedad.

DISTRIBUTION: Af., A., Aust., Eur., N.A., C.A., S.A.

70

Cucumis sativus (and other Cucurbitaceae) + Erwinia trachephila (Smith) Bergey et al. =

ENGLISH	Bacterial Wilt
FRENCH	Bactériose Vasculaire, Flétrissement Bacterien
SPANISH	Podredumbre bacterial
PORTUGUESE	
ITALIAN	Colpo delle curcubitacee, Avvizzimento delle cucurbitacee
RUSSIAN	
SCANDINAVIAN	
GERMAN	
DUTCH	
HUNGARIAN	
BULGARIAN	
TURKISH	
ARABIC	
PERSIAN	
BURMESE	
THAI	Hiaw-Chaow เหี่ยวเฉา
NEPALI	
HINDI	
VIETNAMESE	
CHINESE	
JAPANESE	

This disease begins with the infection of one or two leaves and spreads downward into the petioles and stems until the entire plant has wilted and died. Fruit also wilts and shrivels. If the stem is cut, droplets of ooze can be squeezed out. In advanced stages the sap may be milky in appearance and sticky to the touch, although these two symptoms are not wholly dependable in identifying the disease.

Iste morbo comencia per le infection de un o duo folios e se extende a basso in le petiolos e in le pedunculos usque a tote le planta ha marcescite e es morte. Le fructos anque marcesce e se crispa. Si on sectiona le pedunculo, minute guttas de exsudato pote esser expresse. In stadios avantiate, le succo sembla esser lactose e es glutinose al tacto. Ma iste duo symptomas non es absolutement secur pro le indentification del morbo.

Cette maladie débute par l'infection d'une ou deux feuilles et descend dans le pétiole et la tige jusqu'à ce que la plante entière se flétrisse et meure. Les fruits également se flétrissent et se ratatinent. Si on coupe la tige, on peut en exprimer des gouttelettes d'un exsudat. A une phase avancée, la sève peut paraître laiteuse et collante, mais on ne peut compter seulement sur ces deux symptômes pour diagnostiquer la maladie.

Esta enfermedad comienza con la infección de una o dos hojas y se disemina hacia abajo dentro del pecíolo y tallos hasta que toda la planta se marchita y muere. El fruto también se aja y se marchita. Si se corta un tallo, al apretarlo caen gotitas de fluído. En estados avanzados la savia puede tener apariencia lechosa y pegajosa al tacto, aunque estos síntomas no deben bastar para identificar la enfermedad.

DISTRIBUTION: Af., A., Eur., N.A.

Cucumis sativus + Erysiphe cichoracearum DC. =

ENGLISH	Powdery Mildew
FRENCH	Blanc
SPANISH	Oidium, Oidio, Mal blanco, Cenicilla, Mildiú polvoriento
PORTUGUESE	Oídio
ITALIAN	Nebbia, Mal bianco delle cucurbitacee
RUSSIAN	
SCANDINAVIAN	Agurkmeldug, Gurkmjöldagg
GERMAN	Echter Mehltau der Gurke, Gurkenmehltau
DUTCH	Meeldauw
HUNGARIAN	Lisztharmat
BULGARIAN	брашнеста мана
TURKISH	Külleme
ARABIC	Bayad Dakiky بياض دقيقى
PERSIAN	Sēfēdákē hághēghēyē khēyār سفیدک حقیقی خیار
BURMESE	
THAI	
NEPALI	Kakro Ko pitho-chhare rog काक्राको पिठो छरे रोग , Sete-pat सेतेपात
HINDI	Choorni phaphundi चूरनी फाफुन्दी
VIETNAMESE	
CHINESE	gwua lày bái fěn bìng 胡瓜白粉病
JAPANESE	

Powdery mildew on the foliage and young stems is first evident as a talcum-like growth on the plant surface, especially on the upper surface of the leaves. As the disease advances the spots turn brown and dry, killing the leaves and young stems in severe cases. The white fruiting layer rarely covers the fruit.

Le mal blanc (mildew pulverose) al foliage e al pedunculos se manifesta al initio como un strato similar a talco in le superfacie del planta, specialmente in le superfacie superior del folios. Quando le morbo se avantia, le maculas deveni brun e desiccate, causante le morte del folios e del pedunculos juvene in casos sever. Le fructo es rarmente coperte del structura sporal blanc.

Le blanc sur le feuillage et les jeunes tiges se manifeste d'abord par une végétation ressemblant à du talc à la surface de la plante, particulièrement à la face supérieure des feuilles. A mesure que la maladie progresse, les taches tournent au brun et sèchent; les feuilles et les jeunes tiges meurent dans les cas graves. La couche blanche fructifère du champignon recouvre rarement les fruits.

El mildiú polvoriento sobre el follaje y los tallos jóvenes se nota primero como un crecimiento polvoriento sobre la superficie de las plantas, especialmente sobre el haz de las hojas. A medida que avanza la enfermedad, las manchas se vuelven de color café y se secan, matando las hojas y tallos jóvenes en los casos severos. La capa blanca de cuerpos fructíferos raramente cubre el fruto.

DISTRIBUTION: Af., A., Eur., N.A., C.A., S.A.

Cucurbitaceae + <u>Mycosphaerella</u> <u>citrullina</u> (C. O. Sm.) Grossenb. (<u>Didymella</u> <u>bryoniae</u> (Auersw.) Rehm.) =

ENGLISH	Gummy Stem Blight, Stem-end Rot, Leaf Spot
FRENCH	Pourriture Noire
SPANISH	Tizón gomoso del tallo, Podredumbre del extreme del tallo, Mancha foliar
PORTUGUESE	
ITALIAN	
RUSSIAN	гниль черная микофереллезная
SCANDINAVIAN	
GERMAN	Didymella-Krankheit, Mycosphaerella-Schwarzfäule
DUTCH	
HUNGARIAN	
BULGARIAN	
TURKISH	
ARABIC	
PERSIAN	
BURMESE	
THAI	
NEPALI	
HINDI	
VIETNAMESE	Cháy lá hoăn đốm nhựa trên thân cây
CHINESE	
JAPANESE	Turugare-byô つる枯病

Infection is local at the nodes in the leaf axil, never at the internodes. The edges of the infected areas are oily-green in color, often with resin-colored gummy exudate. The older parts are either dark and gummy or dry and gray and bear brown fruiting bodies.

Infection local occurre in le nodos in le axilla del folios, ma non jammais in le internodos. Le margines del areas infectate es oleose e de color verde, sovente con un exsudate gummose de color de resina. Le partes plus vetule es o obscur e gummose o desiccate e gris. Iste partes porta le brun structuras sporal.

L'infection se limite à l'aisselle des feuilles et n'atteint jamais les entrenoeuds. Le bord des plages infectées est de couleur verte huileuse, souvent avec exsudat gommeux couleur de résine. Les parties âgrées sont ou foncées et gommeuses, ou sèches et grises; elles portent des fructifications brunes.

La infección es local en los nudos y en la axila de la hoja, nunca en los entrenudos. Las orillas de las áreas infectadas son de color verde aceitoso, frecuentemente con exudaciones gomosas de color resinoso. Las partes más viejas son ya sea gomosas y obscuras o secas y de color gris, con cuerpos fructíferos de color café.

DISTRIBUTION: Af., A., Aust., Eur., N.A., C.A., S.A.

Cucurbitaceae + <u>Pseudomonas lachrymans</u> (E. F. Smith & Bryan) Carsner =

ENGLISH	Bacterial Spot
FRENCH	Tache Angulaire, Taches Anguleuses des Feuilles du Concombre, Maladie des Taches Anguleuses
SPANISH	Mancha bacterial
PORTUGUESE	Mancha angular
ITALIAN	Batteriosi dei cetrióli
RUSSIAN	пятнистость угловатая бактериальная
SCANDINAVIAN	
GERMAN	Eckige Blattfleckenkrankheit, Eckige Fleckigkeit, Bakterielle Blattfleckenkrankheit
DUTCH	
HUNGARIAN	
BULGARIAN	бактериален пригор, ъгловати петна
TURKISH	Hıyar köşeli yaprak lekesi
ARABIC	
PERSIAN	Làkêhê zăvêyêyê khêyăr كله زاويهاى مضيار
BURMESE	
THAI	
NEPALI	
HINDI	
VIETNAMESE	
CHINESE	
JAPANESE	Hanten-saikin-byô 斑点細菌病

This disease appears on the leaves, stems and fruit. Spots on the fruit are small, nearly circular and watery. They exude a gummy liquid which later dries and becomes a white residue. Infection spreads to the fruit interior which soon becomes a rotten mass. Spots on the foliage are irregular in shape, angular and water-soaked. Under moist conditions bacteria ooze from the spots in tearlike droplets which later dry and turn white. These water-soaked areas later turn gray, die and tear away into large irregular holes.

Iste morbo occurre in folios, pedunculos, e fructos. In le fructos, le maculas es minime, quasi circular, e aquose. Illos exsuda un liquor gummose, le qual plus tarde se desicca e deveni un residuo blanc. Le infection es extende al interior del fructo, le qual putresce rapidemente. In le foliage, le maculas es irregular de forma, angular, e saturate de aqua. Sub conditiones humide, minute guttas de bacterios es exsudate ab le maculas. Plus tarde, illos se desicca e deveni blanc. Iste areas saturate de aqua plus tarde deveni gris, mori, e se separa, lassante grande perforationes irregular.

Cette maladie se présente sur les feuilles, les tiges et les fruits. Les taches sur les fruits sont petites, presque circulaires et gaufrées. Elles exsudent un liquide gommeux qui plus tard en séchant laisse un résidu blanc. L'infection envahit l'intérieur du fruit, qui ne tarde pas à pourrir. Les taches sur le feuillage sont irrégulières, anguleuses et détrempées. En atmosphère humide, les bactéries exsudent des gouttelettes semblables à des larmes qui se dessèchent par la suite et blanchissent. Ensuite, ces plages détrempées deviennent grises, les tissus meurent et se détachent, laissant de grands trous à contours irréguliers.

Esta enfermedad aparece sobre las hojas, tallos y frutos. Las manchas del fruto son pequeñas, casi circulares y acuosas. Exudan un liquído gomoso que más tarde se seca y se vuelve un residuo blanco. La infección se disemina al interior del fruto el cuál pronto se vuelve una masa podrida. Las manchas en el follaje son de forma irregular, angulares y empapadas de agua. Bajo condiciones húmedas la bacteria sale de las manchas en gotitas como lágrimas que más tarde se sacan y se vuelven blancas. Estas áreas empapadas de agua se agrandan y se vuelven de color gris, mueren y se caen dejando agujeros grandes e irregulares.

DISTRIBUTION: Af., A., Aust., Eur., N.A., C.A., S.A.

74

Cucumis sativus (C. melo and other Cucurbitaceae) + Pseudoperonospora cubensis (Berk. & Curt.) Rostowzew =

ENGLISH   Downy Mildew
FRENCH   Mildiou des Cucurbitacées
SPANISH   Mildiú, Mildiú felpudo, Mildiú velloso, Mildiú lanoso
PORTUGUESE   Mildio
ITALIAN   Peronospora delle cucurbitacee
RUSSIAN   пероноспороз, ложная мучнистая роса
SCANDINAVIAN
GERMAN   Falscher Mehltau
DUTCH
HUNGARIAN
BULGARIAN
TURKISH   Kavun mildiyösü
ARABIC   Bayad Zaghaby   بياض زغبى
PERSIAN   Sēfēdåkê dôroōghēyē khēyår   سفیدک درونی خیار
BURMESE
THAI   Ra-nam-karng   ราน้ำค้าง
NEPALI   Kakro Ko dhuse rog   काक्रोको धुसे रोग
HINDI   Mriduromil phaphundi   मृदुरोमिल फाफुन्दी
VIETNAMESE   Cháy lá và trái non
CHINESE   Hú gwua lùh jìunn bìng   胡瓜露菌病
JAPANESE   Beto-byð   べと病

Symptoms of this disease appear as pale green areas separated by islands of darker green which soon change to yellow angular spots between the leaf veins. These appear first upon the older leaves at the center of the plant. During moist weather the lower surface sometimes becomes covered with a faint purplish layer of the fungus. The spots increase in size and number and soon the entire leaf dies. Fruits are seldom affected, but are dwarfed in growth and have poor flavor.

Symptomas de iste morbo es areas verde clar separate per insulas de verde plus obscur, le quales rapidemente se cambia in jalne maculas angular inter le venas del folios. Istos se manifesta al initio in le folios plus vetule al centro del planta. In tempore humide le superfacie inferior es a vices coperte de un pallide strato fungal purpuree. Le maculas cresce in dimension e deveni plus numerose, e tosto tote le folio mori. Le fructos es rarmente afflicte, ma lor crescimento es impedite e lor gusto es disagradabile.

Les symptômes de cette maladie apparaissent comme des plages vert pâle, séparées par des îlots vert foncé, se changeant rapidement en taches jaunes, anguleuses, entre les nervures de la feuille. Elles se présentent d'abord sur les vieilles feuilles du milieu de la plante. Par temps humide, la face inférieure se couvre parfois d'une couche fongique pourpre pâle. Les taches augmentent en volume et en nombre, et bientôt la feuille entière meurt. Les fruits sont rarement atteints, mais leur croissance est réduite et ils sont peu savoureux.

Los síntomas de esta enfermedad aparecen como áreas de color verde pálido separadas por islas de color verde más obscuro que pronto cambian y se convierten en manchas amarillas angulares entre las venas de las hojas. Estas aparecen primero sobre las hojás más viejas en el centro de la planta. Durante el tiempo húmedo el envés algunas veces se cubre con una capa del hongo, de color levemente purpúrea. Las manchas aumentan en tamaño y en número y pronto la hoja entera muere.. Las frutas casi nunca son afectadas, pero se achaparran y tienen sabor pobre.

DISTRIBUTION: Af., A., Aust., Eur., N.A., C.A., S.A.

Cydonia oblonga + Fabraea maculata (Lév.) Atk. =

ENGLISH	Leaf Blight, Black Spot of Fruit
FRENCH	Entomosporiose, Taches Brunes des Feuilles
SPANISH	Mancha negra de las hojas y frutos, Tizón de la hoja
PORTUGUESE	
ITALIAN	
RUSSIAN	пятнистость бурая, буроватость листьев
SCANDINAVIAN	
GERMAN	Blattbräune, Enthomosporium-Blattfleckenkrankheit
DUTCH	
HUNGARIAN	
BULGARIAN	кафяви петна
TURKISH	
ARABIC	
PERSIAN	
BURMESE	
THAI	
NEPALI	
HINDI	
VIETNAMESE	
CHINESE	
JAPANESE	

The disease is found on leaves, fruit and shoots. On the leaves, the spots appear first as small purple dots. These later extend to circular deep purple or dark brown lesions. When the spots reach their maximum diameter, a small black pimple appears in the center of each spot. When the leaf spots are numerous, extensive defoliation results. On the fruit, the spots are black and become slightly sunken; cracking of the fruit is pronounced. The lesions on the twigs consist of purple or black areas which later form a canker.

Le morbo se trova in folios, fructos, e ramettos nove. In foliage, le maculas se manifesta al initio como minute punctos purpuree. Illos cresce plus tarde in lesiones circular e purpuree o brun obscur. Quando le maculas ha attingite lor diametro maximal, un minute pustula nigre se manifesta al centro de cata macula. Si le maculas in le folios es numerose, defoliation extensive occurre. In fructos, le maculas es nigre e deveni un pauc depresse. Fissuras in le fructos es notabile. In ramettos, le lesiones es areas purpuree o nigre, le quales plus tarde se transforma in cancere.

La maladie se rencontre sur les feuilles, les fruits et les jeunes pousses. Sur les feuilles, les taches se présentent d'abord comme de petits points pourprés. Ceux-ci s'agrandissent ensuite pour devenir des taches rondes pourpres ou brun foncé. Lorsque les taches atteignent leur grosseur maximum, une petite pustule noire apparaît au centre. Quand les taches foliaires sont nombreuses, il s'ensuit une défoliation importante. Sur le fruit, les taches sont noires et deviennent légèrement déprimées; si le fruit est assez gros, il peut s'y former des craquelures. Sur les rameaux, les lésions forment des plages pourprées ou noires qui se transforment en chancres.

Esta enfermedad se encuentra sobre las hojas, frutos y brotes. Sobre las hojas, las manchas aparecen primero como pequeños puntos purpúreos. Estas manchas más tarde crecen hasta convertirse en lesiones circulares de color púrpura encendido o café obscuro. Cuando las manchas alcanzan su diámetro máximo, un pequeño montículo negro aparece en el centro de cada mancha. Cuando las manchas de la hoja son numerosas, resulta una defoliación extensa. Sobre la fruta, las manchas son negras y se vuelven algo hundidas y se pronuncian las rajaduras del fruto. Las lesiones en las ramillas consisten en áreas de color púrpura o negras que más tarde forman un cancro.

DISTRIBUTION: Af., A., Aust., Eur., N.A., C.A., S.A.

76

Daucus carota var. sativa + Cercospora carotae (Pass.) Solh. =

ENGLISH   Leaf Spot, Blight
FRENCH   Brûlure Cercosporéenne, Taches Foliaires, Cercosporose
SPANISH   Mancha cercospora, Mancha foliar, Tizón
PORTUGUESE   Cercosporiose, Mancha de Cercospora
ITALIAN
RUSSIAN   церкоспороз
SCANDINAVIAN
GERMAN   Cercospora-Blattfleckkenkrankheit
DUTCH
HUNGARIAN
BULGARIAN
TURKISH
ARABIC
PERSIAN
BURMESE
THAI
NEPALI
HINDI
VIETNAMESE
CHINESE   Hú lúoh bòh ban dǐan bìng   胡蘿蔔斑點病
JAPANESE   Hanten-byð 斑点病

Lesions form on the leaves, petioles, stems, and floral parts. Nearly circular tan or gray to brown or black spots occur, at first mostly at the lobes. The foliage may shrivel and blacken as the spots increase in number. Elliptic spots occur on the petioles and stems, usually with pale centers and dark margins. Early attacks on the floral parts cause shriveling before the seed is borne, but late attacks may show no outward symptoms.

Lesiones se developpa in le folios, petiolos, pedunculos, e partes floral. Maculas quasi circular, bronzate, gris, burn, o nigre occurre, al initio specialmente, in le lobos. Le foliage es sovente nigrate e crispate quando le maculas deveni plus numerose. Maculas elliptic occurre in le petiolos e in le pedunculos, usualmente con centros pallide e margines obscur. Attaccos precoce in le partes floral face los crispar se ante que le granos se developpa. Attaccos plus tarde sovente monstra nulle symptomas visibile.

Des lésions se forment sur les feuilles, les pétioles, les tiges et les parties florales. Des taches presque circulaires, couleur du tan ou grises à brunes ou noires apparaissent, d'abord principalement sur les lobes. Les feuilles peuvent se ratatiner et noircir à mesure que le nombre de taches augmente. Des taches elliptiques se forment sur les pétioles et les tiges, avec d'ordinaire le centre blanc et les marges foncées. Des attaques précoces sur les parties florales provoquent le ratatinement avant que la graine ne se forme, tandis que les attaques tardives peuvent ne pas laisser paraître de symptômes extérieurs.

Las lesiones se forman sobre las hojas, pecíolos, tallos y partes florales. Aparecen manchas casi circulares de color crema a gris hasta café o negra, primero casi sólo en los lóbulos. El follaje puede ajarse o ennegrecer mientras que las manchas crecen en número. En los pecíolos y tallos aparecen manchas elípticas, generalmente con centros pálidos y orillas obscuras. Ataques tempranos en las partes florales causan que se ajen antes de que la semilla nazca, pero los ataques tardíos pueden no mostrar síntomas externos.

DISTRIBUTION: A., Aust., Eur., N.A.

Daucus carota var. sativa + Macrosporium carotae Ell. & Langl. (Alternaria dauci (Kühn.) Groves & Skolko)=

ENGLISH	Leaf Blight
FRENCH	Brûlure Alternarienne, Brûlure des Feuilles
SPANISH	Tizón tardío, Mancha de la hoja, Lancha
PORTUGUESE	Queima do fôlha, Alternariose, Mancha de alternaria
ITALIAN	
RUSSIAN	альтернариоз преимущественно листьев
SCANDINAVIAN	
GERMAN	Möhrenschwärze, Möhrenverderber, Möhrenblattbrand
DUTCH	
HUNGARIAN	
BULGARIAN	
TURKISH	
ARABIC	
PERSIAN	
BURMESE	
THAI	
NEPALI	
HINDI	
VIETNAMESE	
CHINESE	Hú lúoh bòh hei yèh ku bìng　胡蘿蔔黑葉枯病
JAPANESE	Kurohagare-byð　黒葉枯病

On the foliage, small dark brown to black spots edged with yellow form, at first mostly along the leaf margins. As the spots increase in number, the intervening tissue dies and the entire leaf shrivels and dies. Under moist conditions, this process is so rapid that it resembles frost injury. Sometimes, when the elongated spots occur on the petioles, the entire leaf dies without foliage spots. Damping-off of seedlings, blight of seed stalks and black decay of roots are also caused by this fungus.

In le foliage, minute maculas brun obscur o nigre, con bordos jalne, se forma, al initio specialmente in le margines del folios. Quando le maculas deveni plus numerose, le texito inter illos mori e le folio integre se crispa e mori. Sub conditiones humide, iste mortification occurre si rapidemente que illo resimilia le damno causate per gelo. A vices, quando le maculas elongate se forma in le petiolos, tote le folio mori sin monstrar maculas foliar. Iste fungo causa anque putrefaction del plantulas, alternariosis del pedunculos, e putrefaction nigre del radices.

Il se forme sur les feuilles, d'abord principalement le long de la marge, de petites taches brun foncé a noires bordées de jaune. A mesure que le nombre de taches s'accroît, les tissus adjacents meurent et la feuille entière se recroqueville et meurt. A l'humidité, cela se produit si vite qu'on dirait un coup de froid. Parfois, lorsque des taches allongées se forment sur le pétiole, la feuille entière meurt sans qu'il y ait de tache foliaire. Ce champignon peut aussi causer la fonte des semis, la brûlure des pédoncules et la pourriture noire des racines.

En el follaje se forman pequeñas manchas de color café obscuro a negras, con bordes amarillos, al principio a lo largo de los márgenes de la hoja. A medida que las manchas aumentan en número, el tejido entre las venas muere y toda la hoja se aja y muere. Bajo condiciones húmedas, este proceso es tan rápido que semaja un daño causado por heladas. Algunas veces, cuando las manchas alargadas aparecen en los pecíolos, toda la hoja muere sin manchas foliares. El mal del talluelo de las plantitas, el tizón de los tallos de la semilla y podredumbre negra de las raíces también son causados por este hongo.

DISTRIBUTION: Af., A., Aust., Eur., N.A., C.A., S.A.

78

Fragaria chiloensis + Corynebacterium fascians (Tilford) Dowson =

ENGLISH   Fasciation, Leaf Gall. Cauliflower Disease of Strawberry
FRENCH   Fasciations des Tiges
SPANISH   Fasciación, Agalla foliar, Enfermedad de coliflor de la fresa
PORTUGUESE
ITALIAN   Fasciazione
RUSSIAN   "цветная капуста," фасциация, уродливость
SCANDINAVIAN
GERMAN   Blumenkohlkrankheit, Fasziation, Verbänderung
DUTCH
HUNGARIAN
BULGARIAN
TURKISH
ARABIC
PERSIAN
BURMESE
THAI
NEPALI
HINDI
VIETNAMESE
CHINESE
JAPANESE

Diseased plants may be recognized early by the abnormal character of the crown. The bud is small and poorly developed. The hairs covering the unfolding buds are greatly reduced in number. The development of foliage is retarded. As the leaves unfold and elongate, they become distorted and abnormal. When fully formed, leaves are small and crumpled and often curled. Lightly affected plants may have normal-sized leaves but may show some crinkling. These symptoms are usually associated with the fungus and a nematode.

Plantas morbide es facilemente remarcate de bon hora a causa del character anormal del corona. Le buttones es minute e mal developpate. Le pilos que coperi le buttones emergente es multo minus numerose que normalmente. Developpamento de foliage es impedite. Quando le folios se disrola e deveni elongate, illos deveni distorte e anormal. Quando le folios es totalmente developpate, illos es minute e crispate, sovente severmente. Plantas que es solmente un pauc afflicte pote monstrar folios de dimension normal, ma con corrugation anque. Iste symptomas se associa usualmente con le fungo e con un nematodo.

On peut reconnaître assez tôt les plantes malades à l'aspect anormal de la couronne. Le bourgeon est petit et peu développé. Il y a forte réduction du nombre de poils recouvrant les bourgeons qui s'ouvrent. Le feuillage se développe tardivement. Les feuilles deviennent difformes et anormales dès leur apparition. A maturité, les feuilles sont petites, plissees et souvent enroulées. Les plantes légèrement atteintes peuvent produire des feuilles de taille normale, mais parfois d'un aspect quelque peu ondulé. En outre, le pathogène s'accompagne ordinairement d'un nématode.

Las plantas enfermas pueden ser reconocidas temprano por el carácter anormal de la corona. El capullo es pequeño y pobremente desarrollado. Los pelillos que cubren los capullos sin abrir están bastante reducidos en su número. El desarrollo del follaje se retarda. A medida que las hojas se abren y alargan, se vuelven deformes y anormales. Cuando están completamente formadas, las hojas son pequeñas y encrespadas y frecuentemente enrolladas. Las plantas ligeramente afectadas pueden tener hojas de tamaño normal pero pueden mostrar algo de encrespamiento. Estos síntomas están generalmente asociados con hongos y un nemátodo.

DISTRIBUTION: Eur., N.A.

Fragaria chiloensis (and others) + Cylindrocarpon radicicola Wollenw. (Nectria radicicola
Gerlach & Nilsson) =

ENGLISH	Cortical Root Rot Complex, Black Root
FRENCH	Taches Foliaires
SPANISH	Complejo cortical de podredumbre radicular, Podredumbre negra
PORTUGUESE	
ITALIAN	
RUSSIAN	
SCANDINAVIAN	
GERMAN	
DUTCH	
HUNGARIAN	
BULGARIAN	
TURKISH	
ARABIC	
PERSIAN	
BURMESE	
THAI	
NEPALI	
HINDI	
VIETNAMESE	
CHINESE	
JAPANESE	

A field affected by this disease presents an uneven appearance due to dwarfing of the diseased
plants and to gaps caused by death of the plants. The first evidence of infection is the appearance
of brown areas on the normally white or tan roots. The lesions may be elliptical and on one
side of the root or may extend entirely around the root. Often the small fibrous rootlets are
killed back to the roots from which they arise. The localized lesions soon extend up and down
the root which becomes blackened. The outer bark sloughs off and then the root decays.

Un campo in le qual iste morbo occurre appare irregular a causa del nanismo del plantas morbide e del lacunas, le quales es le resultato del morte del plantas. Le prime symptoma de infection es areas brun in le radices que es normalmente blanc o bronzate. Le lesiones es elliptic; illos pote occurrer o in un latere del radice o extender se circum tote le radice. Sovente
le minute radiculas fibrose se mortifica usque al radices ab le quales illos proveni. Le lesiones
local rapidemente se extende e in alto e in basso per tote le radice que deveni nigrate. Le cortice
exterior se separa e cade e alora le radice putresce.

Un champ atteint de cette maladie apparaît inégal par suite du nanisme des plantes malades et
des vides causés par la mort. L'apparition de lésions brunes sur des racines normalement
blanches ou couleur du tan constitute le premier symptôme d'infection. Les lésions peuvent
être elliptiques, se localiser sur un côté d'une racine ou couvrir toute la racine. Souvent, les
petites radicelles fibreuses sont détruites jusqu'à la racine dont elles portent. Les lésions
localisées ne tardent pas à envahir toute la racine, qui devient noirâtre. La racine se dépou-
ille de son écorce externe, après quoi elle pourrit.

Un campo atacado por esta enfermedad presenta una apariencia dispareja debido al achaparramiento de las plantas enfermas y a brechas causadas por plantas muertas. La primer evidencia de infección es el aparecimiento de áreas de color café en las raíces normalmente blancas
o crema. Las lesiones pueden ser elípticas y estar a un lado de la raíz o pueden extenderse
enteramente alrededor de la raíz. Frecuentemente las pequeñas raicillas fibrosas mueren atrás
hacia las raíces de las cuáles brotan. Las lesiones localizadas se extienden pronto arriba y
abajo de la raíz, la cuál se ennegrece. Las corteza exterior se desprende y luego la raíz
se pudre.

DISTRIBUTION: Af., A., Aust., Eur., N.A., C.A., S.A.

80

Fragaria chiloensis (Fragaria spp.) + Dendrophoma obscurans (Ell. & Ev.) H. W. Anderson =

ENGLISH	Leaf Blight
FRENCH	Brûlure des Feuilles, Taches Angulaires des Feuilles
SPANISH	Tizón de la hoja
PORTUGUESE	Crestamento das fôlhas
ITALIAN	
RUSSIAN	пятнистость коричневая дендрофомозная, пятнистость угловатая
SCANDINAVIAN	
GERMAN	Dendrophoma-Blattfleckenkrankheit, Eckige Blattfleckenkrankheit
DUTCH	
HUNGARIAN	
BULGARIAN	
TURKISH	
ARABIC	
PERSIAN	
BURMESE	
THAI	
NEPALI	
HINDI	
VIETNAMESE	
CHINESE	
JAPANESE	Rinpan-byð 輪斑病

The disease is most conspicuous on the leaves. Usually the spots on the leaflet are limited to one to five. When first observed, the young spots are uniformly reddish-purple and almost circular in outline. If they are near one of the main veins, the spots are elliptical. Later three zones may be observed: an outer purple zone which gradually shades off into the normal green leaf; a light brown zone; and a dark brown area.

Iste morbo es multo remarcabile in le folios. Usualmente le maculas in le foliettos es limitate a un usque a cinque. Al initio, le maculas juvene es uniformemente purpuree-rubiette e quasi circular de forma. Si illos se trova presso un vena principal, le maculas es elliptic. Plus tarde, tres zonas es evidente: un purpuree zona exterior que gradualmente se cambia in le verde normal del folio, un zona brun clar, e un area brun obscur.

La maladie est plus apparente sur les feuilles. D'ordinaire, on ne voit pas plus de cinq taches sur les folioles. Dès leur apparition, les petites taches sont de couleur rouge pourpré et presque rondes. Les taches situées près des nervures principales sont elliptiques. Plus tard, trois zones peuvent apparaître: une zone extérieure pourprée qui se fond graduellement avec le vert normal de la feuille, une zone brun pâle et une zone brun foncé.

La enfermedad es más conspicua en las hojas. Generalmente las manchas en la hojita son limitadas de una a cinco. Cuando se observan primero, las manchas jóvenes son uniformemente rojizas-púrpura y casi circulares en su forma. Si están cerca de una de las venas principales, las manchas son elípticas. Más tarde se pueden observar tres zonas: una zona purpúrea exterior que gradualmente cambia y se vuelve del color verde normal de la hoja; una zona café claro y una café obscuro.

DISTRIBUTION: Af., A., Aust., N.A., S.A.

Fragaria spp. + <u>Diplocarpon</u> <u>earlianum</u> (Ell. & Everh.) Wolf (<u>Gloeosporium</u> <u>fragariae</u>
(Lib.) Mont.) =

ENGLISH   Leaf Scorch, Leaf Spot
FRENCH   Tache Pourpre
SPANISH   Mancha foliar, Quemadura de la hoja
PORTUGUESE   Mancha de Diplocarpon, Podridão do fruto, Antracnose do fruto
ITALIAN
RUSSIAN
SCANDINAVIAN   Jordbærbrunflekk
GERMAN   Rotfleckenkrankheit
DUTCH
HUNGARIAN
BULGARIAN
TURKISH
ARABIC
PERSIAN
BURMESE
THAI
NEPALI
HINDI
VIETNAMESE
CHINESE
JAPANESE

This disease causes lesions on all aboveground parts of the plant except the fruit. The first symptoms are small dark purple spots which are scattered over the upper surface of the leaflets. In the mature condition the spots are large and irregular in outline and never show the white central areas which are characteristic of other leaf spots. Often large areas of the leaflet or the entire leaf take on a reddish or light purple hue quite distinct from that produced by any other disease of the host.

Iste morbo produce lesiones in tote le partes epigee del planta, salvo le fructo. Le prime symptoma es minute maculas purpuree obscur, le quales es disperse per le superfacie superior del foliettos. In su stato maturate le maculas es grande e irregular de forma e non monstra le blanc areas central le quales es un characteristica de altere maladias que se manifesta como maculas foliar. Frequentemente, grande areas del foliettos o de tote le folio monstra un tinta rubiette or purpuree clar le qual es assatis differente del symptomas causate per ulle altere morbo de iste planta-hospite.

Cette maladie cause des lésions sur toutes les parties aériennes de la plante à l'exception du fruit. Les premiers symptômes sont de petites taches pourpre foncé disseminées a la face supérieure des folioles. A maturité, ces taches sont grosses et irrégulières et ne présentent jamais de plages centrales blanches, comme c'est le cas d'autres taches foliaires. Souvent, de grandes plages se forment sur les folioles ou sur toute la feuille, dont la teinte rougeâtre ou pourpre pâle est très distincte des teintes qui resultent des autres maladies du fraisier.

Esta enfermedad causa lesiones en todas las partes aéreas de la planta, excepto en el fruto. Los primeros síntomas son pequeñas manchas de color púrpura obscuro, que están diseminadas sobre el haz de las hojillas. Cuando están maduras las manchas son grandes e irregulares en su forma y nunca muestran las áreas centrales blancas que son características de otras manchas foliares. Frecuentemente las áreas grandes de la hojilla o la hoja entera se torna de color rojizo o púrpura claro, muy distinto del producido por cualquier otra enfermedad de la hospedera.

DISTRIBUTION: Af., A., Aust., Eur., N.A., C.A.,S.A.

Fragaria chiloensis + *Mycosphaerella* fragariae (Tul.) Lindau =

ENGLISH	Leaf Spot
FRENCH	Tache Commune, Tache Brunes des Feuilles, Maladie des Taches Rouges
SPANISH	Mancha de la hoja, Mancha común de la hoja, Mancha foliar
PORTUGUESE	Mancha de Micosferela
ITALIAN	Vaiolatura rossa delle fragole
RUSSIAN	пятнистость белая, рамуляриоз
SCANDINAVIAN	Jordbærbladpletsyge, Ögonfläcksjuka
GERMAN	Weissfleckenkrankheit, Weifleckenkrankheit
DUTCH	Bladvlekkenziekte
HUNGARIAN	Mikoszferellás fehér levélfoltosság
BULGARIAN	бели листни петна
TURKISH	Çilek yaprak lekesi
ARABIC	Tabbakoo Awrak تبقع أوراق
PERSIAN	
BURMESE	
THAI	
NEPALI	
HINDI	
VIETNAMESE	
CHINESE	
JAPANESE	Zyanome-byô じゃのめ病

This leaf spot disease is most frequently evident on the blades, but may appear on the fruit and fruit stems. The lesions are first seen on the upper surface as small, deep purple, some-what indefinite areas. As the spot enlarges, the central area becomes brown, but soon turns to a definite white spot in older leaves or to a light brown in young leaves. On the undersurface the coloring is less intense. Numerous spots cause death of the leaflet.

Iste morbo es evidente le plus frequentemente in le laminas, ma occurre anque in le fructos e in le pedunculos que sustene le fructos. Le lesiones se manifesta al initio in le superfacie superior como minute areas purpuree, un pauc indefinite. Quando le macula cresce, le area central deveni brun e alora tosto un definite macula blanc in le folios plus vetule o brun clar in le folios plus juvene. In le superfacie inferior le coloration non es si intense. Maculas numerose face le fol-iettos morir.

Cette maladie apparaît le plus souvent sur les feuilles, mais on peut aussi l'observer sur les fruits et les pédoncules. On aperçoit d'abord des petites lésions plutôt irrégulières, pourpre foncé, à la surface des feuilles. Dès que la tache s'agrandit, le centre brunit, puis se trans-forme bientôt en une tache blanche bien délimitée sur les vieilles feuilles, ou d'un brun pâle sur les jeunes feuilles. La couleur des taches est moins prononcée à la face inférieure des feuilles. Quand elles sont nombreuses, les taches peuvent détruire les folioles.

Esta enfermedad de manchas foliares es frecuentemente más evidente en las hojas, pero puede aparecer en los frutos y pedunculos. Las lesiones s e ven primero en el haz y son áreas pe-queñas, de color púrpura encendido y algo indefinidas. A medida que la mancha crece, el área central se vuelve café pero pronto se convierte en una definida mancha blanca en las hojas maduras o de color café claro en las hojas jóvenes. En el envés el colorido es menos intenso. Manchas numerosas causan la muerte de las hojillas.

DISTRIBUTION: Af., A., Aust., Eur., N.A., C.A., S.A.

Glycine <u>max</u> + Peronospora <u>manshurica</u> (Naoumff) Syd. =

ENGLISH   Downy Mildew
FRENCH   Mildiou
SPANISH   Mildeo velloso, Mildiú lanoso
PORTUGUESE   Mildio em sementes
ITALIAN
RUSSIAN   пероноспороз, пожная мучнистая роса
SCANDINAVIAN
GERMAN   Falscher Mehltau
DUTCH
HUNGARIAN
BULGARIAN
TURKISH
ARABIC
PERSIAN
BURMESE
THAI
NEPALI
HINDI   Mriduromil phaphundi   मृदुरोमिल फाफुन्दी.
VIETNAMESE
CHINESE   Dà dòw lù jîunn bîng   大豆露菌病
JAPANESE   Beto-byð   べと病

This disease is characterized in its early stages by indefinite, yellowish-green areas on the upper surface of the leaves. In severe cases entire leaflets are discolored. As infection progresses, the diseased areas become grayish-brown to dark brown and are surrounded by yellowish-green margins. Severely infected leaves fall prematurely. The fungus grows within the plant, invades the pods and covers some seeds with a white crust. The gray fungus growth also developes on the lower side of leaves.

Iste morbo se characterisa, in le stadios precoce, per indefinite areas verde-jalnette in le superfacie superior del folios. In casos sever le foliettos integre es anormalmente colorate. Quando le infection se avantia, le areas morbide deveni brun-gris o brun obscur e monstra margines verde-jalnette. Folios severmente afflicte se separa e cade prematurmente. Le fungo cresce intra le planta, invade le siliquas, e coperi alicun granos de un crusta blanc. Le strato fungal gris se developpa anque in le latere inferior del folios.

A ses premières phases, cette maladie se caractérise par des plages indéfinies vert jaunâtre à la face supérieure des feuilles. Dans les cas graves, des folioles entières sont affectées. Quand la maladie progresse, les plages malades deviennent brun grisâtre à brun foncé et sont bordées de marges vert jaunâtre. Les feuilles gravement infectées tombent prématurément. Le champignon croît à l'intérieur de la plante, envahit les gousses et recouvre quelques graines d'une croûte blanche. Du mycélium gris se forme aussi sur la face inférieure des feuilles.

Esta enfermedad se caracteriza al principio por el aparecimiento de áreas indefinidas, de color verde-amarilloso sobre el haz de las hojas. En casos severos la hojita entera se decolora. A medida que la infección progresa, las áreas enfermas se vuelven de color café-grisáceo a café obscuro y están rodeadas por márgenes verde-amarillosos. Las hojas severamente infectadas se caen prematuramente. El hongo crece dentro de la planta, invade las vainas y cubre algunas semillas con una costra blanca. El crecimiento gris del hongo también se desarrolla en el envés de las hojas.

DISTRIBUTION: A., Eur., N.A., C.A.

84

Glycine max + Pseudomonas glycinea Coerper =

ENGLISH   Bacterial Blight
FRENCH   Brûlure Bactérienne, Taches Foliaires Angulaires Bactériennes
SPANISH   Bacteriosis, Tizón bacterial
PORTUGUESE   Crestamento bacteriano
ITALIAN
RUSSIAN   пятнистость угловатая бактериальная, ожог бактериальный
SCANDINAVIAN
GERMAN   Eckige Fleckigkeit, Bakterielle Blattfleckenkrankheit, Bakterienbrand
DUTCH
HUNGARIAN
BULGARIAN
TURKISH
ARABIC
PERSIAN
BURMESE
THAI   Bi-Mai ใบไหม้
NEPALI
HINDI
VIETNAMESE
CHINESE   Dà dòw shìh jiunn shìng lìi ku bìng 大豆細菌性立枯病
JAPANESE   Hanten-saikin-byô 斑点細菌病

This disease produces small, angular wet spots which turn yellow and then brown as the tissue dies. Many small infections sometimes run together. The resulting dead areas may fall out. Heavy infection may cause dropping of lower leaves. Diseased leaves may be severely torn by high wind during rainstorms.

Iste morbo produce minute maculas angular, humide, le quales deveni jalne e alora brun quando le texito mori. Sovente multe infectiones minute coalesce, e le areas morte que es le resultato pote separar se e cader. Infection sever pote facer le folios inferior cader. Folios morbide es a vices lacerate per ventos violente durante tempestas pluviose.

Cette maladie produit de petites taches aqueuses angulaires, qui deviennent jaunes, puis brunes lorsque les tissus meurent. Plusieurs petites infections confluent parfois, ce qui peut entraîner la chute des plages mortes. Une forte infection peut causer la chute des feuilles du bas. Les grands vents d'orage peuvent endommager gravement les feuilles malades.

Esta enfermedad produce pequeñas manchas angulares, acuosas, que se vuelven de color amarillo y luego café a medida que el tejido muere. Muchas infecciones pequeñas se juntan a veces; las áreas muertas que resultan pueden caerse. Las infecciones fuertes pueden causar la caída de las hojas inferiores. Las hojas enfermas pueden ser arrancadas por los vientos fuertes durante las tormentas.

DISTRIBUTION: A., Aust., Eur., N.A.

Glycine max + Septoria glycines Hemmi =

ENGLISH	Brown Spot
FRENCH	Tache Brune, Septoriose Brune
SPANISH	Mancha café
PORTUGUESE	
ITALIAN	
RUSSIAN	пятнистость ржавая, септориоз ржавый
SCANDINAVIAN	
GERMAN	Septoria-Braunfleckenkrankheit
DUTCH	
HUNGARIAN	
BULGARIAN	бактериоза
TURKISH	
ARABIC	
PERSIAN	
BURMESE	
THAI	
NEPALI	
HINDI	
VIETNAMESE	
CHINESE	Dà dòw hor wén bìng 大豆褐紋病
JAPANESE	Katumon-byð 褐絞病

This disease appears early in the growing season on the primary leaves of young plants as angular, reddish-brown lesions which vary greatly in size. Later the disease occurs on mature leaves, stems and pods. Heavily infected leaves gradually turn yellow and fall prematurely. Defoliation pregresses from the base toward the top of the plant. In severe cases the lower half of the stem may be bare of leaves before maturity.

Iste morbo se manifesta in le prime partes del saison in le folios primari de plantas juvene como angular lesiones brun-rubiette, le quales se varia multo in dimension. Plus tarde le morbo se extende al folios, pedunculos, e siliquas maturate. Folios severmente afflicte gradualmente deveni jalne e cade prematurmente. Defoliation se avantia ab le pede del planta usque al partes apical. In casos sever le medietate inferior del pedunculo pote esser denudate de folios ante maturation.

Cette maladie apparaît tôt durant la saison de croissance sur les feuilles primaires des jeunes plantes sous forme de lésions angulaires brun rougeâtre de dimensions très variables. Par la suite, la maladie se montre sur les vieilles feuilles, les tiges et les gousses. Les feuilles fortemente infectées jaunissent graduellement et tombent prématurément. La défoliation progresse du bas de la plante vers le sommet. Dans les cas graves, la moitié inférieure de la tige peut être dépouillée de ses feuilles avant maturité.

Esta enfermedad aparece al principio de la época de crecimiento sobre las hojas primarias de las plantitas jóvenes, como lesiones angulares, café-rojizas que varían mucho en el tamaño. Más tarde la enfermedad aparece en las hojas maduras, tallos y vainas. Las hojas fuertementee infectadas gradualmente se vuelven de color amarillo y se caen prematuramente. La defoliación progresa de la base hacia la punta de la planta. En casos severos la mitad inferior del tallo puede quedar sin hojas antes de madurar.

DISTRIBUTION: A., N.A.

Gossypium spp. + Alternaria macrospora Zimm. =

ENGLISH	Leaf Spot
FRENCH	Taches Foliaires, Alternariose
SPANISH	Mancha foliar
PORTUGUESE	
ITALIAN	
RUSSIAN	пятнистость концентрическая альтернариозная , альтернариоз листьев
SCANDINAVIAN	
GERMAN	Alternaria-Blattfleckenkrankheit
DUTCH	
HUNGARIAN	
BULGARIAN	
TURKISH	
ARABIC	Tabbakoo Awrak, Affan El-loze عـفن اللوز , تبقع أوراق
PERSIAN	Älternarēya pánbeh آلترناریای پنبه
BURMESE	
THAI	Bi-Mai ใบไหม้
NEPALI	
HINDI	
VIETNAMESE	
CHINESE	Mían hwa hei ban bìng 棉花黑斑病
JAPANESE	Kokuhan-byô 黒斑病

This disease is caused by a fungus which produces an abundance of papery, rusty-brown spots of irregular shape. The size of the spots vary from 1/8 to 1/2 inch in diameter. The spots enlarge and become more apparent late in the growing season, forming a series of concentric markings. In many cases the disease appears to be associated with rust injury and potash hunger and often causes severe defoliation.

Le causa de iste morbo es un fungo que produce numerose maculas similar a papiro, burn e (fer) ruginose, de forma irregular. Le dimension del maculas se varia ab 0, 3 cm usque a 1, 25 cm in diametro. Le maculas cresce e deveni plus remarcabile plus tarde in le saison, faciente un serie de marcationes concentric. Frequentemente le morbo sembla associar se con damno caus- ate per (fer) rugine e per carentia potassic, e illo causa defoliation sever.

Le champignon qui cause cette maladie produit une grande quantité de taches brun rouille de forme irrégulière et de consistance de papier. La grandeur des taches varie de 1/8 à 1/2 pouce de diamètre. Les taches s'agrandissent et deviennent plus apparentes tard dans la saison de croissance alors qu'il s'y forme une série de marques concentriques. Dans plusieurs cas, la maladie paraît associée avec la rouille et la carence de potasse. Souvent, elle cause une défoliation grave.

Esta enfermedad es causada por un hongo que produce una abundacia de manchas de textura como de papel, de color café-óxido, de forma irregular. El tamaño de las manchas varía de 1/8 a 1/2 pulgada de diámetro. Las manchas crecen y se vuelven más aparentes tarde durante la época de cultivo, formando una serie de marcas concéntricas. En muchos casos la enferme- dad parece estar asociada con daño de roya y deficiencia de potasio y frecuentemente causa de- foliación severa.

DISTRIBUTION: Af., A., Aust., Eur., W.I., S.A.

Fig. 1. Onion-Purple Blotch; ARS

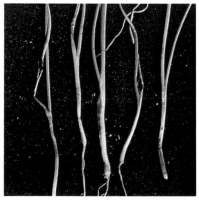

Fig. 2. Onion-Neck Rot; Dept. of Pl. Path., Purdue Univ.

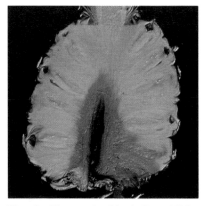

Fig. 3. Onion-Smudge; Arden F. Sherf, Cornell Univ.

Fig. 4. Onion-Downy Mildew; Arden F. Sherf, Cornell Univ.

Fig. 5. Onion-Blight; Dept. of Bot. & Pl. Path, Mich. St. Univ.

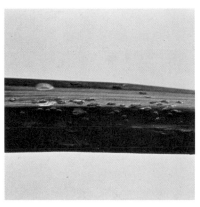

Fig. 6. Garlic-Rust; Charles Y. Yang, Taiwan, Republic of China

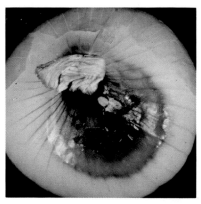

Fig. 7. Onion-White Rot; Peter B. Adams, ARS

Fig. 8. Onion-Smut; Dept. of Bot.& Pl. Path, Mich. St. Univ.

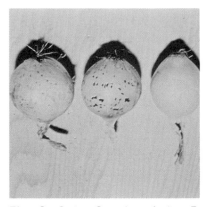

Fig. 9. Pineapple-Black Rot; ARS

PLATE 1

Fig. 10. Pineapple-Heart Rot; Charles Y. Yang, Taiwan, Republic of China

Fig. 11. Celery-Leaf Spot; ARS

Fig. 12. Peanut-Brown Leaf Spot; Lawrence I. Miller, Dept. of Pl. Path. V.P.I.

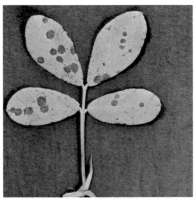

Fig. 13. Peanut-Leaf Spot; Lawrence I. Miller, Dept. of Pl. Path, V.P.I.

Fig. 14. Peanut-Stem Rot; Lawrence I. Miller, Dept. of Pl. Path, V.P.I.

Fig. 15. Peanut-Leaf Rust; Pl. Disease Lab., Ft. Detrick, Md.

Fig. 16. Oat-Head Mold

Fig. 17. Oat-Leaf Spot; ARS

Fig. 18. Oat-Seedling Blight

PLATE 2

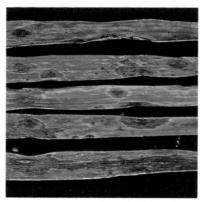

Fig. 19. Oat-Speckled Blotch; W. H. Bragonier

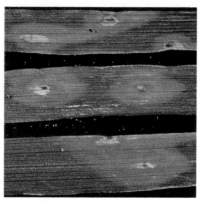

Fig. 20. Oat-Halo Blight; M. D. Simons

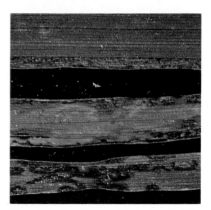

Fig. 21. Oat-Bacterial Stripe; M. D. Simons

Fig. 22. Oat-Leaf Stripe

Fig. 23. Oat-Loose Smut; J. A. Browning

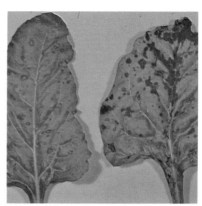

Fig. 24. Beet-Leaf Blight; J. S. McFarlane, ARS

Fig. 25. Beet-Leaf Spot; Arden F. Sherf, Cornell Univ.

Fig. 26. Beet-Powdery Mildew; J. S. McFarlane, ARS

Fig. 27. Beet-Downy Mildew; D. H. Hall, Univ. of Calif.

PLATE 3

Fig. 28. Beet-Leaf Spot; Cornell Univ.

Fig. 29. Beet-Rust; D. H. Hall, Univ. of Calif.

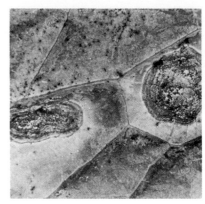

Fig. 30. Cabbage-Gray Leaf Spot; ARS

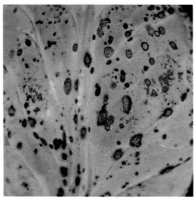

Fig. 31. Cabbage-Black Spot; Pl. Path. Div., Penn St. Univ.

Fig. 32. Turnip-White Spot; Dept. of Pl. Path., Clemson Univ.

Fig. 33. Cabbage-Powdery Mildew

Fig. 34. Cabbage-Yellows; Arden F. Sherf, Cornell Univ.

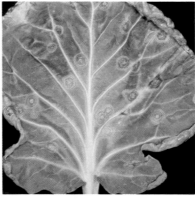

Fig. 35. Cabbage-Ring Spot; D. H. Hall, Univ. of Calif.

Fig. 36. Cabbage-Seedling Disease

PLATE 4

Fig. 37. Cabbage-Downy Mildew; Arden F. Sherf, Cornell Univ.

Fig. 38. Cabbage-Black Leg; Arden F. Sherf, Cornell Univ.

Fig. 39. Cabbage-Club Root; Paul H. Williams, Univ. of Wisc.

Fig. 40. Cauliflower-Leaf Spot; Arden F. Sherf, Cornell Univ.

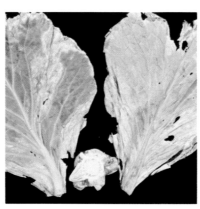

Fig. 41. Cabbage-Black Rot; Arden F. Sherf, Cornell Univ.

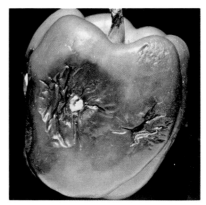

Fig. 42. Pepper-Phythopthora Blight; Pl. Path. Div., Penn. St. Univ.

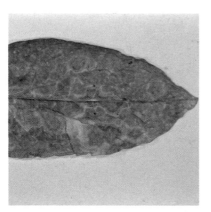

Fig. 43. Safflower-Leaf Spot; Montana St. Univ.

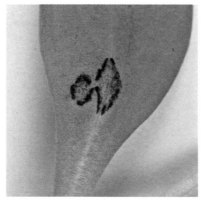

Fig. 44. Safflower-Rust; Dept. of Botany & Pl. Path., U. of Md.

Fig. 45. Chestnut-Blight; R. J. Stipes, Dept. of Pl. Path. V.P.I.

PLATE 5

Fig. 46. Chestnut-Ink Disease

Fig. 47. Lemon-Black Rot; ARS, Orlando, Florida

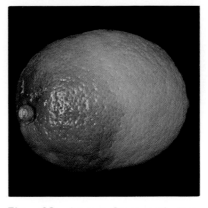

Fig. 48. Lemon-Stem-end Rot; ARS

Fig. 49. Grapefruit-Phomopsis Rot; ARS, Orlando, Florida

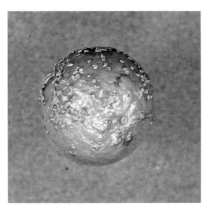

Fig. 50. Orange-Scab; ARS, Orlando, Florida

Fig. 51. Orange-Anthracnose; ARS, Orlando, Florida

Fig. 52. Orange-Black Spot; E. C. Calavan, U. of Calif.

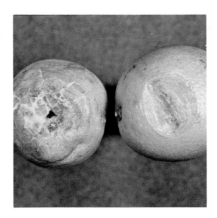

Fig. 53. Orange-Sour Rot; ARS, Orlando, Florida

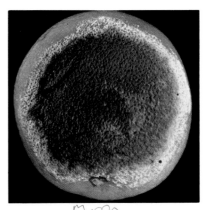

Fig. 54. Orange-Blue Mold Rot; ARS

PLATE 6

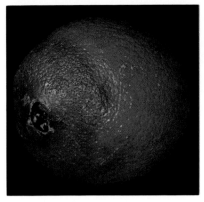

Fig. 55. Orange-Brown Rot; ARS

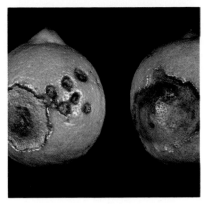

Fig. 56. Lemon-Black Pit; L. J. Klotz, Dept. of Pl. Path., Calif.

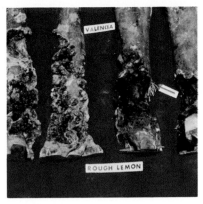

Fig. 57. Orange-Branch Rot; Carlos H. Blazquez, Univ. of Flor.

Fig. 58. Grapefruit-Citrus Canker; L. J. Klotz, Dept. of Pl. Path., Calif.

Fig. 59. Coffee-Brown Spot; Leopoldo Abrego, El Salvador

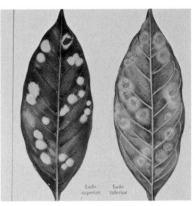

Fig. 60. Coffee-Rust; Frederick L. Wellman; N. C. St. Univ.

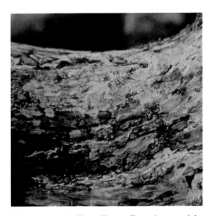

Fig. 61. Coffee-Root Rot; Leopoldo Abrego, El Salvador

Fig. 62. Jute-Anthracnose

Fig. 63. Jute-Leaf Blight

PLATE 7

Fig. 64. Jute-Powdery Mildew   **Fig. 65**. Jute-Bacterial Leaf Spot   **Fig. 66**. Sunn Hemp-Rust

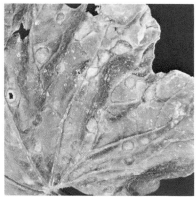

Fig. 67. Cucumber-Leaf Spot; Dept. of Pl. Path., Clemson Univ.

Fig. 68. Cucumber-Scab; ARS

Fig. 69. Cucumber-Anthracnose; ARS

Fig. 70. Cucumber-Bacterial Wilt; R. Provvidenti

Fig. 71. Cucumber-Powdery Mildew; R. Provvidenti

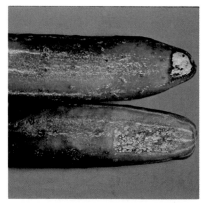

Fig. 72. Cucumber-Gummy Stem Blight; ARS

PLATE 8

Fig. 73. Cucumber-Bacterial Spot; ARS

Fig. 74. Cucumber-Downy Mildew; Dept. of Pl. Path., Clemson Univ.

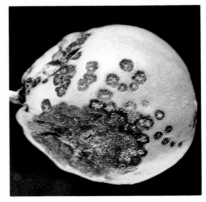

Fig. 75. Quince-Black Spot of Fruit; Pl. Path. Div., Penn. St. U.

Fig. 76. Carrot-Leaf Spot

Fig. 77. Carrot-Leaf Blight; Dept. of Botany & Pl. Path., U. of Md.

Fig. 78. Strawberry-Fasciation; Dept. of Botany & Pl. Path., U. of Md.

Fig. 79. Strawberry-Cortical Root Rot Complex; Pl. Path. Div., Penn. St. Univ.

Fig. 80. Strawberry-Leaf Blight; ARS

Fig. 81. Strawberry-Leaf Scorch; ARS

PLATE 9

Fig. 82. Strawberry-Leaf Spot; ARS

Fig. 83. Soybean-Downy Mildew; R. A. Kilpatrick, ARS

Fig. 84. Soybean-Bacterial Blight; Dept. of Pl. Path., Univ. of Illinois

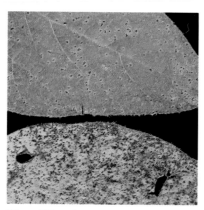

Fig. 85. Soybean-Brown Spot; Dept. of Pl. Path., Purdue Univ.

Fig. 86. Cotton-Leaf Spot; ARS

Fig. 87. Cotton-Leaf Blight; J. C. Wells, N.C. St. Univ.

Fig. 88. Cotton-Boll Rot; J. C. Wells, N.C. St. Univ.

Fig. 89. Cotton-Wilt; J. C. Wells, N.C. St. Univ.

Fig. 90. Cotton-Anthracnose Boll Rot; ARS

PLATE 10

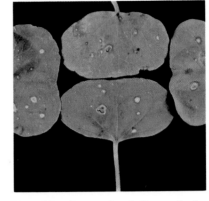

Fig. 91. Cotton-Powdery Mildew

Fig. 92. Cotton-Areolate Mildew

Fig. 93. Cotton-Leaf Spot; J. A. Pinkard

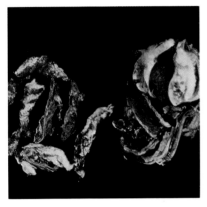

Fig. 94. Cotton-Internal Boll Rot; Lekh Batra, ARS

Fig. 95. Cotton-Rust; ARS

Fig. 96. Cotton-Southwestern Rust; John T. Presley, ARS

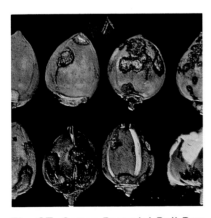

Fig. 97. Cotton-Bacterial Boll Rot; John T. Presley, ARS

Fig. 98. Sunflower-Downy Mildew; R. G. Orellana, ARS

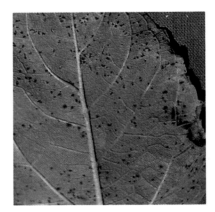

Fig. 99. Sunflower-Rust; Malcolm Shurtleff

PLATE 11

Fig. 100. Sunflower-Leaf Spot

Fig. 101. Rubber-Pink Disease

Fig. 102. Rubber-Mouldy Rot

Fig. 103. Rubber-White Root Rot; Carlos H. Blazquez, U. of Florida

Fig. 104. Rubber-Bird's-eye Spot; Carlos H. Blazquez, U. of Florida

Fig. 105. Rubber-Black Stripe; Carlos H. Blazquez, U. of Florida

Fig. 106. Barley-Leaf Rust; Dept. of Pl. Path., Univ. of Wisc.

Fig. 107. Barley-Net Blotch; Dept. of Pl. Path. Univ. of Wisc.

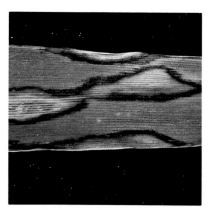

Fig. 108. Barley-Scald; J. A. Browning

Fig. 109. Barley-Speckled Leaf Blotch; Vernyl D. Pederson

PLATE 12

Gossypium spp. (Nicotiana tabacum) + Ascochyta gossypii Syd. =

ENGLISH	Leaf Blight, Stem Canker, Boll Rot, Ascochyta Leaf Spot
FRENCH	Maculicole des Feuilles
SPANISH	Mancha de la hoja, Tizón de la hoja, Mancha ceniza, Cancro del tallo
PORTUGUESE	Manchas das fôlhas
ITALIAN	
RUSSIAN	аскохитоз
SCANDINAVIAN	
GERMAN	Ascochyta-Blattfleckenkrankheit
DUTCH	
HUNGARIAN	
BULGARIAN	
TURKISH	Siyah küf
ARABIC	
PERSIAN	
BURMESE	
THAI	
NEPALI	
HINDI	
VIETNAMESE	
CHINESE	Mían hwa lúen wén bìng 棉花輪紋病
JAPANESE	Rinmon-byô 輪紋病

The fungus attacks all parts of the plant aboveground. The first symptoms are small, round, brownish spots on the leaves of seedlings. The spots enlarge and become water-soaked in appearance, later turning light brown or tan with a reddish-brown border. Spots on the bolls and stems are similar but darker in color. Stem infections advance rapidly up the stems until the leaves droop and die. Early decay and falling out of the centers of the stem spots are characteristic.

Le fungo attacca tote le partes epigee del planta. Le prime symptomas es minute maculas rotunde, brunette, in le folios del plantulas. Le maculas cresce e deveni saturate de aqua, plus tarde cambiante se in un brun clar o bronzate con un margine brun-rubiette. Maculas in le siliquas e in le pedunculos es similar, ma de color plus obscur. Le infectiones del pedunculos se avantia rapidemente in alto usque a facer le folios marcescer e morir. Putrescentia precoce e cadita del centros del maculas in le pedunculos es typic.

Le champignon attaque toutes les parties aériennes de la plante. Les premiers symptômes sont de petites taches rondes, brunâtres, sur les feuilles des semis. Les taches s'agrandissent et paraissent délavées, pour ensuite devenir brun pâle à couleur du tan, avec des bords brun rougeâtre. Sur les capsules et les tiges, les taches sont semblables mais de couleur plus foncée. L'infection gagne rapidement le bout des tiges jusqu'à ce que les feuilles se fanent et meurent. La pourriture précoce et la chute du centre des taches sur la tige sont caractéristiques.

El hongo ataca todas las partes aéreas de la planta. Los primeros síntomas son pequeñas manchas redondas, cafesosas en las hojas de la plántulas. Las manchas se agrandan y se vuelven de apariencia acuosa; más tarde se vuelven de color café claro a crema, con la orilla café rojiza. Las manchas en la cápsula y en los tallos son similares pero de color más obscuro. Las infecciones del tallo avanzan rápidamente hacia los tallos hasta que las hojas se decaen y mueren. La pudrición temprana y caída de los centros de las manchas del tallo son características.

DISTRIBUTION: Af., A., Aust., Eur., N.A., S.A.

88

Gossypium spp. (Arachis hypogaea, Zea mays, Lupinus spp., Sorghum spp., Cicer arietinum) + Aspergillus niger van Teighem =

ENGLISH   Boll Rot of Cotton, Damping-off, Seed Rot, Kernel Rot
FRENCH   Pourriture des Capsules
SPANISH   Pudrición de la fibra, Pudrición de las capsulas, Moho negro de la cebolla
PORTUGUESE
ITALIAN
RUSSIAN   гниль коробочек аспергиллезная
SCANDINAVIAN
GERMAN   Aspergillus-Kapselfaule
DUTCH
HUNGARIAN
BULGARIAN
TURKISH   Koza çürüklügü, Siyah küf
ARABIC   Affan El-loze غن اللوز
PERSIAN
BURMESE
THAI
NEPALI
HINDI
VIETNAMESE
CHINESE
JAPANESE   Kurokabi-byô 黒かび病

This disease begins as a soft, pinkish spot either on the side or near the base of the boll. The color of the older decayed areas turns brown and the original discoloration remains only in the freshly decaying borders. The fungus destroys all parts of the boll, producing spores which become dark and smutty in appearance. Boll rot causes losses by reducing yields, by staining and reducing strength of the lint and by infecting the seed.

Iste morbo se manifesta al initio como un puncto molle e rosate o in le latere o presso le base del siliqua. Le areas putrefacte plus vetule deveni brun e le discoloration original resta solmente in le margines ancora putrescente. Le fungo destrue tote le partes del siliqua, producente sporas le quales deveni obscur e sembla esser carbonose. Iste putrefaction causa perdition de valor economic per le diminution del rendimento, per tintar, e infirmar le fibras de coton, e per infection del granos.

Cette maladie commence par une petite tache molle rosâtre soit sur un côté, soit près de la base de la capsule. Les vieilles plages décomposée deviennent brunes et ce n'est que sur les bords récemment atteints que persiste la teinte originale. Le champignon détruit toute la capsule et produit des spores qui noircissent et ressemblent au charbon. La pourriture des capsules cause des pertes en réduisant les rendements, en altérant la couleur et la force des tissus et en infectant la graine.

Esta enfermedad comienza como una mancha suave y rosadosa ya sea al lado o cerca de la base de la cápsula. El color de las áreas más viejas y podridas se vuelve de color café y la decoloración original permanece sólo en los bordes que comienzan a podrirse. El hongo destruye todas las partes de la cápsula, produciendo esporas que se vuelven obscuras y de apariencia tiznada. La podredumbre de la cápsula causa pérdidas al reducir el rendimiento, al manchar y reducir la fuerza de la fibra y al infectar la semilla.

DISTRIBUTION: Af., A., Aust., Eur., N.A., C.A., S.A.

Gossypium spp., + Fusarium oxysporum f. sp., vasinfectum (Atk.) Sny. & Hans. =

ENGLISH	Wilt
FRENCH	Flétrissement Fusarien, Trachéomycose
SPANISH	Marchitez, Fusariosis, Marchitamiento, Marchitez por fusarium
PORTUGUESE	Murcha de Fusarium, Fusariose
ITALIAN	Avvizzimento del cotone
RUSSIAN	увядание фузариозное
SCANDINAVIAN	
GERMAN	Fusarium-Welke
DUTCH	
HUNGARIAN	
BULGARIAN	
TURKISH	Solganluk hastalığı
ARABIC	Zobol زبــول
PERSIAN	Bôtêh mêrēyê Pánbêh لوته‌میری پنبه
BURMESE	Hnyo-thay roga ဤ၍ဝၖ္ဝၖ္ဝ၁တ်ၖ္ဝၞ၇ၖ - အၱ်ၜၛၙ၎ ဝၞ်တၖ္
THAI	
NEPALI	
HINDI	Mlani मलानी
VIETNAMESE	Thối gộc - Héo cây
CHINESE	
JAPANESE	Tatigare-byô 立枯病

The earliest symptoms of this disease are the yellowing and browning of the leaves at their margins and between the veins. The affected leaves eventually die and fall off, leaving bare stems which soon blacken and die. Early in the season the infected plants appear dwarfed. When an older plant is attacked, the first symptom may be stunting, followed by yellowing, wilting and dropping of leaves. Characteristic symptoms of wilted plants are internal discoloration of stem vascular tissues, short taproots, excessive shedding of leaves and bare stalks.

Le symptomas le plus precoce de iste morbo es que le margines del folios e le areas inter le venas deveni jalne e brun. Folios afflicte finalmente mori e cade, lassante pedunculos denudate, le quales tosto deveni nigrate e morte. In le prime parte del saison, plantas afflicte sembla esser nanos. Si un planta plus vetule es attaccate, le prime symptomas pote esser nanismo, e alora jalnessa, marcescentia, e cadita del folios. Symptomas characteristic de plantas que ha marcescite es anormal coloration interne del texito vascular in le pedunculo, accurtamento del radices primari, defoliation excessive, e pedunculos denudate.

Les symptômes les plus précoces de cette maladie sont le jaunissement et le brunissement des feuilles à la marge et entre les nervures. Les feuilles atteintes finissent par mourir et tomber, laissant des tiges nues qui ne tardent pas à noircir et mourir. Tôt dans la saison, les plantes infectées paraissent rabougries. Lorsque l'attaque porte sur des plantes plus âgeés, le premier symptôme peut être le rabougrissement, suivi du jaunissement, du flétrissement et de la chute des feuilles. Les symptômes caractéristiques des plantes flétries sont l'altération de la couleur des tissus vasculaires internes de la tige, la petitesse des racines pivotantes, la chute excessive des feuilles et la nudité des porte-graines.

Los primeros síntomas de esta enfermedad son el amarillamiento y color café que adquieren las hojas en sus márgenes y entre las venas. Las hojas afectadas mueren eventualmente y se caen, dejando los tallos desnudos, los cuáles pronto se ennegrecen y mueren. Al principio de la estación las plantas infectadas aparecen achaparradas. Cuando una planta adulta es atacada, el primer síntoma puede se el achaparramiento, seguido de amarillamiento, marchitez y caída de las hojas. Los síntomas característicos de las plantas marchitas con la decoloración interna de los tejidos vasculares del tallo, raíz principal corta, caída excesiva de las hojas y tallos desnudos.

DISTRIBUTION: Af., A., Eur., N.A., C.A., S.A.

Gossypium spp. + <u>Glomerella</u> <u>gossypii</u> Edgerton (<u>Colletotrichum</u> <u>gossypii</u> Southworth)=

ENGLISH	Anthracnose, Pink Boll Rot, Seedling Blight
FRENCH	Anthracnose du Cotonnier
SPANISH	Antracnosis, Podredumbre rosada de la cápsula, Tizón de las plántulas
PORTUGUESE	Antracnose, Estiolamento, Tombamento, Mela, Folhas, "damping-off"
ITALIAN	Antracnosi del cotone
RUSSIAN	антракноз
SCANDINAVIAN	
GERMAN	Anthraknose
DUTCH	
HUNGARIAN	
BULGARIAN	
TURKISH	
ARABIC	
PERSIAN	
BURMESE	
THAI	Samore-Nao-Si-Chompu สมอเน่าสีชมพู
NEPALI	
HINDI	Anthracnose ऐनथ्रेक्नाज
VIETNAMESE	
CHINESE	Mían hwa tàn dzǔ bìng 棉花角斑病
JAPANESE	Tanso-byð 炭疽病

This disease occurs primarily on the bolls, although leaves, stems and bracts may also be affected. On the bolls, the first symptoms are small, round, water-soaked spots which later enlarge, become sunken and finally develop reddish borders with pink centers. As the pink area enlarges, it continues to have a well-defined dark border. When the weather is dry, the diseased areas may be grayish in color. Diseased bolls often become darkened and hardened and never open. Frequently the lint from infected bolls is tinted pink and of inferior quality.

Iste morbo occurre principalmente in le siliquas, ma folios, pedunculos, e bracteas es a vice afflicte. In le siliquas le prime symptomas es minute maculas rotunde e saturate de aqua, le quales plus tarde cresce, deveni depresse, e al fin forma margines rubiette con centros rosee. Le area rosee, quando illos cresce, retene su margine obscur ben definite. In tempore sic, le areas morbide pote esser de color aliquanto gris. Siliquas morbide sovente deveni obscurate e dur e non se aperi. Frequentemente le fibras de coton ab siliquas infectate es colorate rosee e es de qualitate inferior.

Cette maladie affecte principalement les capsules, mais aussi les feuilles, les tiges et les bractées. Sur les capsules, les premiers symptômes sont de petites taches rondes et détrempées. Ces taches s'agrandissent ensuite et se dépriment. Finalement, leurs contours deviennent rougeâtres et leur centre rose. Cette bordure sombre bien définie persiste durant l'agrandissement de la zone rose. Par temps sec, les surfaces affectées peuvent être de couleur grisâtre. Les capsules malades noircissent fréquemment, durcissent et n'ouvrent pas. Souvent, les fibres des capsules infectées sont teintées de rose et sont de qualité inférieure.

Esta enfermedad aparece principalmente en las cápsulas, aunque también puede infectas las hojas, tallos y brácteas. En las cápsulas los primeros síntomas son pequeñas manchas redondas, empapadas de agua, que luego crecen, se vuelven hundidas y finalmente desarrollan bordes rojizos con centros rosados. A medida que el área rosada se agranda, continua teniendo un borde obscuro y bien definido. Cuando el tiempo es seco, las áreas enfermas pueden ser grisáceas. Las cápsulas enfermas frecuentemente se obscurecen y endurecen y nunca se abren. Frecuentemente la fibra de las cápsulas enfermas tiene un tinte rosado y es de inferior calidad.

DISTRIBUTION: Af., A., Aust., Eur., N.A., C.A., S.A.

Gossypium spp. + Leveillula taurica (Lév.) Arn. =

ENGLISH	Powdery Mildew
FRENCH	Blanc, Oïdium
SPANISH	Oidium, Mildiú polvoriento
PORTUGUESE	
ITALIAN	
RUSSIAN	мучнистая роса
SCANDINAVIAN	
GERMAN	Echter Mehltau
DUTCH	
HUNGARIAN	
BULGARIAN	
TURKISH	Küllemesi
ARABIC	
PERSIAN	
BURMESE	
THAI	
NEPALI	
HINDI	Choorni phaphundi चूरनी फाफुन्दी
VIETNAMESE	
CHINESE	
JAPANESE	

The cotton plant becomes completely covered with the fungus giving it a white to gray powdery appearance which is typical of many of the powdery mildew diseases. Severe attacks cause defoliation. The foliage may be malformed by drooping or drying out and shriveling. The disease usually appears late in the growing season.

Le planta es totalmente coperte del fungo, le qual causa le blanc o gris apparition pulverose que es typic anque in multe altere mildews que es nominate 'pulverose'. Attaccos sever produce defoliation. Le foliage es a vices deformate a causa de marcescentia, desiccation, o crispation. Usualmente le morbo se manifesta tarde in le saison.

La plante devient entièrement couverte par le champignon, qui forme un revêtement blanc gris pulvérulent, typique des blancs. Dans les cas graves, il y a défoliation. Le feuillage peut être déformé alors que les feuilles s'inclinent vers le sol, se déssèchent et se recroquevillent. La maladie apparaît habituellement tard dans la saison.

La planta de algodón se cubre completamente con el hongo que le da una apariencia polverienta, de color blanco a gris, la cuál es típica de muchas enfermedades de mildiú polveriento. Los ataques severos causan defoliación. El follaje puede estar malformado por decaímiento, al secarse o ajarse. La enfermedad generalmente aparece tarde durante la época de cultivo.

DISTRIBUTION: Af., A., Aust., Eur., N.A., C.A., S.A.

Gossypium spp. + Mycosphaerella areola Ehrlich & Wolf (Ramularia areola Atk.) =

ENGLISH	Areolate Mildew, Frosty Blight
FRENCH	Ramulariose, Taches Blanches Pulverulentes
SPANISH	Mildeo areolado, Tizón escarchado, Mildiú aereolado
PORTUGUESE	
ITALIAN	
RUSSIAN	пятнистость рамуляриозная
SCANDINAVIAN	
GERMAN	Ramularia-Blattflekkenkrankheit
DUTCH	
HUNGARIAN	
BULGARIAN	
TURKISH	
ARABIC	
PERSIAN	
BURMESE	
THAI	
NEPALI	
HINDI	
VIETNAMESE	
CHINESE	Mían hwa bór méi bìng 棉花白黴病
JAPANESE	Sirokabi-byð 白かび病

This disease occurs as angular-shaped spots, white in color though darkening somewhat with age. The spots may become as large as 1/2 inch in diameter although they are usually considerably smaller.

Iste morbo se manifesta como maculas de forma angular, blanc ma post qualque tempore un pauc obscurate. Le maculas pote crescer usque a 1,25 cm de diametro, ma es usualmente multo plus minute.

Cette maladie se manifeste par des taches anguleuses blanches qui deviennent plus foncées avec le temps. Les taches peuvent atteindre 1/2 pouce (12mm) de diamètre, mais elles sont beaucoup plus petites, en général.

La enfermedad aparece como manchas de forma angular, de color blanco, que se van obscureciendo con la edad. Las manchas pueden agrandarse hasta alcanzar 1/2 pulgada de diámetro aunque generalmente son considerablemente más pequeñas.

DISTRIBUTION: Af., A., Eur., N.A., C.A., S.A.

Gossypium spp. + <u>Mycosphaerella gossypina</u> (Atk.) Earle (<u>Cercospora gossypina</u> Cke.) =

ENGLISH	Leaf Spot
FRENCH	Cercosporose, Taches Foliaires
SPANISH	Mancha parda de las hojas, Mancha de las capsulas, Mancha foliar
PORTUGUESE	
ITALIAN	
RUSSIAN	рамуляриоз, церкоспороз
SCANDINAVIAN	
GERMAN	Cercospora-Blattfleckenkrankheit
DUTCH	
HUNGARIAN	
BULGARIAN	
TURKISH	
ARABIC	Tabbakoo Awrak تبغ أوراق
PERSIAN	
BURMESE	
THAI	Bi-Jut ใบจุด
NEPALI	
HINDI	Parn Dhabba परण घव्वा
VIETNAMESE	
CHINESE	Mían hwa yèh shaw bìng 棉花葉燒病
JAPANESE	Hayake-byô 葉燒病

Small, red dots first appear on the leaves. They are usually small, rarely more than 1/4 inch in diameter, and round to irregular in shape with purple borders and white centers. After the spots reach maximum size, the centers fall out.

Minime maculas rubie se manifesta in le folios al initio. Illos es usualmente minute, rarmente plus grande que 0,6 cm de diametro, rotunde e de forma irregular con margines purpuree e centros blanc. Post crescimento maximal, le centros se separa e cade.

Cette maladie débute par de petits points rouges sur les feuilles. Ces points sont ordinaire-ment petits, n'excédant que rarement 1/4 de pouce (5mm) de diamètre. Ils sont ronds ou de forme irrégulière, avec un bord pourpré et un centre blanc. Le centre se détache quand la tache atteint son plein développement.

Primero aparecen en las hojas puntos pequeños y de color rojo. Generalmente son pequeños, escasamente más de 1/4 de pulgada de diámetro y de forma redonda a irregular con bordes de color púrpura y centros blancos. Después que las manchas alcanzan su tamaño máximo, los centros se caen.

DISTRIBUTION: Af., A., Eur., N.A., C.A.

94

Gossypium spp. + Nematospora coryli Peglion =

ENGLISH   Internal Boll Disease
FRENCH
SPANISH   Enfermedad interna de la cápsula
PORTUGUESE
ITALIAN
RUSSIAN
SCANDINAVIAN
GERMAN
DUTCH
HUNGARIAN
BULGARIAN
TURKISH
ARABIC
PERSIAN
BURMESE
THAI
NEPALI
HINDI
VIETNAMESE
CHINESE
JAPANESE

The lint fibers become dirty, yellowish brown and the seedcoat is stained brown in spots. With age, the lint loosens from the seeds and becomes reduced to a papery membrane. Infection results either in premature dropping of the bolls or in drying out of those remaining on the plant. The discoloration of the lint is due to the toxins produced by the fungus.

Le fibras del coton deveni immunde e brun-jalnette e le tegumento es tintate con maculas brun. Post qualque tempore, le fibras se separa ab le granos e deveni un membrana similar a papiro. Infection causa o cadita prematur del siliquas o desiccation de illos que resta attachate al planta. Coloration anormal del fibras es le resultato de toxinas producte per le fungo.

Les fibres sont teintées de brun jaunâtre sale et les enveloppes des graines sont maculées de brun. Subséquemment, les fibres se détachent des graines et se transforment en une membrane mince comme du papier. L'infection s'accompagne de la chute prématurée des capsules ou du dessèchement de celles qui restent attachées à la plante. La coloration des fibres est causée par les toxines provenant du champignon.

Las fibras se vuelven sucias, café-amarillentas y la cubierta de la semilla tiene manchas de color café. Con la edad, la fibra se afloja de las semillas y se ve reducida a una membrana como de papel. La infección resulta ya sea en una caída prematura de las cápsulas o en que las que quedan en la planta se sequen. La decoloración de la fibra se debe a las toxinas producidas por el hongo.

DISTRIBUTION: Af., A., Eur., N.A., C.A., S.A.

Gossypium spp. + Phakopsora desmium (Berk. & Br.) Cumm. (Cerotelium desmium (Berk. & Br.)
Arth.) =

ENGLISH   Rust
FRENCH
SPANISH   Roya, Polvillo
PORTUGUESE   Ferrugem
ITALIAN
RUSSIAN
SCANDINAVIAN
GERMAN
DUTCH
HUNGARIAN
BULGARIAN
TURKISH
ARABIC
PERSIAN
BURMESE
THAI
NEPALI
HINDI
VIETNAMESE
CHINESE   Mían hwa shíou bìng   棉花銹病
JAPANESE   Sabi-byð さび病

This disease is confined to the growing tissues, chiefly the leaves, and spreads from the older
parts of the plant to the new leaves as rapidly as they are formed, causing premature defoliation. The attack is severe on weak plants. The symptoms appear as small orange-colored spots
on the leaf, stem and boll.

Iste morbo se limita al texitos crescente, principalmente al folios, e se extende ab le partes
plus vetule del planta al folios nove tanto rapidemente como illos emerge, producente defoliation prematur. In plantas infirme le attacco es sever. Le symptomas se manifesta como minute
maculas de color orange in le folios, le pedunculo, e le siliquas.

Cette maladie n'attaque que les tissus en croissance, surtout les feuilles; elle se propage à
partir des parties les plus âgées aux jeunes feuilles, au fur et à mesure que celles-ci se
forment et provoque une défoliation précoce. L'attaque est grave chez les plantes faibles.
Les symptômes sont de petites taches orangées sur les feuilles, les tiges et les capsules.

Esta enfermedad está confinada a los tejidos en crecimiento, principalmente las hojas y se
disemina de las partes más viejas de la planta hacia las hojas nuevas tan rápidamente como
éstas se van formando, causando una defoliación prematura. El ataque es severo en plantas
débiles. Los síntomas aparecen como pequeñas manchas de color naranja en las hojas, tallos
y cápsulas.

DISTRIBUTION: Af., A., N.A., C.A., S.A.

Gossypium spp. (Bouteloua & Chloris spp.) + Puccinia stakmanii Presley (Puccinia cacabata Arth. & Holw.) =

ENGLISH   Rust
FRENCH   Rouille
SPANISH   Chahuixtle, Roya
PORTUGUESE   Mancha angular, Mancha bacteriana
ITALIAN
RUSSIAN   ржавчина
SCANDINAVIAN
GERMAN   Rost
DUTCH
HUNGARIAN
BULGARIAN
TURKISH
ARABIC
PERSIAN
BURMESE
THAI   Ra-Snim-Leck   ราสนิมเหล็ก
NEPALI
HINDI
VIETNAMESE
CHINESE
JAPANESE

The first symptoms of rust appear on the leaves as small orange-colored spots on the upper surface. Similar spots occur on the bolls and bracts. Infection leads to reduction of the amount of fiber produced and also fiber strength. Infection may also cause death of the seedlings.

Le prime symptomas del (fer)rugine se manifesta in le folios como minute maculas de color orange in le superfacie superior. Maculas similar occurre anque in le siliquas e in le bracteas. Infection causa diminution del quantitate e del vigor del fibras de coton. Infection pote, in plus, facer le plantulas morir.

Les premiers symptômes de cette rouille sont de petites taches orangées à la face supérieure des feuilles. Des taches similaires apparaissent sur les capsules et les bractées. L'infection entraîne une diminution du nombre et de la solidité des fibres. L'infection peut aussi causer la mort des plantules.

Los primeros síntomas de roya aparecen sobre las hojas como pequeñas manchas de color naranja en el haz. Manchas similares aparecen en las cápsulas y brácteas. La infección lleva a reducir la cantidad de fibra producida y la fuerza de la fibra. La infección puede también causar la muerte de las plántulas.

DISTRIBUTION: N.A., C.A., S.A.

Gossypium spp. + Xanthomonas malvacearum (E. F. Smith) Dowson =

ENGLISH   Angular Leaf Spot, Blackarm, Boll Rot , Bacterial Blight
FRENCH   Bactériose, Taches Anguleuses des Feuilles, Pourriture Bactérienne des Capsules
SPANISH   Brazo negro, Mancha angular de la hoja, Podredumbre de la cápsula,tizón bacterial
PORTUGUESE
ITALIAN   Batteriosi del cotone
RUSSIAN   гоммоз
SCANDINAVIAN
GERMAN   Baumwollbakteriose, Eckige Blattfleckenkrankheit
DUTCH
HUNGARIAN
BULGARIAN   бактериоза
TURKISH   Pamuk köşeli yaprak leke hastalığı
ARABIC   Tabbakoo Awrak Zawi   تبقع أوراق زاوى
PERSIAN   Bäktêrēuzê pånbêh   باكتريوز پنبه
BURMESE
THAI   Bi-Jut-Liam ใบจุดเหลี่ยม,   Samore-Nao สมอเน่า
NEPALI
HINDI   Kala haath Rog   काला हाथ रोग
VIETNAMESE   Đốm gốc
CHINESE   Mían hwa jǎw ban bìng   棉花炭疽病
JAPANESE   Kakuten-byô 角点病

On the leaves, the disease first causes dark green angular spots on the undersurface. Later
these show on the upper surface and become reddish-brown. If the leaf is badly infected, it may
curl up and drop off. On the stems and branches, it causes the affected parts to blacken. On the
bolls, the first signs are small, dark green, water-soaked, roundish spots which gradually
enlarge and turn black in the center. The bolls may become so injured that they fail to open.

In le folios iste morbo produce al initio angular maculas verde obscur al superfacie inferior. Iste
maculas plus tarde se manifesta al superfacie superior e deveni brun-rubiette. Si le folio es
severmente infectate, illo pote crispar se e cader. In le pedunculos e in le ramos, le morbo
causa que le partes afflicte se nigra. In le siliquas, le prime symptomas es minute maculas
quasi circular, verde obscur e saturate de aqua, le quales gradualmente cresce e deveni nigre in
le centro. Le siliquas es a vices tanto severmente lesionate que illos non se aperi.

Sur les feuilles, la maladie se manifeste d'abord à la face inférieure, sous forme de taches
anguleuses vert foncé. Subséquemment, ces taches apparaissent à la face supérieure et deviennent brun rougeâtre. Dans les cas graves, les limbes courbent vers le haut et tombent.
Sur les tiges et les rameaux, les parties atteintes noircissent. Sur les capsules, les premiers
indices sont de petites taches arrondies, détrempées, et vert foncé, qui s'agrandissent graduellement et deviennent noires au centre. Les capsules peuvent être si gravement endommagées
qu'elles n'éclosent pas.

Sobre las hojas, la enfermedad causa primero manchas angulares de color verde obscuro en
el envés. Más tarde éstas aparecen en el haz y se vuelven café-rojizo. Si la hoja está severamente atacada, puede enrollarse y caer. En los tallos y ramas causa el ennegrecimiento de
las partes afectadas. En las cápsulas, los primeros síntomas son manchas pequeñas, verde
obscuro, empapadas de agua, que gradualmente se agrandan y se vuelven negras en el centro.
Las cápsulas pueden dañarse tanto que no llegan a abrirse.

DISTRIBUTION: Af., A., Aust., Eur., N.A., C.A., S.A.

98

Helianthus spp. (other Compositae) + Plasmopara halstedii (Farl.) Berl. & de Toni =

ENGLISH	Downy Mildew
FRENCH	Mildiou
SPANISH	Mildiu, Mildiú lanoso
PORTUGUESE	
ITALIAN	
RUSSIAN	
SCANDINAVIAN	
GERMAN	Falscher Mehltau
DUTCH	
HUNGARIAN	Napraforgó-peronoszpóra
BULGARIAN	мана
TURKISH	Ayçiçeği mildiyösü
ARABIC	
PERSIAN	
BURMESE	
THAI	
NEPALI	
HINDI	
VIETNAMESE	
CHINESE	
JAPANESE	Beto-byð べと病

Diseased plants usually are severely stunted and have pale green areas centered about the mid-ribs of the leaves. The pale parts may be covered on the underside with the white downy growth of the fungus. The disease also causes blackening and sometimes swelling at the base of the stem. These plants are more likely to lodge (fall over) from drought than healthy plants.

Plantas morbide usualmente monstra sever obstruction de crescimento. Il ha areas verde clar a tote le duo lateres del nervatura central del folios. Iste areas clar es a vices coperte, in le superfacie inferior, del blanc strato lanuginose del fungo. Le morbo, in plus, face le pede del pedunculo nigrar se e, a vices, inspissar se. Tal plantas cade al basso per siccitate plus frequentemente que plantas san.

Les plantes infectées sont ordinairement rabougries et les feuilles, surtout le long de la nervure principale, sont vert pâle. Cette partie des feuilles peut être recouverte à la face inférieure par le champignon sous forme d'un duvet blanchâtre. Il y a aussi un noircissement et occasionnellement, un renflement de la base de la tige. Ces plantes sont, durant les sécheresses, plus sujettes à verser que les plantes saines.

Las plantas enfermas son severamente achaparradas y tienen áreas de color verde pálido, centradas alrededor de las venas centrales de las hojas. Las partes pálidas pueden estar cubiertas en la parte de abajo con el crecimiento lanoso blanco del hongo. La enfermedad también causa ennegrecimiento y algunas veces hinchazón en la base del tallo. Estas plantas están más propensas a doblarse por la sequía que las plantas sanas.

DISTRIBUTION: Af., A. Eur., N.A., C.A., S.A., W.I.

Helianthus spp. + Puccinia helianthi Schw. =

ENGLISH	Rust
FRENCH	Rouille du Tournesol
SPANISH	Roya negra, Chahuixtle, Polvillo
PORTUGUESE	Ferrugem
ITALIAN	Ruggine del girasole
RUSSIAN	ржавчина
SCANDINAVIAN	
GERMAN	
DUTCH	
HUNGARIAN	
BULGARIAN	ръжда
TURKISH	Ayçiçeği pası
ARABIC	
PERSIAN	Zångê äftäbgårdän رنگ آفتاب گردان
BURMESE	Than-chi roga ပန်ခ်ိး ဠ်းဃား ဃၥ်လ်ဃ၆သ်
THAI	
NEPALI	
HINDI	
VIETNAMESE	
CHINESE	
JAPANESE	Sabi-byð さび病

This disease occurs on leaves and stems, causing severe damage and defoliation. The first symptoms are rust-colored pustules on the leaves, later as black specks on the stems. Badly rusted leaves die and injury follows on both the flowers and seeds.

Iste morbo occurre in le folios e in le pedunculos, causante damno sever e defoliation. Le prime symptomas es pustulas (fer)ruginose in le folios e, plus tarde, maculas nigre in le pedunculos. Folios que es multo (fer)ruginose mori e le damno se extende e al flores e al granos.

Cette maladie s'attaque aux feuilles et aux tiges, et cause des dégâts importants et la défoliation. Les premiers symptômes sont des pustules couleur rouille sur les feuilles, suivis de points noirs sur les tiges. Les feuilles gravement atteintes meurent et les dégâts n'épargnent ni les fleurs, ni les fruits.

Esta enfermedad aparece sobre hojas y tallos causando daños severos y defoliación. Los primeros síntomas son pústulas de color de óxido sobre las hojas; más tarde aparecen manchas negras en los tallos. Las hojas infectadas severamente mueren y la enfermedad ataca ambas las flores y las semillas.

DISTRIBUTION: Af., A., Aust., Eur., N.A., C.A., S.A.

100

Helianthus spp. + Septoria helianthi Ell. & Kell. =

ENGLISH	Leaf Spot
FRENCH	Tache Septorienne, Taches Foliaires, Septoriose
SPANISH	Mancha foliar
PORTUGUESE	
ITALIAN	
RUSSIAN	пятнистость бурая, септориоз
SCANDINAVIAN	
GERMAN	Septoria-Blattfleckenkrankheit
DUTCH	
HUNGARIAN	
BULGARIAN	листен пригор
TURKISH	
ARABIC	
PERSIAN	
BURMESE	
THAI	
NEPALI	
HINDI	
VIETNAMESE	
CHINESE	
JAPANESE	Kappan-byô 褐斑病

This disease produces yellowish spots over the entire leaf area. Later, when the spots die, they become almost black. These lesions have a polygonal outline because their growth is sharply delimited by the veins. The numerous fruiting bodies which appear are usually on the lower surface of the leaves. Infection usually attacks the cotyledons (primary leaves) spreading later to the developing true leaves. Severe attacks lead to defoliation and loss in yield.

Iste morbo produce maculas jalnette per tote le area del folio. Plus tarde, quando le maculas mori, illos deveni quasi nigre. Iste lesiones es de contorno polygonal perque lor crescimento es stricte delimitate per le venas. Le numerose structuras sporal que se manifesta se trova usualmente in le superfacie inferior del folios. Le infection usualmente attacca le cotyledones (folios primari) e se extende plus tarde al emergente folios secundari. Attaccos sever causa defoliation e diminution del rendimento.

Cette maladie se manifeste par des taches jaunâtres sur toute la surface des feuilles. Plus tard, quand les tissus des taches meurent, elles deviennent presque noires. Ces lésions ont un contour anguleux parce que leur développement est limité par les nervures. De nombreuses fructifications se forment ordinairement à la face inférieure des limbes. La maladie atteint en premier lieu les cotylédons (feuilles primaires) pour s'étendre ensuite aux feuilles en voie de développement. Les attaques graves causent la défoliation et la réduction du rendement.

Esta enfermedad produce manchas amarillentas sobre toda el área de la hoja. Más tarde, cuando las manchas mueren, se vuelven casi negras. Estas lesiones tienen forma poligonal debido a que su crecimiento está bastante limitado por las venas. Los numerosos cuerpos fructíferos que aparecen están generalmente en el envés de las hojas. La infección generalmente ataca los cotiledones (hojas primarias) y luego se disemina hacia las hojas verdaderas que se están desarrollando. El ataque severo causa defoliación y pérdida en el rendimiento.

DISTRIBUTION: Af., A., Aust., Eur., N.A.

Hevea brasiliensis (Citrus spp. and others) + Corticium salmonicolor Berk. & Br. =

ENGLISH   Pink Disease
FRENCH   Maladie Rose
SPANISH   Enfermedad rosada, Mal rosado
PORTUGUESE
ITALIAN
RUSSIAN   "розовая болезнь"
SCANDINAVIAN
GERMAN   Rosakrankheit
DUTCH
HUNGARIAN
BULGARIAN
TURKISH
ARABIC
PERSIAN
BURMESE   Pan-yaung-hmo roga   ကာ �း ၆ယ ၆ ဃယ်ယ် ၄ိ ၄ ကၥၥ
THAI   Si-chom-pu   สีชมพู
NEPALI
HINDI
VIETNAMESE   Đềnh váng hồng
CHINESE
JAPANESE

The first symptom of this disease is the yellowing of the leaves of the parts above the infection. The infection encircles young stems or spreads from basal infections of branches to other larger branches. If complete girdling results, all parts above the canker die. If the fungus is arrested by dry weather, the bark over the affected area cracks, causing the formation of open wounds.

Le prime symptomas de iste morbo es le jalnessa del folios in le partes que se trova super le infection. Le infection incircula ramos juvene o se extende ab le pede de un ramo al altere ramos plus grande. Si le ramos es totalmente cincturate, tote le partes que se trova supere le cancere mori. Si le progresso del fungo es impedite per siccitate, le cortice del area afflicte se finde, lassante vulneres aperte.

Le premier symptôme de cette maladie est le jaunissement des feuilles au-dessus de l'infection. L'infection encercle les jeunes tiges ou s'étend, à partir des infections basales des branches, à d'autres branches plus grosses. Si le chancre ceinture une branche, il tue toutes les parties situées au-dessus. Si un temps sec bloque le développement du champignon, des fissures se produisent dans l'écorce des parties atteintes, provoquant la formation de plaies vives.

El primer síntoma de esta enfermedad es el amarillamiento de las hojas en las plantas, encima de la infección. La infección rodea ramas jóvenes o se propaga desde infecciones en la base de las ramas hacia otras ramas más grandes. Si resulta en un ceñimiento completo, todas las partes encima del cáncer mueren. Si el crecimiento del hongo es detenido por el tiempo seco, la corteza sobre las partes afectadas se raja, dando lugar a la formación de heridas abiertas.

DISTRIBUTION: Af., A., Aust., Eur., N.A., C.A., S.A.

Hevea brasiliensis (Cocos nucifera, Ipomoea batatas, Theobroma cacao) + Endoconidiophora fimbriata (Ell. & Halst.) Davidson (Ceratocystis fimbriata Ellis & Halst. , Sphaeronema fimbriata (E. & H.) Sacc.) =

ENGLISH   Moldy Rot of Rubber, Black Rot of Sweet Potato
FRENCH   Pourridié moisi du tronc du panneau de saignée
SPANISH   Cáncer, Podredumbre Mohosa del Hule, Podredumbre Negra del Camote
PORTUGUESE
ITALIAN
RUSSIAN   гниль плесневидная (раневая гниль при подсочке)
SCANDINAVIAN
GERMAN   Stammschimmelfäule
DUTCH
HUNGARIAN
BULGARIAN
TURKISH
ARABIC
PERSIAN
BURMESE   Hmo-swe roga   သီၢၢတိုအ၆၆တ၆၆ပ၆ဘ၇ုံလးၵၢ:
THAI   Mouldy Rot   โมลคีรอท
NEPALI
HINDI
VIETNAMESE
CHINESE
JAPANESE

The renewing bark of rubber trees is attacked about an inch above the tapping cut by the fungus. The disease starts with small depressed black spots which spread and soon present a grayish bloom. Dampness favors that appearance with the progress of the disease usually starting in lowlands and spreading up the slopes, being favored by moisture. Tapping operation may spread the infection, as may also wind and insects.

Le cortice del gummiero (cauchu) que se renova es attaccate per le fungo circa 2,50 cm plus alto que le puncturation. Al initio se manifesta minute maculas nigre depresse, le quales se extende rapidemente e cresce a un pellicula gris, specialmente sub conditiones humide. Le morbo usualmente comencia in le terra basse e se extende in alto per le declivitates ubi il ha favorabile conditiones de humiditate. Le activitate de puncturation pote extender le infection, e ventos e insectos pote extender lo anque.

Le champignon attaque l'écorce cicatricielle des caoutchoucs, environ un pouce au-dessus de l' incision. La maladie cause d'abord de petites dépressions noires qui s'étendent et présentent bientôt une teinte grisâtre. L'humidité favorise cette apparence suivant le progrès de la maladie, qui débute généralement dans les terres basses pour ensuite remonter les pentes plus sèches par temps humide. L'entaillage peut répandre l'infection, de même que le vent et les insectes.

El hongo ataca la corteza en renovación de árboles de hule más o menos a una pulgada arriba del corte de sangría. La enfermedad comienza con pequeñas manchas negras cóncavas que se extienden y pronto presentan una fructificación grisácea. La humedad favorece esta fructificación y también el avance de la enfermedad que generalmente comienza en las partes más bajas del terreno y progresa pendiente arriba. La operación de sangría, el viento y los insectos pueden diseminar la infección.

DISTRIBUTION: Af., A., Aust., Eur., N.A., C.A., S.A.

Hevea brasiliensis (and others) + Fomes lignosus (Klotzsch) Bres. =

ENGLISH   White Root Rot
FRENCH
SPANISH   Fomes, Podredumbre blanca de la raíz
PORTUGUESE
ITALIAN
RUSSIAN
SCANDINAVIAN
GERMAN
DUTCH
HUNGARIAN
BULGARIAN
TURKISH
ARABIC
PERSIAN
BURMESE
THAI   Rak-Khaow   รากขาว
NEPALI
HINDI
VIETNAMESE   Thối rễ - Mục cây
CHINESE
JAPANESE

This disease causes the foliage to be abnormally thin in all stages of development. The failure of many small twigs of the last new nodes to produce leaves in the upper crown is another characteristic symptom. The internal symptom is soft watery decay. The bark of infected roots may exhibit white, yellowish or reddish strands of the fungus.

Iste morbo face le foliage anormalmente tenue per tote le stadios de developpamento. Un altere symptoma characteristic es que multe minute rametos es incapace a producer folios in le plus alte corona. Le symptoma interne es un molle putrescentia aquose. Le cortice del radices afflicte pote monstrar cordas blanc, jalnette, o rubiette del fungo.

A toutes ses phases, cette maladie réduit considérablement le feuillage des arbres. Un autre symptôme caractéristique est l'absence de feuilles sur un grand nombre de ramilles issues des derniers noeuds de la nouvelle végétation à la cime de la plante. Le symptôme interne est une pourriture aqueuse et molle. Sous l'écorce des racines infectées, on peut voir les lanières blanches, jaunâtres ou rougeâtres du champignon.

Esta enfermedad ocasiona un follaje anormalmente delgado en todas las etapas de su desarrollo. Otro síntoma característico es la falla de muchas ramitas pequeñas de los nudos más nuevos en la copa superior en producir hojas. El síntoma interno es una pudrición húmeda y blanda. La corteza de las raíces infectadas puede mostrar filamentos del hongo, de color blanco, amarillento o rojizo.

DISTRIBUTION: Af., A., S.A.

104

Hevea brasiliensis + Helminthosporium heveae Petch (Drechslera heveae (Petch) M. B. Ellis) =

ENGLISH	Bird's-eye Spot
FRENCH	Maladie à Taches Circulaires des Feuilles
SPANISH	Mancha ojo de pájaro
PORTUGUESE	
ITALIAN	
RUSSIAN	гельминтоспориоз
SCANDINAVIAN	
GERMAN	Helminthosporiose
DUTCH	
HUNGARIAN	
BULGARIAN	
TURKISH	
ARABIC	
PERSIAN	
BURMESE	
THAI	Bi-Jut-Ta-Nok ใบจุดตานก
NEPALI	
HINDI	
VIETNAMESE	
CHINESE	
JAPANESE	

Symptoms are most conspicuous on the leaves, but also occur on green stems. On very young leaves the dark brown spots coalesce and there is leaf distortion and shedding. The upper part of a defoliated stem becomes swollen. On older leaves the initially chlorotic spots develop a straw-colored, papery center with a distinct, red-brown to black margin and a chlorotic halo and finally shotholes form. Fruiting bodies of the fungus on the undersurface of the leaf are in the center of the lesion.

Le symptomas es le plus remarcabile in le folios, ma occurre anque in pedunculos verde. In folios ancora multo juvene le maculas brun obscur coalesce e distortion e defoliation eveni. Le parte le plus superior de un pedunculo defoliate deveni inspissate. In folios plus vetule le maculas que es al initio chlorotic forma un centro similar a papiro, de color paleate, con un distinctive margine brun-rubiette o nigre e un halo chlorotic; al fin cavitates se developpa. Structuras sporal del fungo se trova al centro del lesion in le superfacie inferior del folio.

Les symptômes se voient surtout sur les feuilles, mais on les trouve aussi sur les tiges vertes. Sur les feuilles très jeunes, les taches brun foncé se fusionnent, les feuilles se déforment et tombent. La partie supérieure de la tige défeuillée s'enfle. Sur les vieilles feuilles, les taches d'abord chlorotiques se marquent d'une plage centrale jaune paille, à consistance de papier, avec une bordure distincte allant du brun rougeâtre au noir et un cerne chlorotique, qui se crible à la fin de petites perforations. Les fructifications du champignon se trouvent à la face inférieure de la feuille, au centre de la lésion.

Los síntomas son más conspicuos en las hojas, pero también aparecen en los tallos verdes. En hojas muy jóvenes las manchas café obscuro se fusionan y hay distorsión y caída de las hojas. La parte superior del tallo defoliado se hincha. En hojas muy viejas las manchas inicialmente cloróticas desarrollan un centro semejante al papel, color de paja, con un margen café rojizo a negro y un halo clorótico, finalmente la lesión se cae dejando agujeros. Las estructuras de fructificación del hongo están en el envés de la hoja en el centro de la lesión.

DISTRIBUTION: Af., A., Aust., N.A., C.A., S.A.

Hevea brasiliensis (Theobroma cacao) + Phytophthora palmivora Butl. (Phytophthora faberi Maubl.) =

ENGLISH   Phytophthora Leaf Fall, Pod Rot, Black Stripe
FRENCH   Nécrose de l'écorce, Chancre à Taches
SPANISH   Cáncer del tallo, Cáncer del tronco y muerto de descendente de los brotes, Caída
PORTUGUESE
ITALIAN
RUSSIAN   гниль мокрая фитофторозная
SCANDINAVIAN
GERMAN   Rindennekrose, Fleckenkrebs, Schwarzstreifenkrebs, Phytophthora-Nassfäule
DUTCH
HUNGARIAN
BULGARIAN
TURKISH
ARABIC
PERSIAN
BURMESE   Thee-pok, Ywet-kyway, Gaing-nu-chauk, Ah-net-sinn roga ဖိုက် စော ၁ွ၆သို ၇ ပါလ ၆ဘို ၇
THAI   Nar-Greed-Yang-Nao   หนากรีดยางเนา
NEPALI
HINDI   Canker कँकर
VIETNAMESE   Sưng vết cạo - Khô thân cây
CHINESE
JAPANESE

Symptoms of this disease appear on about all aboveground parts of the trees. The leaves are variously mottled with yellow and purple, particularly along the midrib where grayish spots are also observed. The leaf stalks, which fall along with the leaves, show shrunken discolored spots.

Le symptomas de iste morbo se manifesta in tote le partes epigee del arbores. Le folios es irregularmente tintate de jalne e purpuree, specialmente al nervatura central ubi se monstra anque maculas gris. Le petiolos, le quales cade al mesme tempore que le folios, monstra maculas discolorate e diminute.

Les symptômes de cette maladie apparaissent sur presque toutes les parties aériennes de l' arbre. Les feuilles sont diversement bigarrées de jaune et de pourpre, surtout le long de la nervure médiane où l'on observe aussi de taches grisâtres. Les pétioles qui tombent avec les feuilles montrent des taches ratatinées et décolorées par endroits.

Síntomas de esta enfermedad aparecen en casi todas las partes aéreas de los árboles. Las hojas aparecen con bastante moteado con amarillo y púrpura, particularmente a lo largo de la vena central donde también se observan manchas grisáceas. Los pecíolos que se caen junto con las hojas presentan manchas arrugadas decoloradas.

DISTRIBUTION: A., Eur., N.A., S.A.

106

Hordeum vulgare + Puccinia hordei Otth (Puccinia anomala Rostr.) =

ENGLISH	Leaf Rust
FRENCH	Rouille Naine des Feuilles, Rouille Naine de l'orge
SPANISH	Roya del tallo, Roya enana, Roya de la hoja
PORTUGUESE	
ITALIAN	Ruggine bruno dell'orzo
RUSSIAN	ржавчина карликовая
SCANDINAVIAN	Bygrust, Kornrost, Brunrost på korn
GERMAN	Zwergrost der Gerste, Braunrost
DUTCH	Bruine Roest, Dwergroest op Gerst
HUNGARIAN	
BULGARIAN	кафяни листни петна
TURKISH	Kahverengi pas
ARABIC	Sadaa Awrak صدا أوراق
PERSIAN	Zångě jð زنگ جو
BURMESE	
THAI	
HINDI	Parn Ratua परण रतुवा
NEPALI	
VIETNAMESE	
CHINESE	
JAPANESE	Kosabi-byð 小さび病

The characteristic symptoms of this disease are small, round, yellowish-brown pustules on the leaves. Severely infected leaves turn yellow and usually die. Infection occurs mostly late in the growing season, first on the lower leaves and then progressively upward toward the plant tip. Fields of severely infected plants have a yellow appearance. A winter pustule stage is produced as plants near maturity. The pustules, slate gray in color, tend to coalesce and form gray patches on the leaves.

Le symptoma characteristic de iste morbo es minute pustulas rotunde, de color brun-jalnette, in le folios. Folios que es severmente afflicte deveni jalne e usualmente mori. Infection generalmente comencia tarde in le saison, al initio in le folios inferior e alora extendente se gradualmente in alto usque al extremitate apical. Campos de plantas severmente afflicte ha un apparition jalne. Quando le plantas se approxima al maturation, un periodo de pustulas hibernal eveni. Le pustulas de color ardesiose propende a coalescer e formar areas gris in le folios.

La présence sur les feuilles de petites pustules circulaires brun jaunâtre est un symptôme propre à cette maladie. Les feuilles gravement infectées jaunissent et, généralement, meurent. L'infection survient le plus souvent tard dans la saison, d'abord sur les feuilles inférieures, pour monter graduellement vers le sommet de la plante. Une infection grave colore le champ en jaune. Avec la maturité des plantes apparaît un stade d'hiver. Devenues gris ardoise, les pustules ont tendance à se fondre pour former des plages grises sur les feuilles.

Los síntomas característicos de esta enfermedad son pequeñas pústulas redondas, amarillo-cafesosas sobre las hojas. Las hojas severamente infectadas se vuelven amarillas y generalmente mueren. La infección ocurre por lo general tarde en la estación, primero en las hojas inferiores y progresivamente sube hacia la punta de la planta. Los campos con plantas severamente infectadas tienen apariencia amarilla. Cuando las plantas se acercan a la madurez se produce la etapa de pústulas de invierno. Las pústulas, de color gris pizarra, tienden a fusionarse y forman parches grises en las hojas.

DISTRIBUTION: Af., A., Aust., Eur., N.A., S.A.

Hordeum vulgare (Triticum aestivum and other Gramineae) + Pyrenophora teres Dreschl. (Helminthosporium teres Sacc.) =

ENGLISH	Net Blotch
FRENCH	Helminthosporiose, Maladie des Stries Réticulaires
SPANISH	Mancha en red, Helmitosporiosis, Mancha reticular
PORTUGUESE	
ITALIAN	Macchie reticolate dell'orzo, Elmintosporiosi
RUSSIAN	Гельминтоспориоз, пятнистость сетчатая
SCANDINAVIAN	Byggbrunflekk
GERMAN	Netzfleckenkrankheit, Blattfleckenkrankheit, Braunfleckigkeit
DUTCH	Vlekkenziekte bij Gerst
HUNGARIAN	
BULGARIAN	
TURKISH	
ARABIC	Tabbakoo Shabaki تبقع شبكي
PERSIAN	
BURMESE	
THAI	
HINDI	Jaal Daag जाल दाग
NEPALI	
VIETNAMESE	
CHINESE	
JAPANESE	Amihan-byô 網斑病

This disease can be seen on seedling leaves as brown spots or blotches, usually at the tips. These blotches have a netted appearance which is caused by longitudinal and transverse lines of dark brown pigment within a lighter brown area of the spot. The spots enlarge to form narrow streaks. Sometimes, when infection is severe, they coalesce and form long brown stripes with irregular margins. Infection occurs on young leaves of plants until the plant reaches maturity. Small brown streaks without the netted appearance develop on the chaff. Infected kernels may have dark spots at their bases.

Iste morbo se manifesta como maculas o pustulas brun in le folios del plantulas, usualmente al extremitates. Iste pustulas pare esser reticulate a causa del lineas e longitudinal e transversal de pigmento brun obscur inter un area plus clar del macula brun. Le maculas cresce e evolve in stricte venas de color. A vices, quando infection sever occurre, illos coalesce e se developpa in longe bandas brun con margines irregular. Infection attacca le folios juvene del plantas usque a maturation. Minute bandas brun, sin reticulation, se monstra in le palea. Granos afflicte pote monstrar maculas obscur al base.

Cette maladie se reconnait sur les semis par la présence de taches ou d'éclaboussures brunes, généralement au sommet des feuilles. Ces éclaboussures paraissent réticulées à cause des lignes longitudinales et transversales de pigment brun foncé dans une plage d'un brun plus pâle. Les taches s'agrandissent pour former des rayures étroites. Dans certains cas d'infection très grave, ces rayures se soudent pour former de longues bandes brunes aux marges irrégulières. L'infection se produit sur les jeunes feuilles jusqu'à la maturité de la plante. Sur le chaume, il se forme de petites rayures brunes sans apparence réticulée. Les grains infectés peuvent présenter des points foncés à leur base.

Esta enfermedad puede verse en hojas de plántulas como manchas café o borrones, generalmente en las puntas. Estas manchas tienen una apariencia de malla causada por líneas longitudinales y transversales de pigmento café obscuro dentro del área café más claro de la mancha. Las manchas crecen para formar rayas angostas. Algunas veces, cuando la infección es severa, las rayas se unen y forman largas franjas café con márgenes irregulares. La infección ocurre en las hojas jóvenes hasta que las plantas puedan alcanzar la madurez. Rayas pequeñas, color café sin la apariencia reticular se desarrollan en las envolturas del grano. Los granos infectados pueden tener manchas obscuras en sus bases.

DISTRIBUTION: Af., A., Eur., N.A., S.A.

108

Hordeum vulgare (Secale spp. and other Gramineae) + Rhynchosporium secalis (Oud.) Davis =

ENGLISH   Scald
FRENCH   Tache Pâle
SPANISH   Quemaduras foliares, Escaldadura, Mancha café,Rincosporiasis
PORTUGUESE   Escaldadura
ITALIAN
RUSSIAN
SCANDINAVIAN   Grå øyeflekk, Byggens skoldpletsyge
GERMAN   Blattfleckenkrankheit
DUTCH
HUNGARIAN
BULGARIAN   листен пригор
TURKISH   Cavdar yaprak lekesi
ARABIC
PERSIAN
BURMESE
THAI
HINDI
NEPALI
VIETNAMESE
CHINESE
JAPANESE   Kumogata-byð   雲形病

The first symptom on the leaf blades and sheaths is the appearance of oval water-soaked blotches which later change from a bluish-green to brown and finally to a bleached straw color with brown margins. Sometimes the spots have a zonate appearance. On the tips of small, glumes, adjacent to the flower,   inconspicuous medium brown scald spots sometimes occur.

Le prime symptoma in le laminas e vaginas del folios es oval pustulas saturate de aqua, le quales plus tarde se cambia ab color verde-blau a brun e, al fin, a un blanchite color paleate con margines brun. A vices le maculas es zonate. In le extremitates de folios minute e pallide presso le flor se manifesta a vices maculas de escaldatura, assatis brun e non remarcabile.

Le premier symptôme est l'apparition, sur les limbes et les gaines des feuilles, d'éclabous-sures détrempées, qui virent plus tard du vert bleuâtre au brun et enfin à une couleur paille délavée, avec bordures brunes. Quelquefois, les taches paraissent zonées. A l'extrémité des petites feuilles pâles près de la fleur, on trouve parfois de minuscules taches d'échaudage, d'un brun moyen, peu apparentes.

El primer síntoma en la superficie de las hojas y vainas es la aparición de manchas ovales acuosas que posteriormente cambian de un verde azuloso a café y finalmente a un color paja decolorado, con márgenes café. A veces las manchas tienen un aspecto zoneado. En las puntas de hojas pequeñas y livianas cerca de la espiga hay algunas veces manchas poco visibles, de color café claro.

DISTRIBUTION: Af., A., Aust., Eur., N.A., S.A.

Hordeum spp. + Septoria passerinii Sacc. =

ENGLISH   Speckled Leaf Blotch
FRENCH   Tache Septorienne, Septoriose, Taches Foliaires
SPANISH   Tizón de las hojas, Mancha moteada de la hoja
PORTUGUESE
ITALIAN
RUSSIAN   септориоз
SCANDINAVIAN
GERMAN   Septoria-Braunfleckigkeit
DUTCH
HUNGARIAN
BULGARIAN
TURKISH   Arpa yaprak lekesi
ARABIC
PERSIAN
BURMESE
THAI
HINDI
NEPALI
VIETNAMESE
CHINESE
JAPANESE

Characteristic symptoms of this disease are elongated, yellowish-brown lesions with indefinite margins on leaves. Very small, dark brown fruiting bodies develop in rows between the veins in the straw-colored blotches. Defoliation, low yields and formation of light kernels result.

Symptomas characteristic de iste morbo es elongate lesiones brun-jalnette con margines indefinite in le folios. Minuscule structuras sporal brun obscur se forma in lineas inter le venas in le pustulas paleate. Le resultato es defoliation, diminution de rendimento, e legieressa de granos.

Comme symptôme caractéristique, cette maladie produit sur les feuilles de longues lésions brun jaunâtre aux bordures indéfinies. De toutes petites fructifications brun foncé se développent en rangées entre les nervures, dans les éclaboussures couleur paille. Il s'en suit une défoliation, une diminution des rendements et la formation de grains légers.

Los síntomas característicos de esta enfermedad son lesiones alargadas, color café amarillento, con márgenes definidos en las hojas. Entre las venas en las manchas color paja se forman en filas estructuras fructíferas muy pequeñas, de color café obscuro. Hay defoliación, bajos rendimientos y formación de granos livianos.

DISTRIBUTION: Af., Aust., Eur., N.A.

Hordeum vulgare + Ustilago hordei (Pers.) Lagerh. =

ENGLISH   Covered Smut
FRENCH   Charbon Couvert, Charbon Vêtu
SPANISH   Carbón vestido, Carbón cubierto
PORTUGUESE
ITALIAN   Carbone, Carbone coperto dell'orzo
RUSSIAN   головня каменная, головня твердая
SCANDINAVIAN   Dækket bygbrand, Hårdsot
GERMAN   Gedeckter Brand Gerste, Gerstenhartbrand, Hartbrand, Gedeckter Brand
DUTCH   Steenbrand, Gerstesteen Brand
HUNGARIAN
BULGARIAN   покрита главня
TURKISH   Kapalı rastık
ARABIC   Taphahom moghata   تفحم مغطى
PERSIAN   Sēyāhàkê jô   سیاهک جو
BURMESE
THAI
NEPALI
HINDI   Avrit Kand   अबरित कंद
VIETNAMESE
CHINESE
JAPANESE   Katakuroho-byô   堅黒穂病

The smut masses which are characteristic of this disease are enclosed in a thin grayish-white membrane which remains intact until harvest. The dark brown to black spore masses sometimes replace the kernels (seed) and floral parts. Smutty heads are a complete loss.

Le massas carbonose le quales es characteristic de iste morbo se contine in un tenue membrana blanc-gris que resta coperte usque al rendimento. Le massas de structuras sporal brun obscur o nigre a vices reimplacia le granos e le partes floral. Spicas que es afflicte con carbon suffre perdition total.

Les masses charbonneuses qui caractérisent cette maladie sont enrobées dans une membrane d'un blanc grisâtre qui demeure intacte jusqu'à la moisson. Les masses de spores, brun foncé ou noires, remplacent parfois les grains et les parties florales. Les épis charbonneux sont une perte totale.

Las masas de carbón que son características de esta enfermedad están dentro de una delgada membrana blanco-grisácea que permanece intacta hasta la cosecha. Las masas de esporas color café obscuro a negro algunas veces substituyen a los granos (semillas) y partes florales. Las espigas con carbón son una pérdida total.

DISTRIBUTION: Af., A., Aust., Eur., N.A., C.A., S.A.

Hordeum vulgare (Triticum aestivum, Avena sativa) + Ustilago nuda (Jens.) Rostr. =

ENGLISH	Loose Smut
FRENCH	Charbon Nu, Charbon de l'orge
SPANISH	Carbón desnudo, Carbón volador, Carbón descubierto
PORTUGUESE	
ITALIAN	Carbone volante, Carbone nudo dell'orzo
RUSSIAN	головня пыльная
SCANDINAVIAN	Nøgen bygbrand, Kornflygsot, Flygsot på korn
GERMAN	Flugbrand Gerste, Gerstenflugbrand, Gerstestuifbrand, Naakte Gerstebrand
DUTCH	Stuifbrand, Gerstestuifbrand, Naakte Gerstebrand
HUNGARIAN	
BULGARIAN	кафява праховита
TURKISH	Baş rastığı
ARABIC	Taphahom saaeb تـفحم سايب
PERSIAN	Sēyähàkè jǒ سياهک آشکار جو
BURMESE	
THAI	
NEPALI	Jau Ko kalopoke जउको कालो पोके , Jau Ka nedha rog जउका नेभा रोग
HINDI	Shaith Kand सैथ कंद
VIETNAMESE	
CHINESE	
JAPANESE	Hadakakuroho-byǒ 裸黒穂病

This disease is first evident when the head emerges from the growing shoot, which may occur earlier than with healthy heads. The heads are replaced by a mass of medium brown to dark brown powdery spores which are enclosed for a time with delicate, silvery membranes. Spores are released when the membranes rupture, leaving only naked spikes. Infected seed may be reduced in size and lighter than healthy seed.

Iste morbo non se manifesta ante que le spica emerge ab le planta crescente, e iste emergentia pote occurrer ante le tempore normal pro plantas san. Pro le spicas se substitue un massa de pulverose sporas assatis brun o brun obscur, le quales se contine durante qualque tempore inter delicate membranas argentee. Quando le membrana es rupte, le sporas es liberate, lassante spicas denudate. Granos infectate es a vices plus minute e plus legier que granos san.

Cette maladie se révèle d'abord quand la tête émerge des pousses atteintes, ce qui peut se produire avant la sortie des épis sains. Les têtes sont remplacées par des masses de spores poudreuses, brun moyen à brun foncé, enrobées pendant un certain temps de délicates membranes argentées. La ruptures des membranes libère les spores, ne laissant que les épis dégarnis. La semence infectée peut être plus petite et plus légère que la semence saine.

Esta enfermedad se nota primero cuando la espiga emerge del tallo en crecimiento, lo cuál puede ocurrir más temprano que con espigas sanas. Las espigas son reemplazadas por una masa polvorienta de esporas color café claro a café obscuro que por un tiempo están cubiertas por delicadas membranas plateadas. Las esporas se liberan al romperse las membranas, quedando solamente las espigas desnudas. La semilla infectada puede ser reducida en tamaño y puede ser más liviana que una semilla sana.

DISTRIBUTION: Af., A., Aust., Eur., N.A., C.A., S.A.

Ipomoea batatas + Cercospora bataticola Cif. & Bruner (Cercospora batatae Zimm.) =

ENGLISH	Leaf Spot
FRENCH	Taches Foliaires
SPANISH	Mancha de la hoja
PORTUGUESE	
ITALIAN	
RUSSIAN	церкоспороз
SCANDINAVIAN	
GERMAN	Cercospora-Blattfleckenkrankheit
DUTCH	
HUNGARIAN	
BULGARIAN	
TURKISH	
ARABIC	
PERSIAN	
BURMESE	
THAI	Bi-Jut ใบจุด
NEPALI	
HINDI	Parn Dhabba परण घब्बा
VIETNAMESE	
CHINESE	Gan shǔh yèh sèh bìng 甘藷葉澁病
JAPANESE	Kakuhan-byô 角斑病

The spots caused by this fungus are circular to irregular, up to 1/3 inch in diameter, with pale brown to dingy-gray centers and dark purple borders. The fungus fruiting bodies occur mostly on the lower surface of the leaves.

Le causa de iste morbo es un fungo que produce maculas, circular o irregular de forma, le dimension usque 0,4 cm, con centros brun clar o gris immunde e margines purpuree obscur. Le structoras sporal del fungo se manifesta generalmente in le latere inferior del folios.

Les taches qui sont le résultat de ce champignon sont circularies à irregulières, mésurant jusqu'à un tiers d'un pouce en diamètre avec un centre de couleur marron pâle à grise sale et avec une bordure de couleur pourpre foncé. Les organes de fructification du champignon se produit le plus souvent sur la partie inférieure de la surface des feuilles.

Las manchas causadas por estos hongos varían de circulares a irregulares, hasta un tercio de una pulgada de diámetro cuyo centro tira de un pardo pálido a un tono grisáceo con un borde morado obscuro. Los cuerpos fructíferos del hongo aparece con mayor frecuencia en la superficie inferior de las hojas.

DISTRIBUTION: A., Eur., N.A.

Ipomoea _batatas_ + _Elsinoe_ _batatas_ Jenk. & Viégas =

ENGLISH	Scab, Spot Anthracnose
FRENCH	Tavelure des Tiges et des Feuilles
SPANISH	Roña o sarna, Mancha antracnosa
PORTUGUESE	
ITALIAN	
RUSSIAN	парша стеблей и листьев
SCANDINAVIAN	
GERMAN	Stengel-und Blattschorf
DUTCH	
HUNGARIAN	
BULGARIAN	
TURKISH	
ARABIC	
PERSIAN	
BURMESE	
THAI	
NEPALI	
HINDI	
VIETNAMESE	
CHINESE	Gan shǔh sùh iá bìng 甘藷縮芽病
JAPANESE	Syukuga-byô 縮芽病

The spots are small, circular to elliptical, tan or brown, sometimes with purple borders and have depressed centers. They occur on leaf veins, petioles, leaf blades, and some of the youngest stems. The affected parts are usually stunted and deformed. Sometimes the stems grow upright instead of lying flat as is normal for healthy plants.

Le maculas es minute, circular o elliptic, bronzate o brun, a vices con margines purpuree e centros depresse. Illos se trova in le venas de folios, in petiolos, in le laminas de folios, e in alicunes del pedunculos le plus juvene. Usualmente le partes afflicte monstra obstruction de crescimento e deformation. A vices le pedunculos cresce vertical in loco de horizontal como es normal in plantas san.

Les taches sont petites, circulaires ou elliptiques, beiges ou brunes, avec quelquefois des bordures violettes. Leur centre est déprimé. Elles se forment sur les nervures des feuilles, les pétioles, les limbes, et sur les plus jeunes tiges. Les parties atteintes sont générale-ment rabougries et difformes. Quelquefois, les tiges poussent verticalement, au lieu de rester à plat comme c'est le cas pour les plantes saines.

Las manchas son pequeñas, circulares a elípticas, de color crema o café, a veces con bordes púrpura y tienen los centros hundidos. Aparecen en las venas, pecíolos y sobre la superficie de las hojas y sobre algunos de los tallos más tiernos. Las partes afectadas están general-mente achaparradas y deformes. A veces los tallos crecen erectos en vez de acostados como es normal en plantas sanas.

DISTRIBUTION: Af., Aust., S.A.

Ipomoea batatas + Fusarium oxysporum (Wr.) Snyd. & Hans. f. batatas Wr. =

ENGLISH	Soft Rot
FRENCH	Flétrissement Fusarien, Fusariose
SPANISH	Podredumbre blanda
PORTUGUESE	
ITALIAN	
RUSSIAN	**увядание фузариозное, фузариоз**
SCANDINAVIAN	
GERMAN	Fusarium-Welke
DUTCH	
HUNGARIAN	
BULGARIAN	
TURKISH	
ARABIC	
PERSIAN	
BURMESE	
THAI	
NEPALI	
HINDI	
VIETNAMESE	Thối nhũn
CHINESE	Gan shǔh màn guóh bìng 甘藷蔓割病
JAPANESE	Turuware-byð つる割病

The first symptom of this disease is a slight yellowing of the vines, followed by a more pronounced discoloration, puckering of the foliage and finally wilting of the plant. The petioles of the older leaves drop off when they are infected, leaving those near the tip of the vine. These are usually the first to show yellowing. When the young vines are infected, many short stems grow at the center of the hill, giving a rosette appearance. If the skin of an infected stem were peeled away, the exposed tissue would be darkly discolored.

Le prime symptoma de iste morbo es un pauc de jalnessa in le vites e, plus tarde, un discoloration plus marcate, crispation del foliage, e, al fin, marcescentia del planta. Le petiolos del folios plus vetule, si illos es infectate, se separa e cade, lassante illos presso le extremitate del vite. Iste ultimes es usualmente le primes que monstra jalnessa. Quando le vites juvene es infectate, multe curte pedunculos emerge al centro de cata gruppamento de plantas e resimila a un rosetta. Si on remove le pelle de un pedunculo infectate, le texito discoperte monstra un anormal coloration obscur.

Le premier symptôme de cette maladie est un léger jaunissement de la partie aérienne, suivi d'une décoloration plus prononcée, d'un plissement du feuillage et enfin du flétrissement de la plante. Les pétioles des vieilles feuilles tombent sous l'effet de l'infection, mais il en reste au sommet de la tige. C'est d'ordinaire là que débute le jaunissement. Quand l'infection attaque une jeune plante, plusieurs petites tiges poussent au centre de la butte, lui donnant l'apparence d'une rosette. Si on pèle une tige infectée, on y trouve des tissus plus foncés que les tissus sains.

El primer síntoma de esta enfermedad es un ligero amarillamiento de las guías, seguido por una decoloración más pronunciada, arrugamiento del follaje y finalmente marchitez de la planta. Los pecíolos de las hojas más viejas se caen al infectarse, dejando aquéllos cerca de la punta de la guía. Estos son generalmente los que primero muestran amarillamiento. Cuando las enredaderas jóvenes estan infectadas, en el centro de la planta crecen muchos tallos pequeños, dando una apariencia de roseta. Si se quita la epidermis de un tallo infectado, el tejido expuesto mostraría una decoloración obscura.

DISTRIBUTION: Af., Eur., N.A.

Ipomoea batatas + Monilochaetes infuscans Ell. & Halst. =

ENGLISH	Scurf
FRENCH	Tavelure des Tubercules
SPANISH	Scurf, Costra
PORTUGUESE	
ITALIAN	
RUSSIAN	парша клубней
SCANDINAVIAN	
GERMAN	Knollenschorf
DUTCH	
HUNGARIAN	
BULGARIAN	
TURKISH	
ARABIC	
PERSIAN	
BURMESE	
THAI	
NEPALI	
HINDI	
VIETNAMESE	Hà vỏ
CHINESE	Gan shǔh hei jỳh bìng 甘藷黑痣病
JAPANESE	Kuroaza-byô 黒あざ病

Lesions start as small brown specks on the roots. When many patches of spots exist, they coalesce and form a uniform rusting of a part or the entire surface of the potato. Although the disease rarely damages the root, it does cause the skin to shrivel from loss of moisture. In severe cases the potato may crack and decay.

Le lesiones es al initio minute maculas brun in le radices. Quando il ha grande areas de maculas, illos coalesce e produce un (fer)rugine uniforme de un parte o de tote le superfacie del patata. Ben que le morbo rarmente face mal al radice, illo causa que le pelle se cripsa a causa del desiccation. In casos sever, le patata pote finder se e putrescer.

Les lésions commencent par de petits points bruns sur les racines. Quand plusieurs plages sont ainsi tachetées, elles se fondent pour donner une couleur rouille à une partie de la patate ou à toute sa surface. Quoique la maladie endommage rarement la racine, elle fait plisser la peau par manque d'humidité. Dans les cas graves, le tubercule peut se fissurer et pourrir.

Las lesiones comienzan como pequeñas motitas de color café en las raíces. Cuando existen muchos parches de manchas, se fusionan y forman una herrumbre uniforme de parte o toda la superficie del tubérculo. Aunque la enfermedad rara vez daña la raíz, si causa que la epidermis se arrugue por falta de humedad. En casos severos el tubérculo se raja y se pudre.

DISTRIBUTION: Af., A., Aust., Eur., N.A., S.A.

116

Juglans spp. + Gnomonia leptostyla (Fr.) Ces. & De Not. =

ENGLISH	Anthracnose, Leaf Spot, Leaf Blotch
FRENCH	Tache des Feuilles, Anthracnose des Feuilles
SPANISH	Antracnosis, Mancha de la hoja, Parche de la hoja
PORTUGUESE	
ITALIAN	Nebbia del noce, Antracnosi, Vaiolo del noce
RUSSIAN	**антракноз листьев, пятнистость** бурая
SCANDINAVIAN	
GERMAN	Marssonina-Krankheit, Blattfleckenkrankheit, Gnomonia-Blattfleckenkrankheit
DUTCH	Vlekkenziekte
HUNGARIAN	Gnomóniás barnafoltosság
BULGARIAN	
TURKISH	Ceviz antraknozu
ARABIC	
PERSIAN	
BURMESE	
THAI	
NEPALI	
HINDI	
VIETNAMESE	
CHINESE	
JAPANESE	

This disease first appears as a small, inconspicuous brown spot on the leaves. The shape of the spots varies from circular to greatly elongated. The most distinguishing characteristic of this leaf spot is that affected areas are confined to a narrow space between the lateral veinlets, forming along narrow, dead areas.

Iste morbo se manifesta al initio como minute maculas, brun e non remarcabile, in le folios. Le forma del maculas se varia ab circular usque a grandemente elongate. Le characteristica le plus distinctive de iste macula foliar es que le areas afflicte se contine in un spatio stricte inter le venulas lateral, causante morte areas elongate e stricte.

La maladie se montre d'abord sous forme de taches brunes peu apparentes sur les feuilles. La forme de ces taches varie de circulaire à très allongée. Le caractère le plus particulier de ces taches sur les feuilles est qu'elles se limitent à une bande mince entre les petites nervures latérales, formant d'étroites lisières de tissu mort.

Esta enfermedad aparece primero como pequeñas manchas inconspicuas, color café, en las hojas. La forma de las manchas varía desde circular hasta grandemente alargadas. La característica más singular de esta mancha foliar es que la áreas afectadas están confinadas a un espacio angosto entre las venitas laterales, formando angostas áreas muertas.

DISTRIBUTION: Af., A., Eur., N.A., S.A.

Juglans spp. + Xanthomonas juglandis (Pierce) Dowson =

ENGLISH	Bacterial Blight
FRENCH	Brûlure Bactérienne, Maladie Bactérienne du Noyer
SPANISH	Bacteriosis, Tizón bacterial, Mal seco, Peste negra, Tizón bacteriano
PORTUGUESE	
ITALIAN	Mal secco del noce
RUSSIAN	ожог бактериальный
SCANDINAVIAN	
GERMAN	Bakterienbrand der Walnuss, Bakterienkrankheit
DUTCH	
HUNGARIAN	
BULGARIAN	бактериоза
TURKISH	Ceviz bakteriyal yanıklığı
ARABIC	
PERSIAN	Bäktêrêuzê gêrdoo باکتریوزگردو
BURMESE	
THAI	
NEPALI	
HINDI	
VIETNAMESE	
CHINESE	
JAPANESE	Kassyokuhuhai-byô 褐色腐敗病

This disease appears as reddish-brown spots on young leaves, or as black, slightly depressed spots on young shoots and nuts. Nuts infected early may be killed and mummified before much growth has occured. If infection occurs later, it may remain dormant inside the kernel and then become active about the time the nuts are fully developed. The nuts then drop off. Post-bloom infection shows as black spots on the hull. This type of infection may or may not penetrate the shell.

Iste morbo se manifesta como maculas brun-rubiette in le folios juvene o como nigre maculas aliquanto depresse in le ramettos juvene e in le nuces. Si le nuces es afflicte in le prime parte del saison, illos pote morir e mumificar se sin troppo de crescentia. Si le infection occurre plus tarde, illo pote restar latente intra le grano e reactivar se quando le nuces es totalmente maturate. Le nuces, alora, se separa e cade. Le infection que occurre post florescentia se manifesta como maculas nigre in le scalias. Tal infection pote o penetrar le scalia o non.

Cette maladie se manifeste sous forme de taches brun rougeâtre sur les jeunes feuilles, ou de taches noires légèrement déprimées sur les jeunes pousses et sur les noix. Les noix infectées au début de leur développement peuvent mourir et se momifier avant de grossir pour la peine. Si l'infection se produit plus tard, la maladie peut rester à l'état dormant dans l'amande pour se réveiller quand les noix ont atteint leur plein développement. Alors, les noix tombent. L'infection postflorale se manifeste sous forme de points noirs sur la noix. Ce genre d'infection peut ou non traverser les parois de la noix.

Esta enfermedad aparece como manchas café-rojizo en hojas tiernas, o como manchas negras ligeramente aplastadas en brotes tiernos y nueces. Las nueces infectadas temprano pueden morir y se momifican antes de que hayan crecido. Si la infección ocurre posteriormente, puede permanecer latente dentro de la semilla, volviéndose activa al tiempo que las nueces alcanzan su pleno desarrollo. Entonces se caen las nueces. La infección después de la floración se muestra como manchas negras en la parte carnosa. Este tipo de infección puede o no penetrar la cáscara dura interna.

DISTRIBUTION: Af., A., Aust., Eur., N.A., C.A., S.A.

118

Lactuca sativa (Cichorium endivia, C. intybus, Sonchus spp.) + Bremia lactucae Regel =

ENGLISH	Downy Mildew
FRENCH	Mildiou, Blanc
SPANISH	Mildiu, Mildiu felpudo, Mildiú velloso, Mildiú lanoso
PORTUGUESE	Míldio
ITALIAN	Peronospora delle composite, Marciume dell'insalata
RUSSIAN	**пероноспороз, ложная мучнистая роса**
SCANDINAVIAN	Salatskimmel, Salladbladmögel, Salatbladskimmel
GERMAN	Falscher Mehltau
DUTCH	Valse Meeldauw
HUNGARIAN	Peronoszpóra
BULGARIAN	
TURKISH	Marul mildiyösü
ARABIC	Bayad Zaghaby بياض زغبي
PERSIAN	Sefedake doroogheye kahoo سفیدک دروغی کاهو
BURMESE	
THAI	
NEPALI	
HINDI	
VIETNAMESE	
CHINESE	
JAPANESE	Beto-byo　べと病

The initial symptoms of this disease are light green or pale yellow areas on the upper surface of the leaves. On the reverse side and opposite the discolored spots, a downy white growth will be observed. The lesions may join and in later stages turn brown. Eventually, the entire leaf turns yellow. Severe infection causes stunting of the plants,. and they fail to head. The disease may continue to develop after harvesting.

Le symptomas initial de iste morbo es areas verde clar o jalne pallide al superfacie superior del folios. Al reverso sub le maculas con coloration anormal se manifesta un lanuginositate blanc. Le lesiones pote coalescer e, plus tarde, devenir brun. Al fin tote le folio deveni jalne. Infection sever impedi le developpamento de plantas e le partes apical non se forma. Le morbo a vices progrede post le rendimento.

Les premiers symptômes de cette maladie sont des plages vert gai ou jaune pâle à la face supérieure des feuilles. Sous la feuille à l'opposite des taches on remarque un duvet blanc. Les lésions peuvent se joindre et, dans les stades plus avancés, virent au brun. Eventuellement, toute la feuille jaunit. Dans la cas d'infection grave, la plante ne peut se développer normalement ni produire sa pomme. La maladie peut poursuivre son évolution après la récolte.

Los síntomas iniciales de esta enfermedad son áreas de color verde pálido o amarillo pálido en el haz de las hojas. En el envés y opuesto a las manchas se observa un crecimiento blanco lanoso. Las lesiones pueden unirse y en etapas más avanzadas se vuelve café. Eventualmente toda la hoja se vuelve amarilla. La infección severa causa achaparramiento de las plantas, las cuáles no llegan a formar cabeza.La enfermedad puede continuar desarrollándose después de la cosecha.

DISTRIBUTION: Af., A., Aust., Eur., N.A., C.A., S.A.

<u>Lactuca</u> <u>sativa</u> + <u>Pseudomonas</u> <u>marginalis</u> (Brown) Stevens =

ENGLISH	Marginal Leaf Blight
FRENCH	Brûlure Marginale, Bactériose de la Laitue
SPANISH	Tizón marginal de la hoja
PORTUGUESE	
ITALIAN	Marciume molle della lattuga
RUSSIAN	ожог краевой
SCANDINAVIAN	
GERMAN	Kansas-Salatkrankheit
DUTCH	
HUNGARIAN	
BULGARIAN	
TURKISH	Marul bakteriyel cürüklügü
ARABIC	Affan Tary Aswad عـفن طرى اسود
PERSIAN	
BURMESE	
THAI	
NEPALI	
HINDI	
VIETNAMESE	
CHINESE	
JAPANESE	Herigare-byô 縁枯病

This disease usually starts as a decay at the margin of the leaf, progressing downward until, in many instances, the whole leaf is involved. The first visual symptom is a wilting of the tissue followed by browning or blackening, or in dry weather, by a pale papery texture. The disease usually attacks leaves of similar age, not necessarily the oldest leaves first however, and then moves either toward the center or to the outer leaves.

Iste morbo usualmente comencia como un putrescentia al margine del folio e progrede a basso usque a involver, frequentemente, tote le folio. Le prime symptoma evidente es marcescentia del texito que alora deveni brun o nigre o, in tempore sic, como papiro pallide. Usualmente le morbo attacca folios del mesme etate, ma non semper primo le folios le plus vetule, e alora progrede al folios central o exterior.

La maladie commence généralement par une pourriture à la marge de la feuille, d'où elle se propage aux tissus sous-jacents, de façon, souvent, à envahir toute la feuille. Le premier symptôme perceptible est le flétrissement des tissus suivi d'un brunissement ou d'un noircissement ou, par temps sec, par une teinte pâle et une consistance de papier. La maladie s'attaque généralement à des feuilles du même âge, pas nécessairement aux plus vieilles feuilles d'abord, et s'en va de là, soit vers les feuilles du centre, soit vers celles de la périphérie.

Esta enfermedad generalmente comienza con un deterioro en el margen de la hoja que progresa hacia abajo hasta que, en muchos casos, ataca toda la hoja. El primer síntoma visual es un marchitamiento del tejido, seguido de un obscurecimiento café o negro; en tiempo seco el tejido tiene textura de papel y color pálido. La enfermedad no ataca necesariamente las hojas más viejas primero, sino que ataca hojas de edad similar y luego se mueve ya sea hacia las hojas del centro o hacia las de afuera.

DISTRIBUTION: Af., A., Aust., Eur., N.A., C.A., S.A.

120

Lactuca sativa + Septoria lactucae Pass. =

ENGLISH   Leaf Spot
FRENCH   Tache Septorienne, Taches Foliaires, Septoriose
SPANISH   Mancha foliar, Mancha de la hoja
PORTUGUESE   Crestamento das fôlhas, Septoriose
ITALIAN
RUSSIAN   септориоз
SCANDINAVIAN
GERMAN   Septoria-Blattfleckenkrankheit
DUTCH
HUNGARIAN
BULGARIAN   мана
TURKISH
ARABIC
PERSIAN   Septōrēyāyē kāhōō   سپتوریای کاهو
BURMESE
THAI
NEPALI
HINDI
VIETNAMESE
CHINESE   Uo jìuh ban dǐan bìng   萵苣斑點病
JAPANESE   Hanten-byô   斑点病

The symptoms of this disease are lesions which begins as small yellow areas that enlarge
rapidly and become irregular in shape and greenish in color. They then acquire a large number
of black fungus fruiting bodies. The dead tissue may drop out, leaving holes or ragged edges.
Usually the oldest leaves are the first to be attacked. These finally wither and die.

Le symptomas de iste morbo es lesiones le quales comencia como minute areas jalne que
cresce rapidemente e deveni de forma irregular e de color aliquanto verde. Alora le numerose
structuras sporal nigre se manifesta. Le texito morte sovente cade, lassante perforationes
o margines zigzag. Usualmente le folios le plus vetule es attaccate ante le alteres. Al fin
illos marcesce e mori.

Les symptômes de cette maladie sont des lésions débutant par de petites plages jaunes qui
s'étendent rapidement en prenant une forme irrégulière et une teinte verdâtre. Il s'y forme
alors un grand nombre de fructifications noires. Les tissus morts peuvent tomber, laissant
des trous ou des bordures déchirées. En général, les plus vieilles feuilles sont les premières
infectées. Elles finissent par se flétrir et mourir.

Los síntomas de esta enfermedad son lesiones que comienzan como pequeñas áreas amarillas
que se agrandan rápidamente tomando forma irregular y color verdoso. Entonces adquieren un
gran número de cuerpos fructíferos negros del hongo. El tejido muerto puede caerse, dejando
agujeros o bordes rasgados. Generalmente las hojas más viejas son atacada primero. Estas
finalmente se marchitan y mueren.

DISTRIBUTION: Af., A., Aust., Eur., N.A., C.A., S.A.

Linum usitatissimum + Colletotrichum lini (Westerdijk) Toch (Colletotrichum linicola Pethybr. & Laff.) =

ENGLISH	Anthracnose, Seedling Blight
FRENCH	Anthracnose, Brûlure des Semis
SPANISH	Antracnosis, Tizón del semillero
PORTUGUESE	
ITALIAN	Antracnosi, Cancro del lino
RUSSIAN	**антракноз, пожелтение проростков, мраморность стеблей**
SCANDINAVIAN	
GERMAN	Brennfleckenkrankheit, Sämlingssterben, Keimlings-Anthraknose, Wurzelbrand
DUTCH	Wegvallen, Vlaskanker, Regenvlekkanker
HUNGARIAN	
BULGARIAN	**антракноза**
TURKISH	
ARABIC	
PERSIAN	
BURMESE	
THAI	
NEPALI	
HINDI	
VIETNAMESE	
CHINESE	ya má tàn dzǔ bìng 亞蔴炭疽病
JAPANESE	tanso-byð 炭そ病

This disease causes cankers on the primary leaves. These are circular, zonated, sunken brown spots that spread under cool and moist conditions. Seedling blight occurs either before or after emergence, in the latter case, usually as a stem canker at the soil line. Leaf spots and stem cankers are common during the growing season, especially under conditions of high moisture. Brown spots form on the seed.

Iste maladia produce canceres in le folios primari, le quales es circular, zonate, depresse maculas brun que se extende sub conditiones frigide e humide. Necrosis del plantulas occurre o ante o post emergentia. Si le ultime, usualmente con un cancere in le pedunculo presso le terreno. Maculas foliar e canceres del pedunculo es commun durante le periodo de crescentia, specialmente sub conditiones de humiditate excessive. Maculas brun se forma in le granos.

Cette maladie provoque des chancres sur les feuilles primaires. Ce sont des taches brunes, circulaires, zonées et déprimées, qui s'étendent par temps frais et humide. La brûlure des semis se produit soit avant, soit après l'émergence, dans ce dernier cas, généralement sous forme de chancre de la tige, au ras du sol. Les taches des feuilles et les chancres de la tige sont communs durant la saison de végétation, surtout si le temps est très humide. Il se forme aussi des taches brunes sur les graines.

Esta enfermedad causa cancros en las hojas primarias. Estas son manchas color café, circulares, zoneadas, hundidas, que se aparecen bajo condiciones húmedas y frescas. El tizón del semillero ocurre antes o después de emerger las plantitas, en este último caso generalmente se produce un cancro del tallo a nivel del suelo. Las manchas de la hoja y los cancros del tallo son comunes durante la época de crecimiento, especialmente bajo condiciones de mucha humedad. En la semilla se forman manchas color café.

DISTRIBUTION: Af., A., Eur., N.A., S.A.

122

Linum usitatissmum + Fusarium oxysporum Schlect ex Fr. f. sp. lini (Bolley)Snyder & Hansen =

ENGLISH   Wilt
FRENCH   Flétrissement
SPANISH   Pudrición radicular, Marchitamiento, Marchitez, Fusariosis
PORTUGUESE   Murcha
ITALIAN   Avvizzimento delle piante di lino
RUSSIAN   увядание фузариозное
SCANDINAVIAN
GERMAN   Flachewelke, Welkekrankheit des Leines, Leinmüdigkeit
DUTCH   Wegvallen, Verwelkingsziekte
HUNGARIAN
BULGARIAN
TURKISH   Keten solgunluğu
ARABIC
PERSIAN
BURMESE
THAI
NEPALI
HINDI
VIETNAMESE
CHINESE   ya má lì ku bìng   亞蔴立枯病
JAPANESE   tatigare-byô   立枯病

Plants are attacked at any stage in their development. Although this disease is primarily a wilt, seedling blight occurs when seedlings are grown at high temperatures. In typical wilt, the leaves turn yellow or grayish yellow, the top leaves thicken, growth stops and the plants die and turn light brown. Frequently the plant is only stunted, in which case the leaves turn yellow and fall prematurely, or the stem dies and new, apparently healthy lateral branches develop. A late infection or a weak attack may be evidenced by premature ripening.

Plantas es attaccate per tote le stadios de lor crescimento. Ben que iste maladia es essential-mente un marcescentia, necrosis de plantulas occurre in tempore calide. Typicamente, como in marcescentia, le folios deveni jalne o jalne-gris; le folios le plus apical es inspissate; le crescimento es obstructe; e le plantas mori e deveni brun clar. Si le planta suffre solmente obstruction de crescimento, como occurre frequentemente, le folios deveni jalne e cade pre-maturmente, o le pedunculo mori e nove ramos lateral, le quales sembla esser san, se forma. Un infection tarde o un attacco legier a vices se manifesta per maturation ante le tempore ordinari.

Les plantes sont attaquées à n'importe quel stade. Bien que la maladie soit avant tout une flé-trissure, il peut se produire une brûlure des semis lorsqu'on cultive les plantules à des températures élevées. Dans la flétrissure typique, les feuilles deviennent jaunes ou jaune grisâtre, les feuilles du haut s'épaississent, la croissance s'arrête et les plantes meurent et brunissent légèrement. Souvent, la plante n'est que rabougrie; en ce cas, les feuilles jaunis-sent et tombent prématurément, ou bien la tige meurt et il se forme de nouvelles branches latérales apparemment saines. Une maturation précoce permet de déceler une infection tardive ou une attaque légère.

Las plantas son atacadas en cualquier etapa de su desarrollo. Aunque esta enfermedad es más que nada una marchitez, el tizón del semillero ocurre cuando las plántulas crecen a altas temperaturas. En el marchitamiento típico, las hojas se vuelven amarillas o amarillo grisá-ceas, las de arriba se engruesan, se detiene el crecimiento y las plantas mueren tomando un color café claro. Frecuentemente la planta sólo detiene su desarrollo, en cuyo caso las hojas se vuelven armaillas y se caen prematuramente o el tallo muere y se desarrollan ramas laterales aparentemente sanas. Una infección tardía o ataque débil puede evidenciarse por una maduración prematura.

DISTRIBUTION: Af., A., Aust., Eur., N.A., S.A.

Linum usitatissimum + Melampsora lini (Ehrenb.) Lév. =

ENGLISH	Rust
FRENCH	Rouille du Lin
SPANISH	Roya, Polvillo
PORTUGUESE	Ferrugem
ITALIAN	Ruggine del lino
RUSSIAN	"мухосед", "присуха", ржавчина
SCANDINAVIAN	
GERMAN	Flachsrost, Leinrost
DUTCH	Zwartstip (Roest), Vlasroest
HUNGARIAN	
BULGARIAN	ръжда
TURKISH	
ARABIC	Sadaa صدا
PERSIAN	Zånge kàtän زنگ کتان
BURMESE	
THAI	
NEPALI	
HINDI	Ratua रतुवा
VIETNAMESE	
CHINESE	
JAPANESE	sabi-byô さび病

The disease symptoms are characterized by light yellow to orange-yellow fungus fruiting bodies on leaves and stems early in the growing season, followed by reddish-yellow ones on leaves and stems during the growing season, and later to brown or black ones chiefly on the stems. The fiber is also weakened and disfigured.

Le symptomas characteristic de iste maladia es fungal structuras sporal in folios e in pedunculos, jalne clar o jalne-orange, in le prime parte del periodo de crescentia; plus tarde, illos deveni jalne-rubiette e al fin, principalmente in le pedunculos, brun o nigre. Le fibras anque es infirmate e deformate.

La maladie est caractérisée par des fructifications jaune pâle à jaune orange sur les feuilles et les tiges en début de saison. En pleine saison, ces fructifications tournent au jaune rougeâtre. Plus tard, les fructifications seront brunes ou noires, surtout sur les tiges. Les fibres sont affaiblies et gâtées.

Los síntomas de la enfermedad se caracterizan por estrucutras de fructificación amarillo pálido a amarillo-anaranjado del hongo en las hojas y tallos temprano en la estación, seguidas por unas amarillo-rojizas en las hojas y tallos durante la época de crecimiento y más tarde unas café o negro, principalmente en los tallos. La fibra también se debilita y desfigura.

DISTRIBUTION: Af., A., Aust., Eur., N.A., S.A.

124

Linum usitatissimum (_Linum_ spp.) + _Mycosphaerella linorum_ (Wollenw.) Garcia Rada =

ENGLISH Pasmo, Rust-blotch
FRENCH Pasmo
SPANISH Pasmo, Mancha de herrumbre
PORTUGUESE Septoriose, Pasmo
ITALIAN
RUSSIAN пасмо
SCANDINAVIAN
GERMAN Pasmo-Krankheit, Phlyctaena-Blattfleckenkrankheit
DUTCH
HUNGARIAN
BULGARIAN
TURKISH Yaprak döken
ARABIC
PERSIAN
BURMESE
THAI
NEPALI
HINDI
VIETNAMESE
CHINESE
JAPANESE katumon-byô 褐紋病

The symptoms of this disease are striking and easily recognized, especially during the latter part of the growing season. Lesions develop first on the primary leaves and then on the lower leaves of the seedlings. The lesions are generally circular in outline and vary in color from green-ish-yellow to dark brown. Later the stem lesions develop, first as small elongated spots which then enlarge and coalesce, extending around the stem as well as longitudinally. The infected areas alternate with green tissue until infection becomes severe, then the stems brown as the plants are defoliated in the spots where the disease is severe. Lesions also occur on the bolls.

Le symptomas de iste morbo es multo remarcabile e facilemente distinctive, specialmente in le secunde medietate del periodo de crescentia. Lesiones se forma, al initio, in le folios pri-mari e, plus tarde, in le folios inferior del plantulas. Le lesiones es generalmente de contorno circular e se varia ab le color jalne-verde a brun obscur. Plus tarde, le lesiones del pedun-culos se manifesta, al initio como minute maculas elongate, le quales cresce e coalesce, ex-tendente se e circum le pedunculo e in le longitude. Le areas afflicte se alterna con texito verde e le infection progrede usque a devenir sever. Alora le pedunculos deveni brun e le plantas suffre defoliation in le areas ubi le morbo es sever. Lesiones occurre anque in le siliquas.

Les symptômes de cette maladie sautent aux yeux, surtout en fin de saison. Des lésions se for-ment d'abord sur les feuilles primaires et par la suite sur les feuilles de la base des plantules. Les lésions sont généralement circulaires. Leur couleur varie du jaune verdâtre au brun foncé. Il se forme ensuite des lésions sur la tige, d'abord sous forme de petites taches allongées, qui, plus tard, s'agrandissent et se fusionnent, s'étendant aussi bien autour que le long de la tige. Les plages infectées alternent avec les plages vertes jusqu'à ce que l'infection s'aggrave; alors, les tiges brunissent et les plantes perdent leurs feuilles aux endroits gravement atteints. Des lésions se forment aussi sur les capsules.

Los síntomas de la enfermedad son llamativos y fáciles de reconocer, especialmente durante la última parte de la época del cultivo. Las lesiones se desarrollan primero en las hojas primarias y más tarde en las hojas inferiores de las plántulas. Las lesiones son general-mente de contorno circular y varían en color desde un amarillo verdoso hasta café obscuro. Despúes se desarrollan las lesiones del tallo, comenzando como pequeñas manchas alargadas que se agrandan y se unen, extendiéndos alrededor del tallo, así como longitudinalmente. Las áreas infectadas alternan con tejido verde hasta que la infección se vuelve severa; entonces los tallos se torman color café al defoliarse las plantas en las manchas donde la enfermedad es severa. También ocurren lesiones en las cápsulas.

DISTRIBUTION: Af., A., Aust., Eur., N.A., S.A.

Linum usitatissimum (Gossypium spp.) + Polysora lini Lafferty =

ENGLISH	Browning, Stem-break
FRENCH	Oxychromose, Brunissure des Tiges, Polysporose
SPANISH	Bronceado, Ruptura del tallo
PORTUGUESE	
ITALIAN	
RUSSIAN	**пятнистость бурая стеблей, полиспороз, хрупкость стеблей**
SCANDINAVIAN	
GERMAN	Flachsbräune, Flachsstengelbruch, Bräune und Stengelbrand
DUTCH	Wegvallen, Verbruinen
HUNGARIAN	
BULGARIAN	
TURKISH	Esmer leke
ARABIC	
PERSIAN	
BURMESE	
THAI	
NEPALI	
HINDI	
VIETNAMESE	
CHINESE	
JAPANESE	kappen-byô 褐変病

The symptoms of browning are conspicuous on the seedlings and on plants approaching maturity. This disease is largely seedborn, and consequently the primary light gray to brown circular lesions with darkened margins develop on the primary leaves. From the lesions the infection spreads to the nodes where a canker is formed which causes the stem to break. Circular, gray to brown lesions develop on the leaves and stems late in the season.

Le symptomas de brunimento es remarcabile in le plantulas e in le plantas presso maturation. Iste morbo es grandemente transportate per le semines, e, per consequente, le prime lesiones circular, gris clar verso brun con margines obscur, se forma in le folios primari. Ab le lesiones le infection se extende in le nodos ubi se forma un cancere le qual face le pedunculo finder se. Circular lesiones gris verso brun se forma in le folios e in le pedunculos plus tarde in le saison.

Les symptômes du brunissement sont frappants sur les semis et sur les plantes qui approchent de la maturité. Cette maladie se propage surtout par la semence et, en conséquence, les lésions primaires, circulaires, dont la couleur varie du gris pâle au brun, avec des bordures foncées, se forment sur les feuilles primaires. A partir de ces lésions, l'infection s'étend aux noeuds où se forme un chancre qui fait casser la tige. En fin de saison, des lésions circulaires, dont la couleur va du gris au brun, se forment sur les feuilles et les tiges.

Los síntomas del bronceado son conspicuos en plántulas y en plantas que se aproximan a la madurez. Esta enfermedad en su mayoría es acarreada en la semilla y por consiguiente las primeras lesiones que son circulares, de color gris pálido a café, con los bordes obscuros se desarrollan en las hojas primarias. La infección se extiende desde las lesiones hacia los nudos, donde se forma un cancro que hace que el tallo se quiebre. Tarde en la época de crecimiento, se desarrollan lesiones circulares, color gris a café, en las hojas y tallos.

DISTRIBUTION: Af. , A., Aust. , N. A.

126

Lycopersicon esculentum + Alternaria tomato (Cke.) Brinkman =

ENGLISH	Nailhead Spot of Fruit and Stems
FRENCH	Alternariose, Tête de Clou
SPANISH	Viruela del fruto y ramillas, Mancha cabeza de clavo de frutos y tallos
PORTUGUESE	
ITALIAN	
RUSSIAN	альтернариоз
SCANDINAVIAN	
GERMAN	Alternaria-Fleckigkeit
DUTCH	
HUNGARIAN	
BULGARIAN	
TURKISH	
ARABIC	
PERSIAN	
BURMESE	
THAI	
NEPALI	Golheda Ko dade rog गोलभेंडाको दढे रोग
HINDI	
VIETNAMESE	
CHINESE	
JAPANESE	Kokuhan-byô 黒斑病

This disease is characterized by brown, irregular spots with concentric rings in a target pattern on the lower leaves. On old fruits, it sometimes may cause black, leathery, sunken spots near the stem end of the fruit. The disease also causes defoliation of the lower portions of the plant and exposes the ripening fruits to the direct sun. This may causes sunscald and poor color.

Iste morbo se characterisa per brun maculas irregular con anellos concentric in le folios inferior. In fructos vetule illo a vices produce nigre maculas coriacee e depresse presso le stirpe del fructo. Le morbo causa anque defoliation del portiones inferior del planta e discoperi le fructos maturante se directemente al sol. Isto pote causar escaldatura e discoloration.

Cette maladie est caractérisée par des taches brunes irrégulières avec des cercles concentriques évoquant une cible sur les feuilles du bas. Sur les fruits avancés, elle peut parfois causer des taches noires, deprimées et coriaces, près du calice. La maladie peut aussi causer la défoliation du bas de la plante et exposer aux rayons directs du soleil les fruits en train de mûrir, ce qui peut provoquer l'échaudage et détériorer la couleur.

Esta enfermedad se caracteriza por manchas irregulares en las hojas inferiorres, de color café, con círculos concéntricos con forma de blanco para tiro. En frutos viejos, hacia el extremo del pédunculo, puede causar manchas hundidas, correosas, de color negro. Le enfermedad también causa defoliación de la parte inferior de la planta, exponiendo los frutos en maduración al sol directo. Esto puede provocar quemaduras de sol y colorido pobre.

DISTRIBUTION: A., Aust., Eur., N.A.

Lycopersicon esculentum + Cladosporium fulvum Cooke (Fulvia fulva (Cooke) Ciferri) =

ENGLISH	Leaf Mold
FRENCH	Moisissure Olive, Cladosporiose de la Tomate, Maladie Fauve
SPANISH	Moho gris de las hojas, Pudrición de la parte terminal del fruto
PORTUGUESE	
ITALIAN	Bladticchiolatura, Ticchiolatura, Macchie nere del pomodoro
RUSSIAN	**пятнистость бурая кладоспориозная, кладоспориоз**
SCANDINAVIAN	Fløjelsplet, Sammetsfläcksjuka
GERMAN	Samtklecken, Braun-oder Samtflecken, Samtfleckigkeit
DUTCH	Bladziekte, Meeldauw, Bruine Vlekkenziekte
HUNGARIAN	
BULGARIAN	бактериално изсъхване
TURKISH	Domates yaprak küfü
ARABIC	
PERSIAN	Klădôspŏrēyŏmē gŏjeh fárángē کلادوسپوریم گوجه فرنگی
BURMESE	
THAI	Bi-Jut ใบจุด
NEPALI	
HINDI	Parn phaphundi पर्ण फाफुन्दी
VIETNAMESE	Cháy trái
CHINESE	Fan chýe yêh méi bìng 蕃茄葉黴病
JAPANESE	Hakabi-byŏ 葉かび病

Early symptoms include light yellow spots on the upper side of the leaf. On the corresponding lower side, greenish or purple mold soon appears. Finally the leaf tissue dies and the leaf curls up and drops. The mold is seldom found on the stem and is difficult to detect on the fruit.

Le symptomas initial include maculas jalne clar in le latere superior del folio. Al correspondente latere inferior mucor quasi verde o purpuree rapidemente se manifesta. Al fin le texito del folio mori e le folio se crispa e cade. Le mucor rarmente se trova in le stirpe e es a pena visibile in le fructo.

Les symptômes initiaux comportent des taches jaune pâle à la face supérieure de la feuille. Vis-à-vis à la face inférieure, apparaît bientôt une moisissure verdâtre ou pourprée. Finalement, les tissus atteints meurent, la feuille s'enroule et tombe. On ne voit pas souvent de moisissure sur la tige et il n'est pas facile d'en déceler sur le fruit.

Los primeros síntomas son manchas amarillo pálido en la superficie superior de la hoja. En la parte correspondiente del envés, aparece pronto un moho verdoso o púrpura. Finalmente el tejido de la hoja muere, la hoja se enrolla y cae. El moho es rara vez encontrado en el **tallo** y difícil de detectar en el fruto.

DISTRIBUTION: Af., A., Aust., Eur., N.A., C.A., S.A.

128

Lycopersicon esculentum + Colletotrichum phomoides (Sacc.) Chester (Corticium solani (Prill. & Del.) Bourd. & Galz.) =

ENGLISH	Anthracnose
FRENCH	Anthracnose des Fruits Mûrs
SPANISH	Antracnosis
PORTUGUESE	Antracnose
ITALIAN	
RUSSIAN	антракноз плодов
SCANDINAVIAN	
GERMAN	Anthraknose, Reifefäule
DUTCH	
HUNGARIAN	
BULGARIAN	
TURKISH	
ARABIC	Zobol Tary ذبول طري
PERSIAN	
BURMESE	
THAI	
NEPALI	
HINDI	
VIETNAMESE	
CHINESE	Fan chýe tàn dzǔ bìng 蕃茄炭疽病
JAPANESE	Tanso-byô 炭そ病

In the early stages this disease causes lesions to appear as small, circular, sunken spots in the skin of the fruit much as though the surface had been dented with a blunt pencil. As these spots increase in size, the central portion becomes dark from the presence of the black fungus beneath the skin. Eventually the entire fruit becomes affected, often developing watery rot.

In le stadios initial iste morbo produce lesiones le quales se manifesta como minute maculas circular e depresse in le pelle del fructo. Le superfacie sembla monstrar le impressiones de un obtuse stilo de graphite. Quando iste maculas cresce, le area central deveni obscurate a causa del fungo nigre sub le pelle. Al fin le fructo integre es afflicte e sovente un putrefaction aquose occurre.

A ses premières phases, cette maladie provoque l'apparition de lésions sous forme de petites taches rondes, déprimées, sur la pelure du fruit, dont la surface devient comme marquée des coups d'un crayon épointé. Puis, les taches s'agrandissent et leur centre devient sombre à cause du champignon noir présent sous la peau. Eventuellement, tout le fruit est envahi et il s'ensuit souvent une pourriture aqueuse.

En sus primeras etapas esta enfermedad causa pequeñas manchas circulares hundidas en la cáscara del fruto, como si la superficie hubiera sido abollada con un lápiz sin punta. A medida que estas manchas aumentan de tamaño, la parte central se obcurece por la presencia del hongo negro debajo de la piel. Eventualmente todo el fruto resulta afectado, desarrollándose a menudo una pudrición acuosa.

DISTRIBUTION: Af., A., Aust., Eur., N.A., C.A.

Lycopersicon esculentum + Corynebacterium michigenense (E. F. Smith) Jensen =

ENGLISH	Bacterial Canker of Fruit and Stems, Bird's-eye of Fruit
FRENCH	Chancre Bactérien, Bactériose vasculaire, Flétrissement Bactérien
SPANISH	Marchitamiento, Cancro bacteriano, Ojo de pájaro, Lancha bacteriana
PORTUGUESE	Cancro bacteriano
ITALIAN	Avvizzimento batteriaceo del pomodoro
RUSSIAN	рак бактериальный
SCANDINAVIAN	
GERMAN	Bakterienwelke, Bakterienkrebs, Stengelerkrankung, Bakterielle Welke
DUTCH	
HUNGARIAN	
BULGARIAN	бактериално изсъхване
TURKISH	Domates bakteri solgunluğu
ARABIC	
PERSIAN	
BURMESE	
THAI	
NEPALI	
HINDI	
VIETNAMESE	
CHINESE	
JAPANESE	Kaiyo-byô　かいよう病

The first symptoms of this disease are wilting, rolling and browning of the leaves on one side of the plant. The pith may disappear or be discolored. In severe cases the whole plant dies. Sometimes the disease causes splitting of the stems and yellowish or reddish-brown cavitites in the pith. Under certain conditions fruit spots occur. These are white at first, later with light brown roughened centers surrounded by a white halo .

Le prime symptomas de iste morbo es marcescentia, inrolamento, e brunimento del folios in un latere del planta. Le medulla pote extinguer se o suffrer coloration anormal. In casos sever, tote le planta mori. A vices le morbo causa fissura del pedunculos e produce cavitates quasi jalne o brun-rubiette in le medulla. Sub alicun conditiones maculas occurre in le fructo. Istos es al initio blanc e plus tarde monstra asperate centros brun clar cincte per un halo blanc.

Les premiers symptômes de cette maladie sont le flétrissement, l'enroulement et le brunissement des feuilles sur un côté de la plante. La moëlle peut disparaître ou se décolorer. Dans les cas graves, la plante entière meurt. Quelquefois, la maladie fait fendre la tige en deux et provoque la formation de cavités jaunâtres ou brun rougeâtre. Dans certaines conditions, des taches apparaissent sur le fruit. D'abord blanches, ces taches deviennent brun pâle et rugueuses au centre et sont entourées d'un halo blanc.

Los primeros síntomas de esta enfermedad son la marchitez, enrollamiento y bronceado de las hojas de un lado de la planta. La médula del tallo puede desaparecer o decolorarse. En caso severos muere toda la planta. Algunas veces la enfermedad causa que los tallos se rajen, así como cavidades amarillas o café rejizo en la médula. Bajo ciertas condiciones aparecen manchas en el fruto. Estas son blancas primero y luego tienen centros color café claro rodeadas de un halo blanco.

DISTRIBUTION: Af., A., Aust., Eur., N.A., S.A.

Lycopersicon esculentum + Didymella lycopersici Kleb. =

ENGLISH	Stem Rot
FRENCH	Chancre de la Tomate
SPANISH	Podredumbre del tallo
PORTUGUESE	
ITALIAN	Marciume del fusto, Cancro
RUSSIAN	рак дидимеллезный
SCANDINAVIAN	Tomatsyge, Tomatkräfta, Tomatstengelsyke, Koldbrand
GERMAN	Stengelfäule, Tomatenstengelfäule, Didymella-Krebs
DUTCH	Kanker, Tomatenkanker
HUNGARIAN	
BULGARIAN	раковини, дръжково гниене
TURKISH	Domates sap yarası
ARABIC	
PERSIAN	
BURMESE	
THAI	
NEPALI	
HINDI	
VIETNAMESE	
CHINESE	
JAPANESE	Kukigare-byô 茎枯病

The disease causes a stem and fruit rot. Secondary cankers may later develop higher up on the stem. The plant collapses and dies. The soft, outer diseased tissue contains the fungus fruiting bodies which extrude slimy pink masses. Infection can occur on roots, leaves and flowers.

Le morbo causa putrefaction del fructos e lor pedunculos. Canceres secundari pote formar se plus tarde plus alto in le pedunculos. Le planta suffre collapso e mori. Le molle texito exterior morbide contine le structuras sporal del fungo, le quales exsuda rosee massas viscose. Infection pote occurrer in radices, folios, e flores.

La présente maladie cause une pourriture de la tige et du fruit. Plus tard peuvent se former des chancres secondaires plus haut sur la tige. La plante s'affaisse et meurt. Les tissus superficiels malades, flasques, portent les fructifications du champignon, qui sortent en masses roses gluantes. L'infection peut atteindre les racines, les feuilles et les fleurs.

La enfermedad causa una podredumbre del tallo y del fruto. Los cancros secundarios pueden desarrollarse más tarde más arriba en el tallo. La planta sufre un colapso y muere. El tejido blando enfermo más externo contiene las estructuras de fructificación del hongo, que expulsan masas viscosas color rosado. La infección puede ocurrir en raíces, hojas y flores.

DISTRIBUTION: Af., A., Aust., Eur., N.A., C.A., S.A.

Lycopersicon esculentum + Fusarium oxysporum Schlect. f. lycopersici (Sacc.) Wr. =

ENGLISH   Fusarium wilt
FRENCH   Flétrissure Fusarienne, Fusariose
SPANISH   Marchitez por Fusarium, Fusariosis, Marchitez radical
PORTUGUESE   Fusariose, Murcha de Fusarium
ITALIAN   Tracheomicosi
RUSSIAN   фузапиоз, увядание фузариозное
SCANDINAVIAN   Fusarium sovesyge
GERMAN   Fusarium-Welke der Tomate, Verwelkingsziekte
DUTCH   Verwelkingsziekte
HUNGARIAN
BULGARIAN
TURKISH   Fusarium solgunluğu
ARABIC   Zobol   ذ بــول
PERSIAN
BURMESE
THAI   Bi-Hiaw   ใบเหี่ยว
NEPALI
HINDI   Fusarium Mlani   फयोज्ँरयम मलानी
VIETNAMESE   Bệnh mạch quản - Héo cây
CHINESE   Fan chýe wǔei diao bìng   蕃茄萎凋病
JAPANESE   Ityo-byô   萎ちょう病

This disease causes a true vascular wilt, the first symptom of which is a yellowing of the lower, older leaves on a single stem. These leaves soon die, as do others, progressing from the base of the stem outward. Soon the top portions are affected; they, too, yellow, wilt and die. Internal examination of the stem will disclose a brown discoloration in the vascular area, generally extending the full length of the stem.

Iste morbo causa un typic marcescentia vascular cuje prime symptoma es jalnessa del inferior folios plus vetule de un pedunculo. Iste folios mori rapidemente e le alteres, ab le pede del pedunculo in alto, mori etiam. Tosto le partes alte es affecte; illos deveni jalne, marcesce, e mori. Examination interne del pedunculo revela coloration anormal brun in le area vascular, generalmente extendente se per tote le longitude del pedunculo.

Cette maladie cause un véritable flétrissure vasculaire, dont le premier symptôme est le jaunissement des vieilles feuilles du bas, seulement sur une tige. Bientôt, ces feuilles meurent, puis d'autres, et cela continue depuis la base de la tige vers le haut. La partie supérieure ne tarde pas à être atteinte; là encore, il y a jaunissement, flétrissement et mortalité. L' examen   de l'intérieur de la tige révèle un brunissement de la région vasculaire, qui s'étend généralement à toute la longueur de la tige.

Esta enfermedad causa un verdadera marchitez vascular cuyo primer síntoma es un amarillamiento de las hojas inferiores más viejas de un solo tallo. Estas hojas mueren pronto, así como otras, progresivamente desde la base dal tallo hacia afuera. Pronto se ven afectadas las partes de arriba, las cuáles también se amarillan, se marchitan y mueren. El examen interno del tallo revela una decoloración café del área vascular, generalmente extendiéndose a todo lo largo del tallo.

DISTRIBUTION: Af., A., Aust., Eur., N.A., C.A., S.A.

132

Lycopersicon esculentum (Coffea spp., Gossypium spp.) + Myrothecium roridum Tode ex Fr. =

ENGLISH	Ring Rot of Fruit
FRENCH	
SPANISH	Podredumbre anular del fruto
PORTUGUESE	
ITALIAN	
RUSSIAN	
SCANDINAVIAN	
GERMAN	
DUTCH	
HUNGARIAN	
BULGARIAN	
TURKISH	
ARABIC	
PERSIAN	
BURMESE	
THAI	
NEPALI	
HINDI	
VIETNAMESE	
CHINESE	
JAPANESE	

This disease causes a rot of fruit. The fruit spots are dark brown to black, roughly circular, flattened to slightly depressed. The fruiting of the fungus is often in concentric zones. The fungus spores collect in black ink-like drops on the surface. The masses later dry down into flattened discs. At first the invaded fruit tissue is rather tough and can be removed as a core.

Iste morbo produce putrefaction del fructo. Le maculas in le fructos es brun obscur verso nigre, quasi circular, e platte verso un pauc depresse. Le structuras sporal del fungo sovente se trova in zonas concentric. Le sporas fungal se gruppa in guttas nigre obscur al superfacie. Le massas plus tarde se desicca e deveni discos platte. Al initio le texito invase del fructo es assatis dur e on pote remover lo como un corde.

La présente maladie cause une pourriture du fruit. Les taches du fruit sont brun foncé à noires, grossièrement circulaires, aplaties ou légèrement déprimées. Les fructifications du champignon sont souvent disposées en zones concentriques. Les spores du champignon s'assemblent souvent à la surface en gouttes noires comme de l'encre. Par la suite, les masses s'assèchent pour former des disques aplatis. Au début, les tissus envahis du fruit sont plutôt durs et peuvent s'enlever comme un cornillon.

Esta enfermedad causa una podredumbre del fruto. Las manchas del fruto son café obscuro a negro, toscamente ciculares, aplastadas a ligeramente hundidas. La fructificación del hongo está frecuentemente en zonas concéntricas. Las esporas del hongo se congregan en gotas como de tinta negra en la superficie. Posteriormente las masas se secan formando discos aplastados. Al principio el tejido del fruto que ha sido invadido es algo resistente y puede ser removido.

DISTRIBUTION: Af., A., Aust., Eur., N.A., C.A., S.A.

Lycopersicon esculentum + Phoma destructiva Plowr. =

ENGLISH   Phoma Rot, Black Spot of Stems, Fruits and Leaves
FRENCH   Pourriture Phoméene, Pourriture des Fruits
SPANISH   Mancha negra de las hojas y tallos, Podredumbre Phoma
PORTUGUESE   Mancha de Phoma, Podridão de Phoma
ITALIAN   Marciume del pomodoro, Carie
RUSSIAN   гниль плодов фомозная, фомоз
SCANDINAVIAN   Svartfläcksjuka hos tomater
GERMAN   Schwarzfleckenkrankheit der Tomaten, Stielfäule, Phoma-Fruchtfäule
DUTCH
HUNGARIAN
BULGARIAN
TURKISH
ARABIC
PERSIAN
BURMESE
THAI
NEPALI
HINDI
VIETNAMESE
CHINESE   Fan chỷe gwǔo fǔ bìng   蕃茄果腐病
JAPANESE   Migusare -byồ   実腐病

This disease occurs on the fruit, stems and leaves. The spots are subcircular to irregular in outline, minute to large, and dark brown in color. They may or may not show concentric rings.
One large spot or a few small ones on a single leaflet may cause it to turn yellow and drop.
Ordinarily, infection begins on the lowest leaves and gradually progresses upward until most of
the vine is defoliated. The spots on the stem are similar but are somewhat sunken. With
severe infection the plant is killed.

Iste morbo occurre in le fructos, in le pedunculos, e in le folios. Le maculas es de contorno
subcircular verso irregular, minute verso grande, e de color brun obscur. Illos pote monstrar
anellos concentric o non. Un macula grande o alicun maculas minute pote causar que un folietto
deveni jalne e cade. Usualmente, le infection comencia in le folios le plus basse e progrede
per grados in alto usque a causar defoliation del major parte del vite. In pedunculos le maculas
es similar, ma aliquanto depresse. Infection sever face le planta morir.

Cette maladie se rencontre sur le fruit, les tiges et les feuilles. Les taches sont de presque
rondes à irrégulières, de minuscules à grosses, et de couleur brun foncé. Des cercles concentriques peuvent être présents ou non. Une grosse tache ou plusieurs petites sur une même
foliole peut la faire jaunir et tomber. L'infection commence d'ordinaire sur les feuilles du
bas pour progresser graduellement vers le haut jusqu'à ce que la tige soit presque complètement effeuillée. Les taches sur la tige sont semblables mais quelque peu déprimées. Une grave
infection tue la plante.

Esta enfermedad ataca los frutos, tallos y hojas. Las manchas son de contorno subcircular a
irregular, varían desde infimas a grandes y son de color café obscuro. Una mancha grande o
algunas pequeñas en una sola hojilla pueden hacer que ésta se vuelva amarilla y se caiga.
Comunmente la infección comienza en las hojas inferiores y progresa gradualmente hacia arriba,
hasta que casi toda la planta se defolia. Las manchas en el tallo son similares, pero algo
hundidas. La planta muere si la infección es severa.

DISTRIBUTION: A., Aust., Eur., N.A., C.A., S.A.

134

Lycopersicon esculentum + Septoria lycopersici Speg. =

ENGLISH   Leaf Spot
FRENCH   Tache Septorienne, Septoriose, Taches Foliaires
SPANISH   Septoriosis, Viruela, Mancha zonal de la hoja, Mancha foliar
PORTUGUESE   Septoriose, Mancha de Septoria
ITALIAN   Septoria del pomodoro, Macchie delle foglie, Nebbia
RUSSIAN   пятнистость белая, септориоз
SCANDINAVIAN   Bladpletsyge på tomat m. fl.
GERMAN   Blattfleckenkrankheit
DUTCH   Vlekkenziekte der Tomaten
HUNGARIAN   Fehér levélfoltosság
BULGARIAN   бели листни петна
TURKISH   Domates septoria yaprak lekesi
ARABIC
PERSIAN
BURMESE
THAI
NEPALI   Golbeda Ko kalothople rog   गोल भेंडाको कालो थोप्लेरोग
HINDI
VIETNAMESE   Đốm lá
CHINESE
JAPANESE   Sirahosi-byô   白星病

This disease may cause the plant to be attacked at any stage of its growth -- from that of the seedling to the time that the plant dies. The small circular spots first are observed as water-soaked areas on the undersurface of the lower leaves. As the spots enlarge they develop dark brown margins and sunken white or gray centers. The spots vary greatly in size and, if numerous, the leaflet dies. The centers of the spots on the upper surface are marked characteristically by the presence of minute black fruiting bodies. Similar spots sometimes form on the flower, stem and rarely on the fruit.

Iste morbo pote attaccar le planta per tote le stadios de su crescimento -- ab le plantula usque al morte del planta maturate. Le minute maculas circular se manifesta al initio como areas aquose al superfacie inferior del folios inferior. Quando le maculas cresce, illos monstra margines brun obscur e depresse centros blanc o gris. Le maculas se varia de grande mesura in dimension, e, si illos es numerose, le folietto mori. Le centros del maculas in le superfacie superior se characterisa per minute structuras sporal nigre. Maculas similar a vices se forma in le flores, le pedunculos, e (rarmente) le fructos.

Cette maladie peut attaquer la plante à n'importe quel moment de sa croissance, depuis le semis jusqu'à la mort. Des petites taches circulaires se présentent d'abord comme de petites plages détrempées à la face inférieure des feuilles du bas. A mesure qu'elles s'agrandissent, les taches s'entourent de marges brunes et leur centre devient blanc ou gris et déprimé. La grandeur des taches varie beaucoup. Lorsqu'elles sont nombreuses, la foliole meurt. Le centre des taches à la face supérieure est marqué de façon caractéristique par la présence de minuscules fructifications noires. Des taches similaires se forment parfois sur la fleur, sur la tige et, plus rarement, sur le fruit.

Esta enfermedad puede atacar la planta en cualquier etapa de su crecimiento, desde el semillero hasta que la planta muere. Las pequeñas manchas circulares se observan primero como áreas aguajosas en el envés de las hojas inferiores. A medida que las manchas crecen, van desarrollando márgenes color café obscuro, con centros blancos o grises hundidos. Las manchas varían mucho de tamaño y cuando son numerosas la hojilla muere. En la superficie superior de las hojas el centro de las manchas está característicamente marcado por la presencia de minúsculas estructuras de fructificación de color negro. A veces se forman manchas similares en la flor, tallo y rara vez en el fruto.

DISTRIBUTION: Af., A., Aust., Eur., N.A., C.A., S.A.

Lycopersicon esculentum (Capsicum spp. and others) + Xanthomonas vesicatoria (Doidge)
Dowson =

ENGLISH   Bacterial Spot of Fruit and Leaves, Stem Canker
FRENCH   Tache Bactérienne
SPANISH   Marchitez de hojas, Mancha bacterial, Cancro del tallo
PORTUGUESE   Mancha bacteriane
ITALIAN
RUSSIAN   пятнпстость черная бактериальная
SCANDINAVIAN
GERMAN   Bakterielle-Schwarzfleckenkrankheit
DUTCH
HUNGARIAN   Baktériumos varasodás
BULGARIAN   бактериално струпясване
TURKISH   Domates bakteriyel yaprak kara lekesi
ARABIC
PERSIAN
BURMESE
THAI
NEPALI
HINDI   Jeewanu Parn daag   जीवनु परण दाग
VIETNAMESE
CHINESE   Fan chýe shìh jîunn kwèi yáng bìng   蕃茄細菌潰瘍病
JAPANESE   Hanten-saikin-byô   斑点細菌病

The disease pathogen causes a superficial corky scab of the fruits with irregular lobed margins and water-soaked halos, a leaf spot or blight and a canker of the stems. The first symptoms usually appear on the leaves as minute water-soaked spots, later they become angular, turn black and have a greasy appearance. The lesions never become large. The most noticeable symptom is on the green fruit.

Le pathogeno de iste morbo produce un crusta superficial similar a corco in le fructos con margines irregular e lobate con halos aquose, un macula o necrosis foliar, e un cancere in le pedunculos. Le prime symptoma usualmente se manifesta como minute maculas aquose in le folios. Plus tarde illos deveni angular e nigre e sembla esser oleose. Le lesiones non jammais deveni grande. Le symptoma le plus remarcabile occurre in le fructo verde.

Le pathogène de cette maladie produit sur les fruits une galle liégeuse superficielle à marges irrégulières lobées avec des aréoles délavées, ainsi qu'une tache ou brûlure des feuilles et un chancre des tiges. Les premiers symptômes apparaissent d'ordinaire sur les feuilles sous forme de miniscules taches délavées qui, par la suite, deviennent anguleuses, noires et prennent un aspect graisseux. Les lésions ne deviennent jamais grosses. Le symptôme le plus apparent se trouve sur le fruit vert.

El patógeno de la enfermedad causa en los frutos una costracorchosa superficial, con márgenes irregulares lobulados y halo aguajoso; una mancha foliar o tizón y un cancro en los tallos. El primer síntoma generalmente aparece en las hojas como minúsculas manchas aguajosas que después se vuelven angulares, color negro y tienen apariencia grasosa. Las lesiones nunca se hacen grandes. El síntoma más notorio está en el fruto verde.

DISTRIBUTION: Af., A., Aust., Eur.,   N.A., C.A., S.A.

136

Malus sylvestris = Malus pumila (and others) + Armillaria mellea (Vehl. ex Fr.) Kummer =

ENGLISH	Mushroom Root Rot
FRENCH	Pourridié - agaric, Pourridié des Racines
SPANISH	Pudrición radicular, Pudrición de la base del tronco y raíz
PORTUGUESE	
ITALIAN	
RUSSIAN	гниль корней армилляриеллезная
SCANDINAVIAN	
GERMAN	Armillariella-Wurzelfäule
DUTCH	
HUNGARIAN	
BULGARIAN	чума
TURKISH	
ARABIC	
PERSIAN	Pōōsedēgēyē rēshēyē ghārchē ásálléyē sēb پوسیدگی ریشهٔ قارچ عسلی سیب
BURMESE	
THAI	
NEPALI	
HINDI	
VIETNAMESE	
CHINESE	
JAPANESE	Naratake-byô ならたけ病

This disease is characterized by the presence of dead areas in the bark, on or around which dark brown or black threads of the fungus may be found. These threads are about the size of lead in a pencil or slightly smaller. Often they form a branched system around the roots, with some strands penetrating the decayed bark and some extending into the surrounding soil. Thick, creamy-white, fan-like sheets of the fungus growth are often found between the bark and the wood. Honey-colored groups of the mushroom-like growths are definite evidence of this disease, but they often don't appear until most of the roots have been destroyed and sometimes not at all.

Iste morbo se characterisa per areas morte in le cortice, in o circum le quales hyphas brun obscur o nigre del fungo se trova. Iste hyphas es del mesme dimension como le graphite in un stilo o un pauc plus minute. Frequentemente illos es un systema ramificate circum le radices con alicun filos penetrante le cortice putrefacte e alteres extendente se in le terra vicin. Inspissate stratos del fungo, blanc como crema e similar a ventilatores, sovente se trova inter le cortice e le ligno. Gruppamentos mellate de iste tumores que es similar a champignons forna evidentia conclusive de iste morbo, ma frequentemente illos se manifesta solmente post que le major parte del radices esseva putrefacte e a vices non del toto se manifesta.

Cette maladie est caractérisee par la présence de plages nécrotiques dans l'écorce. Sur ces plages mortes ou à leur voisinage peuvent se trouver des cordons bruns ou noirs du champignon. Ces cordons ont approximativement la grosseur de la mine d'un crayon ou sont un peu plus petits. Ils forment souvent des ramifications autour des racines; certains cordons pénètrent dans l' écorce cariée alors que d'autres s'en vont dans le sol avoisinant. Des nappes épaisses, blanc crémeux, en éventail, de mycélium se trouvent souvent entre l'écorce et le bois. Des touffes du champignon couleur de miel sont le signe manifeste de cette maladie, mais il arrive souvent qu'ils ne se montrent pas du tout ou n'apparaissent qu'une fois les racines en majorité détruites.

Esta enfermedad se caracteriza pór la presencia de áreas muertas en la corteza, sobre o alrededor de las cuáles pueden encontrarse las hebras color café obscuro o negras del hongo. Estas hebras son más o menos del tamaño de la mina de un lápiz o ligeramente más delgadas. A menudo éstas forman un sistema ramificado alrededor de las raíces, con algunos cordones penetrando la corteza podrida y algunos extendiéndose al suelo cercano. A menudo, entre la corteza y la madera, se encuentra el crecimiento del hongo como lámina gruesa en forma de abanico, de color blanco cremoso. La evidencia definitiva de esta enfermedad son los grupos de crecimiento como hongos o setas de color miel, pero los cuáles no aparecen sino hasta que la mayoría de las raíces han sido destruídas y a veces no aparecen del todo.
DISTRIBUTION: Af., A., Aust., Eur., N.A., C.A., S.A.

<u>Malus sylvestris</u> = <u>Malus pumila</u> (<u>Prunus cerasus</u>, <u>P.</u> <u>domestica</u> and others) + <u>Bactererium</u> <u>tumefaciens</u> E. F. Sm. & Towns. =

ENGLISH	Crown Gall
FRENCH	
SPANISH	Agalla de la corona
PORTUGUESE	Galha da corða
ITALIAN	Rogna, Tumori del colleto
RUSSIAN	рак корней бактериальный, зобоватость
SCANDINAVIAN	Rodhalsgalle, Rotkräfta, Krongalle
GERMAN	Wurzelkropf
DUTCH	Wortelknobbels
HUNGARIAN	
BULGARIAN	бактериален рак
TURKISH	Kök kanseri
ARABIC	Tadaron Tagi تدرن تاجی
PERSIAN	Sàràtānê gēyāhē سرطان گیاهی
BURMESE	
THAI	
NEPALI	
HINDI	
VIETNAMESE	
CHINESE	
JAPANESE	Konto-gansyu-byð 根頭がんしゅ病

Two types of abnormal growth characterize this disease: (1) tumorlike enlargements (galls) which form at or near the groundline, at graft unions and on roots and (2) overgrowth of stems and roots. The rough, dark, corky galls vary in size from 1/4 inch in diameter to 50 to 100 pounds. The galls persist throughout the growing season and develop a covering of bark and an interior woody structure. In severe attacks, the trees may lose vigor and later die.

Duo generes de crescimento anormal es characteristic de iste morbo: (1) gallas o tumores que se forma in o presso le terreno, in uniones de insertar, e in radices; e (2) hypertrophia del truncos e del radices. Le asperate gallas obscur e similar a corco se varia in grandor ab un diametro de 0,6 cm usque a un peso de 20-45 kg. Le gallas resta per tote le saison e deveni coperte de cortice; un interior structura lignose anque se developpa. In attaccos sever le arbores pote perder lor vigor e plus tarde morir.

Deux types de croissance anormale caractérisent cette maladie: (1) des excroissances ressemblant à des tumeurs (galles) qui se forment au ras du sol, aux points de jonction des greffes et sur les racines et (2) la surcroissance des tiges et des racines. Les galles, qui sont rugueuses, foncées et liégeuses, peuvent varier en grosseur depuis 1/4" de diamètre jusqu'à peser 50 à 100 livres. Les galles persistent durant toute la saison de croissance; elles sont recouvertes d'écorce et leurs tissus sont liégeux à l'intérieur. Dans les cas graves, les arbres peuvent perdre leur vigueur et mourir.

Esta enfermedad se caracteriza por dos tipos de crecimiento anormal: 1) Protuberancias parecidas a tumores (agallas) que se forman a nivel o cerca del suelo, en uniones de injertos o en raíces y 2) Crecimiento exhuberante de tallos y raíces. Las agallas, ásperas, obscuras, corchosas, varían en tamaño desde 1/4" de diámetro hasta 50 a 100 libras. Las agallas persisten durante toda la época de crecimiento y desarrollan una cubierta sobre la corteza y un interior de estructura leñosa. En ataques severos, los árboles pierden vigor y posteriormente se meuren.

DISTRIBUTION: Af., A., Aust., Eur., N.A., C.A., S.A.

138

<u>Malus</u> <u>sylvestris</u> = <u>Malus</u> <u>pumila</u> (<u>Ribes</u> spp.) + <u>Botryosphaeria</u> <u>ribis</u> Gros. & Dug. =

ENGLISH   Botryosphaeria Canker, Fruit rot of Apple, Cane Blight
FRENCH
SPANISH   Muerte apical de las ramillas, Tizón de la caña
PORTUGUESE   Cancro de papel, Cancro Botryosphaeria
ITALIAN
RUSSIAN
SCANDINAVIAN
GERMAN
DUTCH
HUNGARIAN
BULGARIAN
TURKISH
ARABIC
PERSIAN
BURMESE
THAI
NEPALI
HINDI   Sheersarambi Kshyay   शिरसराम्बी द्यय
VIETNAMESE
CHINESE
JAPANESE   Dôgusare-byô   胴腐病

The first symptoms on twigs are small circular spots which become depressed and watery. The following spring, a few pimplelike elevations appear on the lesion's surface. New cankers may exude liquid in midsummer and become quite large. Large limbs may be girdled by coalesced cankers. The easy sloughing off of the bark's outer covering and the slimy character of the underlying tissue are characteristic. Rings of fruiting bodies can be seen on the exposed surface. External and internal rots of fruit can be found. The external rot lesions are circular and slightly depressed, with the rotted flesh cup-shaped and mushy.

Le prime symptomas in ramettos es minute maculas circular le quales deveni depresse e aquose. In le primavera sequente, alicun elevationes similar a pustulas se manifesta in le superfacie del lesion. Nove canceres a vices exsuda liquor in le medio del estate e deveni assatis grande. Grande ramos es a vices imbraciate per canceres coalescente. Le separation spontanee del exterior integumento del cortice, similar a papiro, e le natura oleose del texito subjacente es characteristic. Anellos de structuras sporal es evidente al superfacie exposite. Putrefactiones e externe e interne de fructos pote occurrer. Le externe lesiones del putrefection es circular e un pauc depresse; le pulpa putrefacte es in le forma de un tassa e molle.

Les premiers symptômes se présentent sur les rameaux sous forme de petites taches circulaires qui deviennent déprimées et aqueuses. Le printemps suivant, quelques petites pustules font irruption à la surface des lésions. De nouveaux chancres peuvent exsuder un liquide au milieu de l'été et devenir très gros. De grosses branches peuvent être encerclées par des chancres fusionnés. La facilité avec laquelle on peut détacher la couche externe à consistance de papier de l'écorce ainsi que la viscosité des tissus sous-jacents sont des points caractéristiques. On peut observer des cercles de fructifications à la surface exposée. Il peut y avoir pourriture externe et interne du fruit. La pourriture externe comporte des lésions circulaires et légèrement déprimées, la chair pourrie du fruit, d'aspect détrempé, se présentant sous forme de soucoupe.

Los primeros síntomas en las ramitas son manchas circulares, pequeñas, que se vuelven hundidas y acuosas. En la siguiente primavera, en la superficie de la lesión aparecen unas pocas protuberancias como pústulas. Nuevos cancres pueden exudar líquidos a mediados del verano y hacerse bastante grandes. Las ramas grandes peuden ser ceñidas por cancres fusionados. La facilidad para desprender la corteza papelosa más externa y el carácter viscoso del tejido que está debajo, son características de la enfermedad. Anillos de estructuras fructíferas pueden verse en la superficie expuesta. Puede encontrarse podredumbre externa e interna de la fruta. Las lesiones externas de podredumbre son circulares y ligeramente deprimidas, con la carne podrida con forma de copa y pulposa.
DISTRIBUTION: Af., A., Aust., N.A., S.A.

Malus sylvestris = Malus pumila (Pyrus communis and others) + Erwinia amylovora (Burrill) Winslow et. al. =

ENGLISH	Fire Blight, Pear Blight
FRENCH	Brûlure bactérienne, Feu Bactérien
SPANISH	Lancha bacterial, Tizón de fuego
PORTUGUESE	
ITALIAN	
RUSSIAN	ожог бактериальный семечковых пород американский
SCANDINAVIAN	Ildsot
GERMAN	Feuerbrand, Bakterienfeuerbrand, Apfelbrand
DUTCH	Bakteriënbrand
HUNGARIAN	
BULGARIAN	
TURKISH	Ateş yanıklığı
ARABIC	
PERSIAN	
BURMESE	
THAI	
NEPALI	
HINDI	
VIETNAMESE	
CHINESE	
JAPANESE	Kasyô-byô 火傷病

This disease attacks all parts of the tree and fruit. Blossoms wilt, turn dark brown and die. The fungus spreads downward into the branchlets, twigs and leaves. When the branches are affected, bark lesions (cankers) which are water-soaked and reddish in color later become brown and dry. Infected shoots turn brown or black. If the tree trunk is invaded, the tree may die. Diseased, immature fruit turn an oily dark green and later turn black.

Iste morbo attacca tote le partes del arbore e fructos. Flores marcesce, deveni brun obscur, e mori. Le fungo se extende in basso in le ramos, ramettos, e folios. Quando le ramos es infectate, lesiones o canceres aquose e rubiette, ma plus tarde brun e desiccate, se forma in le cortice. Partes crescente que es infectate deveni brun o nigre. Si le trunco del arbore es invase, le arbore pote morir. Morbide fructos immatur deveni oleose verde obscur e plus tarde nigre.

Cette maladie attaque toutes les parties de l'arbre et le fruit. Les fleurs se flétrissent, deviennent brun foncé et meurent. Le champignon descend et pénètre dans les branchettes, les ramilles et les feuilles. Quand les branches sont atteintes, il s'y forme des lésions de l'écorce (chancres) d'aspect détrempé et de couleur rougeâtre qui par la suite deviennent brunes et seches. Les pousses infectées brunissent ou noircissent. Lorsque le tronc est envahi, l'arbre peut morir. Les fruits non mûrs atteints de cette malade deviennent vert foncé d'aspect huileux et finissent par noircir.

Esta enfermedad ataca todas las partes del árbol y fruta. Las flores se marchitan, se vuelven café obscuro y mueren. El hongo se propaga hacia abajo hacia las ramitas y hojas. Cuando las ramas se ven afectadas, las lesiones de la corteza (cancros) que son aguanosas y de color rojizo más tarde se vuelven café y secas. Los brotes infectados se vuelven color café o negro. Si el tronco del árbol es invadido, el árbol puede morir. La fruta inmadure enferma se vuelve un verde obscuro grasoso y después negra.

DISTRIBUTION: A., Aust., Eur., N.A.

<u>Malus</u> <u>sylvestris</u> = <u>Malus</u> <u>pumila</u> + <u>Glomerella</u> <u>cingulata</u> (Atk.) Spauld. & Schrenk. (Gloeosporium
<u>fructigenum</u> Berh.) =

ENGLISH	Bitter Rot, Gloeosporium Rot
FRENCH	Pourriture Amère des Fruits
SPANISH	Podredumbre amarga, Pudrición del fruto, Podredumbre Gloesporium
PORTUGUESE	Podridão amarga
ITALIAN	Marciume amaro delle mele
RUSSIAN	гнилв плодов горькая, гниль плодов "спелая"
SCANDINAVIAN	Gloeosporium, Bitterröta Gloeosporiumröta
GERMAN	Bitterfäule, Fruchtfäule an Apfel
DUTCH	Bitterrot van Appels
HUNGARIAN	
BULGARIAN	горчиво гниене
TURKISH	Acı cürüklük
ARABIC	
PERSIAN	
BURMESE	
THAI	
NEPALI	
HINDI	Tikt Viglan टिक्त विगलान
VIETNAMESE	
CHINESE	
JAPANESE	Tanso-byô 炭そ病

On half- to fully-grown fruit, rot may start as small circular, light brown areas. These enlarge
rapidly and become somewhat sunken in the center. The number of spots may vary from a few
to many but usually only a small number develop into advanced stage. Usually more than one
spot develops on a fruit, and by uniting and extensions of these spots the entire fruit is rotted.
Premature ripening and dropping of the fruit is common. Sometimes the mummified fruit remains attached to the twigs all winter.

In fructos medio e completemente maturate, le putrefaction pote comenciar como minute areas
circular brun clar. Istos cresce rapidemente e deveni aliquanto depresse in le centro. Le
numero del maculas pote variar se ab pauc usque a multe, ma usualmente pauc de illos se
developpa in le stadio avantiate. Usualmente alicun maculas se forma in un fructo e per extension e union de iste maculas tote le fructo es putrefacte. Maturation ante le tempore ordinari
e separation e cadita del fructo es commun. A vices le fructo mumificate resta attachate al ramettos per tote le hiberno.

Sur des fruits à demi formés ou de grosseur normale, la pourriture peut se présenter au départ
sous forme de petites plages circulaires brun pâle. Ces plages s'agrandissent rapidement et
deviennent quelque peu déprimées au centre. Les taches peuvent être rares ou nombreuses,
mais, en général, seulement un petit nombre parviennent au stade avancé. D'ordinaire, il y a
plus d'une tache par fruit; la fusion et le grossissement de ces taches font que tout le fruit
pourrit. La prématuration et la chute du fruit sont communes. Parfois, le fruit se momifie
et reste sur l'arbre tout l'hiver.

En la fruta medio o completamente desarrollada la podredumbre puede empezar con pequeñas
áreas circulares color café claro. Estas se agrandan rápidamente y se vuelven algo hundidas
en el centro. El número de manchas puede variar desde unas pocas hasta muchas, pero generalmente sólo un pequeño número progresa hasta llegar a la etapa avanzada. Generalmente se
desarrolla más de una mancha en una fruta y al unirse o extenderse las manchas, la fruta
se pudre por completo. Es común la maduración prematura y la caída de la fruta. Algunas
veces la fruta momificada permanece unida a las ramitas todo el invierno.

DISTRIBUTION: A., Aust., Eur., N.A., C.A., S.A.

Malus sylvestris = Malus pumila (Pyrus communis, Cydonia oblonga) + Mycosphaerella sentina (Fr.) Schroet. =

ENGLISH	Leaf Spot
FRENCH	Taches Foliaires, Septoriose
SPANISH	Mancha de la hoja
PORTUGUESE	
ITALIAN	
RUSSIAN	**септориоз**
SCANDINAVIAN	
GERMAN	Septoria-Blattfleckenkrankheit
DUTCH	
HUNGARIAN	
BULGARIAN	
TURKISH	Elma yaprak lekesi
ARABIC	
PERSIAN	
BURMESE	
THAI	
NEPALI	
HINDI	
VIETNAMESE	
CHINESE	
JAPANESE	

This disease causes a leaf spot. The mature spots are easily recognized by their grayish-white centers and well-defined margins. When the spots first appear, they are evident on the upper surface as small, brownish areas. The centers turn to a grayish color with a dark brown border. On the gray center, the black scattered fruiting bodies of the fungus appear. With severe infection defoliation occurs.

Iste morbo produce un macula foliar. Le maculas maturate es facilemente recognoscite per lor centros blanc-gris e lor ben definite margines. Quando le maculas se manifesta, illos es evidente in le superfacie superior como minute areas quasi brun. Le centros deveni quasi gris con un margine brun obscur. In le centro gris se manifesta le disperse nigre structuras sporal del fungo. In infection sever defoliation occurre.

Cette maladie cause une tache des feuilles. A maturité, les taches sont faciles à reconnaître par leur centre blanc grisâtre et leurs marges bien définies. Dès que les taches apparaissent, elles se manifestent à la face supérieure sous forme de petites plages brunâtres. Leur centre devient grisâtre et les contours se colorent en surgissent çà et là, les fructifications noires du champignon. Les infections graves provoquent la chute des feuilles.

Esta enfermedad causa una mancha en la hoja. Las manchas maduras se reconocen fácilmente por sus centros blanco grisáceos y márgenes bien definidos. Cuando las manchas aparecen por primera vez son evidentes en el haz de la hoja como pequeñas áreas obscuras. Los centros se vuelven de color grisáceo, con un borde café obscuro. En el centro gris aparecen esparcidas las estructuras de fructificación del hongo, de color negro. Con una infección severa de la enfermedad ocurre la defoliación.

DISTRIBUTION: Af., A., Eur., N.A.

142

Malus sylvestris = Malus pumila (Pyrus communis) + Nectria galligena Bres. =

ENGLISH	Canker
FRENCH	Chancre Européen, Chancre des Arbres Fruitiers à Pepins
SPANISH	Chancro europeo, Cáncer, Cancro
PORTUGUESE	Podridão radicular
ITALIAN	Cancro
RUSSIAN	рак стволов и ветвей нектриозный
SCANDINAVIAN	Kræft, Fruktträdskräfta, Lövträdskräfta, Frukttrekreft
GERMAN	Obstbaumkrebs, Apfelbaumkrebs, Nectriakrebs, Krebs
DUTCH	Kanker, Appelkanker, Neusrot Appelkanker
HUNGARIAN	
BULGARIAN	
TURKISH	Elma dal kanseri
ARABIC	
PERSIAN	Shänkrê sêb شانكر سيب
BURMESE	
THAI	
NEPALI	
HINDI	
VIETNAMESE	
CHINESE	
JAPANESE	Gansyu-byô がんしゅ病

The symptoms of this disease in an advanced stage are easy to distinguish from other disease on this tree. Removal of the bark shows a series of callous folds surrounding a central cavity which extends into the wood. The cankers are usually centered around wounds or in the crotches of the limbs. The folds are black and may number from one to several. Early infection starts as small circular brown areas usually around a leaf scar. Later the central area becomes sunken, with the surrounding healthy bark.

Le symptomas de iste morbo in le stadio avantiate es facilemente distinctive ab altere maladias que afflige iste arbore. Si on remove le cortice, on vide un serie de plicas callose circum un cavitate central le qual se extende a in le ligno. Le canceres usualmente se trova circum vulneres o in le bifurcationes del ramos. Le plicas es nigre e pote numerar se ab un singule usque a satis multes. Infection precoce se manifesta como minute areas circular brun, usualmente circum un cicatrice foliar. Plus tarde le area central deveni depresse con le margines elevate super le circumjacente cortice salubre.

Lorsque la maladie en est à ses dernièrs phases, il est facile, sur ces hôtes, d'en distinguer les symptômes de ceux d'autres maladies. L'enlèvement de l'écorce laisse voir une série de replis calleux entourant une cavité centrale qui se prolonge dans le bois. Les chancres se produisent généralement près des blessures ou des fourches. Les replis sont noirs. Il peut y en avoir un ou plusieurs. L'infection se présente au départ sous forme de petites plages circulaires brunes, généralement au voisinage d'une cicatrice de feuille. Par la suite, la partie centrale devient déprimée, es marges s'élevant au-dessus de l'écorce saine avoisinante.

Los síntomas de esta enfermedad en una etapa avanzada son fáciles de distinguir de alguna otra enfermedad que tenga el árbol. Al remover la corteza se ve una serie de pliegues callosos rodeando una cavidad central que se extiende hasta la madera. Los cánceres están generalmente centrados alrededor de heridas o en los nudos de las ramas. Los pliegues son negros y puede haber uno o varios. La infección al principio comienza con pequeñas áreas circulares de color café, generalmente alrededor de alguna cicatriz que tenga la hoja. Más tarde el área central se vuelve hundida con los bordes levantados por encima de la corteza sana de alrededor.

DISTRIBUTION: Af., A., Aust., Eur., N.A., S.A.

<u>Malus</u> <u>sylvestris</u> = <u>Malus</u> <u>pumila</u> + <u>Penicillium</u> <u>expansum</u> Link ex F. S. Gray =

ENGLISH	Blue Mold of Apple, Soft Rot
FRENCH	Moisissure Bleue, Pourriture des Fruits
SPANISH	Podredumbre blanda, Pudrición húmeda de frutos, Moho azul
PORTUGUESE	
ITALIAN	Muffa azzurra delle mele
RUSSIAN	гниль плодов пенициллезная
SCANDINAVIAN	
GERMAN	Grünfäule, Penicillium-Fruchtfäule
DUTCH	
HUNGARIAN	
BULGARIAN	меко гниене
TURKISH	
ARABIC	
PERSIAN	
BURMESE	
THAI	
NEPALI	
HINDI	
VIETNAMESE	
CHINESE	
JAPANESE	Aokabi-byô 青かび病

This disease can usually be distinguished in early stages by the light color and soft, watery texture of the rotted fruit areas. The taste is characteristic, being disagreeably moldy. Once the rot is started it advances rapidly when temperature conditions are favorable. The whole fruit may be rotted in less than two weeks after infection. The rotted area does not become sunken; it is usually yellow or light brown in color, very soft in texture and not wrinkled.

Iste morbo usualmente es distincte in le prime stadios per le color clar e le textura molle e aquose del areas putrefacte del fructo. Le sapor, disagradabilemente mucide, es characteristic. Quando le putrefaction comencia, illo se avantia rapidemente si le conditiones de temperatura es favorabile. Tote le fructo pote putrescer in minus que duo septimanas post le infection. Le area putrefacte non deveni depresse. Illo es usualmente de color jalne o brun clar, de textura multo molle, e non corrugate.

A ses premières phases, cette maladie peut généralement se distinguer par la pâleur et la texture molle et aqueuse des plages pourries du fruit. Le goût désagréable de moisi est caractéristique. Une fois installée, la pourriture progresse rapidement quand les conditions sont favorables. Tout le fruit peut pourrir en moins de deux semaines. La partie pourrie n'est pas déprimée; elle est généralement jaune ou brun pâle, très molle et sans rides.

Esta enfermedad, en etapas tempranas, puede generalmente reconocerse por el color claro y la textura blanda, acuosa de las áreas podridas de la fruta. El sabor es característico, siendo desagradablemente enmohecido. Una vez que la podredumbre comienza avanza rápidamente cuando las condiciones de temperatura son favorables. Toda la fruta puede podrirse en menos de dos semanas después de la infección. El área podrida no se hunde; generalmente es de color amarillo o café claro, de textura muy blanda y no se arruga.

DISTRIBUTION: Af., A., Aust., Eur., N.A., C.A., S.A.

Malus sylvestris = Malus pumila (Pyrus communis, Citrus spp., Humulus lupulus, Syringa spp., Rubus spp. etc.) + Phytophthora cactorum (Leb. & Cohn) Schroet. =

ENGLISH	Collar Rot of Apple and Pear, Die-back, Root Rot
FRENCH	Pourridié du Collet, Pourriture des Fruits
SPANISH	Pudrición del cuello, Podredumbre del cuello del manzano y del peral
PORTUGUESE	
ITALIAN	
RUSSIAN	гниль корневой шейки фитофторозная, фитофтороз
SCANDINAVIAN	
GERMAN	Kragenfäule, Phythopthora-fruchtfäule
DUTCH	
HUNGARIAN	
BULGARIAN	
TURKISH	
ARABIC	
PERSIAN	
BURMESE	
THAI	
NEPALI	
HINDI	
VIETNAMESE	
CHINESE	
JAPANESE	Eki-byô 疫病

Although this is primarily a trunk disease, cankers occasionally appear in the crotches of the branches. Usually, the cankered areas are near the groundline and appear most frequently on trees five years or older. Young cankers are difficult to detect, but after killing of the bark, a moist exudate may be noticed over the invaded area. In this stage the inner bark is slimy. The canker is irregular in outline but usually oval. Rapid extension of the cankered area both in a lateral and a vertical direction occurs, often resulting in girdling of the tree in one season.

Ben que isto es primarimente un morbo del trunco, canceres a vices se manifesta in le bifurcationes del ramos. Usualmente le areas con canceres se trova presso le terreno e, le plus frequentemente, occurre in arbores que ha cinque annos o plus. Canceres juvene es a pena observabile, ma post le morte del cortice on vide un exsudato humide in le area invase. In iste stadio le cortice interior es oleose. Le canceres es de contorno irregular ma usualmente oval. Le area cancerose se extende rapidemente in directiones e lateral e vertical de maniera que le arbore es frequentemente imbraciate in un sol saison.

Bien que cette maladie s'attaque principalement au tronc, on peut trouver occasionnellement des chancres a l'enfourchure des branches. D'ordinaire, les parties chancreuses sont situées au ras du sol et s'observent le plus souvent sur des arbres de cinq ans ou plus. Les jeunes chancres sont difficiles à dépister, mais après la mort de l'écorce on peut déceler un exsudat moite au-dessus de la partie envahie. A ce stade, l'écorce interne est visqueuse. Le chancre a des contours irréguliers, mais il est ordinairement oval. Le chancre progresse rapidement en hauteur et en largeur, ce qui entraîne souvent l'encerclement de l'arbre en une saison.

Aunque ésta es primordialmente una enfermedad del tronco, ocasionalmente aparecen cancros en las uniones de las ramas. Generalmente las áreas cancrosas están cerca del suelo y aparecen con más frecuencia en árboles de cinco años o más. Los cancros jóvenes son difíciles de detectar, pero después de la muerte de la corteza puede verse un exudado húmedo sobre el área invadida. En esta etapa la corteza interior es viscosa. El cancro es de contorno irregular, pero generalmente es ovalado. El área cancrosa aumenta rápidamente, tanto en dirección lateral como vertical, resultando con frecuencia un ceñimento del árbol en una sola estación.

DISTRIBUTION: Af., A., Aust., Eur., N.A., C.A., S.A.

Malus sylvestris = Malus pumila (Pyrus communis, Mespilus germanica, Cydonia oblonga) +
Podosphaera leucotricha (Ell. & Everh.) Salm. =

ENGLISH   Powdery Mildew
FRENCH   Blanc, Oïdium du Pommier
SPANISH   Oidium, Oidio, Blanco, Mildiú polvoriento
PORTUGUESE   Oidio
ITALIAN   Mal bianco del melo, Nebbia del melo
RUSSIAN   мучнистая роса
SCANDINAVIAN   Aeblemeldug, Äpplemjöldagg, Eplemeldugg
GERMAN   Apfelmehltau
DUTCH   Meeldauw, Appelmeeldauw
HUNGARIAN   Almafalisztharmat
BULGARIAN   брашнеста мана
TURKISH   Elma küllemési
ARABIC
PERSIAN   S. hàghēghēyê sĕb   سفیک حغبق سیب
BURMESE
THAI
NEPALI   Syau ko pitho-chhare rog   स्याउको पिठो करें रोग
HINDI   Choorni phapahundi   चूरनी फाफुन्दी
VIETNAMESE
CHINESE
JAPANESE   Udonko-byô   うどんこ病

This disease causes the twigs, leaves and blossoms to be covered with a white to pearly-gray, velvety mold. The twig is dwarfed. Infected leaves are narrower than normal, and they roll inward. Later the older infections turn a darker gray-tan color and may be imbedded with small black fruiting bodies of the fungus. Infected blossoms may fail to set fruit. Infected fruit develops russeting or etching.

Iste morbo causa que le ramettos, folios, e flores es coperte de un mucor como villuto blanc verso gris. Le ramettos es nanos. Folios infectate es plus stricte que normal folios e illos se inrola. Plus tarde le infectiones plus vetule deveni de color bronzate-gris plus obscur e pote manifestar minute structuras sporal nigre del fungo. Flores infectate pote non esser fructificate. Fructos infectate monstra brunimento o gravure.

Cette maladie fait se couvrir les rameaux, les feuilles et les fleurs d'une moisissure veloutée blanche à gris perle. La ramille se rabougrit. Les fruilles infectées sont plus étroites que la normale et s'enroulent vers l'intérieur. Par la suite, les infections d'un certain âge deviennent gris sale et peuvent abriter les petites fructifications noires du champignon. Les fleurs infectées peuvent ne pas porter fruit. Le fruit infecté peut devenir roussâtre ou rongé.

Esta enfermedad causa que las ramitas, hojas y flores se cubran de un moho aterciopelado de color blanco o gris perla. El crecimiento de la ramita se atrofia. Las hojas infectadas son más angostas que las normales y se enrollan hacia dentro. Más tarde las infecciones antiguas se vuelven de un color gris crema más obscuro y pueden contener pequeñas estructuras de fructificación negras del hongo. Las flores infectadas pueden fracasar en fijar el fruto. El fruto infectado se desarrolla en una manzana rojiza de piel áspera o arrugada.

DISTRIBUTION: Af., A., Aust., Eur., N.A., S.A.

Malus sylvestris = Malus pumila (Prunus cerasus, P. domestica, Pyrus communis and others) +
Sclerotinia fructigena Aderh. & Ruhl. (Sclerotinia cinerea (Bon.) Shr.) =

ENGLISH   Brown Rot, Spur Canker
FRENCH   Moniliose, Rot-brun des Arbres Fruitiers, Pourriture Brune
SPANISH   Podredumbre café, Cancro de espuela
PORTUGUESE
ITALIAN   Muffa della frutta, Marciume nero
RUSSIAN   гниль плодов монилиальная, монилиоз, Плодовая гниль
SCANDINAVIAN   Gul monilia, Kärnfruktmögel, Gul frugtskimmel
GERMAN   Polsterschimmel, Monilia-Fruchtfäule, Monilia-Spitzendurre
DUTCH   Moniliaziekte der Pitvruchten, Monilia-rot
HUNGARIAN
BULGARIAN
TURKISH   Meyve monilyası
ARABIC
PERSIAN
BURMESE
THAI
NEPALI
HINDI
VIETNAMESE
CHINESE
JAPANESE   Haibosi-byô   灰星病

This fungus disease causes blossom and twig blight, fruit rot and cankers. On the blossom, it may
attack the floral parts. The infected tissue turn to a gray or light brown color. Later the fruit
becomes invaded. Cankers may develop on twigs, branches as a result of blossom blight. These
cankers are usually elliptical at first, rather definite in outline, sunken and brown in color.
During rainy weather gum exudes on the cankered surface.

Iste fungo produce necrosis del flores e del ramettos e putrefaction e canceres in le fructo.
In le flores, illo pote attaccar le partes floral. Le texito infectate deveni de color gris o brun
clar. Plus tarde le fructo es invase. Canceres pote formar se in ramettos e ramos a causa
del necrosis del flores. Iste canceres es usualmente elliptic al initio, assatis definite in contorno, depresse, e de color brun. In tempore pluviose gumma se exsuda ab le superfacie cancerose.

Cette maladie fongueuse cause une brûlure des fleurs et des rameaux, une pourriture du fruit,
et produit des chancres. Elle peut se limiter aux parties florales. Les tissus infectés deviennent gris ou brun pâle. Ensuite, le fruit est envahi. Des chancres peuvent se former sur les
ramilles et les branches par suite de la brûlure des fleurs. Ces chancres sont en général
elliptiques d'abord, leurs contours sont assez bien définis, ils sont déprimés et de couleur
brune. Par temps pluvieux, de la gomme exsude à la surface des chancres.

Esta enfermedad fungosa causa tizón en la flor y en las ramitas, podredumbre del fruto y cancros. En la floración puede atacar las partes florales. El tejido infectado se vuelve color gris
o café claro. Más tarde la fruta es invadida. Se pueden desarrollar cancros en ramitas y
ramas a consecuencia del tizón de la flor. Estos cancros al principio son generalmente elípti-
cos, con contorno bien definido, hundidos y de color café. En tiempo lluvioso exuda goma de
la superficie cancrosa.

DISTRIBUTION: Af., A., Eur., S.A.

Malus sylvestris = Malus pumila + Sphaeropsis malorum Berk. (Physalospora obtusa (Schw.) Cke.) =

ENGLISH	Black Rot, Frogeye Leaf Spot
FRENCH	Taches Foliaires Noires, "Black Rot" du Pommier
SPANISH	Podredumbre negra, Mancha ocular, Viruela, Mancha de ojo de sapo de la hoja
PORTUGUESE	Podridão preta, Olho de rã, Cancro de Physalospora
ITALIAN	Ticchiolatura delle foglie, Cancro dei rami, Imbrunimento
RUSSIAN	гниль плодов черная физалоспорозная, рак плодов черный
SCANDINAVIAN	
GERMAN	Froschaugenkrankheit der Apfelblätter, Physalospora Schwarzfäule am Apfel
DUTCH	Kikvorsenoogziekte
HUNGARIAN	Fekete rák
BULGARIAN	черно гниене
TURKISH	
ARABIC	
PERSIAN	
BURMESE	
THAI	
NEPALI	
HINDI	Kala Viglan काला विगलान
VIETNAMESE	
CHINESE	
JAPANESE	Kurogusare-byô 黒腐病

The first evident symptoms of this disease are brown spots on the fruit. These spots may remain brown or turn black as the lesions develop. Frequently a series of alternating concentric rings are formed, the darker bands being a deep mahogany brown to black. The rotted fruit finally turn black, but the tissue stays firm and leathery. On the leaf the first evidence of the disease is the appearance of a few to many small purple specks. These spots enlarge and the margin remains purple, while the central area is brown.

Le prime symptomas evidente de iste morbo es maculas brun in le fructo. Iste maculas pote restar brun o devenir nigre quando le lesiones se developpa. Frequentemente un serie de anellos concentric alternative se forma -- le anellos plus obscur essente de color de mahagoni o nigre. Le fructos putrefacte al fin deveni nigre, ma le texito resta dur e coriacee. In le folios le prime symptomas de iste morbo es qualque minute maculas purpuree. Iste maculas cresce e le margine resta purpuree ma le area central deveni brun.

Les premiers symptômes apparents de cette maladie sont des taches brunes sur le fruit. Ces taches peuvent rester brunes ou noircir quand les lésions évoluent. Fréquemment, on peut voir alterner des cercles concentriques, les bandes les plus sombres étant acajou brun foncé à noir. Le fruit pourri finit par noircir, mais les tissus restent fermes et coriaces. Sur la feuille, la maladie apparaît d'abord sous forme de mouchetures pourprées plus ou moins nombreuses. Ces petites taches grossissent et la marge demeure pourprée, alors que la plage centrale est brune.

Los primeros síntomas evidentes de la enfermedad son las manchas café en la fruta. Estas manchas pueden permanecer de color café o volverse negras a medida que las lesiones progresan. Frecuentemente se forman series de anillos concéntricos alternados, siendo las bandas más obscuras de un color café caoba obscuro. La fruta podrida finalmente se vuelve negra, pero el tejido permanece firme y coriáceo. En la hoja la primera evidencia de la enfermedad es la aparición de algunos a muchos pequeños puntos de color púrpura. Estos puntos se agrandan, permaneciendo el margen de color purpúra, mientras al área central es café.

DISTRIBUTION: Af., A., Eur., N.A., S.A.

148

Malus sylvestris = Malus pumila + Venturia inaequalis (Cooke) Wint. =

ENGLISH	Scab
FRENCH	Travelure
SPANISH	Sarna, Roña, Talón de indio
PORTUGUESE	Sarna
ITALIAN	Ticchiolatura del melo, Brusone del melo
RUSSIAN	парша
SCANDINAVIAN	Aeblekurv, Äppleskorv, Päronskorv, Körsbärsskorv, Epleskurv
GERMAN	Schorf, Apfelschorf, Lagerschorf
DUTCH	Schurft, Appelschurft
HUNGARIAN	Almafa varasodás
BULGARIAN	струпясване
TURKISH	Elma kara lekesi
ARABIC	Garab جرب
PERSIAN	
BURMESE	
THAI	
NEPALI	Syau ko dade rog स्याउको दडे रोग
HINDI	
VIETNAMESE	
CHINESE	
JAPANESE	Kurohosi-byð 黒星病

This disease attacks leaves, fruit and less frequently, the twigs. Indefinite olive-green spots appear on both sides of the leaf. Later, the lesions are outlined more sharply and covered by a greenish-black velvety growth. Severe infections result in dwarfed, blistered leaves with marginal dead areas. On the developed fruit, the lesions are well-defined, usually round, olive-green with a velvety coating. With age the lesions turn darker, scabby and often crack. Heavily infected fruit are misshapen and drop prematurely. If infection occurs late in the season, black, smooth, shiny, pinpoint lesions may develop on the fruit in storage.

Iste morbo attacca folios, fructos, e, minus frequentemente, ramettos. Indefinite maculas olivacee se manifesta in tote le duo lateres del folio. Plus tarde le contorno de lesiones deveni plus definite e coperte de un strato como villuto nigre-verde. Infectiones sever produce folios nanos con ampullas e con areas morte in le margines. In le fructo maturate le lesiones es multo definite, usualmente rotunde, verde como olivas, e con un strato como villuto. Post qualque tempore le lesiones deveni plus obscur, monstra cicatrices, e sovente se finde, Fructos severmente infectate es mal formate e cade ante le tempore ordinari. Si infection occurre in le ultime parte del saison, nigre lesiones plan, fulgide, e multo minute pote formar se in le fructos in immagasinage.

Cette maladie attaque les feuilles, les fruits et, moins fréquemment, les rameaux. Des taches mal définies de couleur vert olive, apparaissent sur les deux faces de la feuille. Plus tard, les lésions deviennent mieux définies et se couvrent d'une végétation veloutée, verdâtre a noire. Des infections graves rendent les feuilles rabougries, boursouflées, avec des plages mortes à la marge. Sur les pommes, les lésions sont bien définies, généralement rondes, vert olive, avec un revêtement velouté. Avec le temps, les lésions deviennent plus sombres, scabieuses, et souvent se crevassent. Les pommes gravement atteintes sont difformes et tombent prématurément. Quand l' infection se produit tard dans la saison, il peut se former sur le fruit en entrepôt des lésions en tête d'épingle, noires, satinées et luisantes.

Esta enfermedad ataca hojas, fruto y con menos frecuencia, las ramitas. En ambos lados de la hoja aparecen manchas indefinidas color verde olivo. Más tarde las lesiones tienen contornos más definidos y están cubiertas por un crecimiento aterciopelado, de color negro-verdoso. Las infecciones severas dan como resultado hojas enanas, con ampollas y áreas muertas marginales. En la fruta desarrollada las lesiones son bien definidas, generalmente redondas, color verde olivo y con una cubierta aterciopelada. Con la edad las lesiones se vuelven más obscuras, roñosas y a menudo se rajan. La fruta severamente infectada es deforme y se cae prematuramente. Si la infección ocurre tarde en la estación, en la fruta almacenada pueden formarse lesione como cabeza de alfiler, brillantes, lisas, de color negro.

DISTRIBUTION: Af., A., Aust., Eur., N.A., C.A., S.A.

Medicago sativa + Ascochyta imperfecta Peck (Phoma herbarum West. var. medicaginis West ex Rab.) =

ENGLISH	Black Stem
FRENCH	Tige noire, Taches Foliaires, Ascochytose
SPANISH	Tallo negro, Lanchamiento
PORTUGUESE	
ITALIAN	
RUSSIAN	аскохитоз
SCANDINAVIAN	Sneglebælgens stængelsvamp
GERMAN	Ascochyta-Blattfleckenkrankheit
DUTCH	
HUNGARIAN	
BULGARIAN	
TURKISH	
ARABIC	
PERSIAN	
BURMESE	
THAI	
NEPALI	
HINDI	
VIETNAMESE	
CHINESE	
JAPANESE	Kukigare-byô 茎枯病

The disease appears as dark brown to black lesions on the stems. When the disease is severe, young shoots are blackened, and killed and stems are girdled by the lesions. The brown spots on the leaves are small, irregular in shape and coalesce to form blackened areas. The infected leaves turn yellow and wither before they drop. Brown spots appear on the pods under cool growing conditions.

Iste morbo se manifesta como lesiones brun obscur verso nigre in le pedunculos. Quando le morbo es sever, plantulas es nigrate e mortificate e pedunculos es imbraciate per le lesiones. Le maculas brun in le folios es minute, de contorno irregular, e illos coalesce in areas nigrate. Le folios infectate deveni jalne e se crispa ante que illos cade. Maculas brun se manifesta in le siliquas in tempore fresc.

La maladie se présente sous forme de lésions brun foncé à noires sur les tiges. Quand la maladie est grave, les jeunes pousses noircissent et meurent et les lésions encerclent les tiges. Les taches brunes sur les feuilles sont petites, de forme irrégulière, et se fusionnent pour former des plages noircies. Les feuilles infectées jaunissent et se flétrissent avant de tomber. Des taches brune apparaissent sur les gousses par temps frais.

La enfermedad aparece como lesiones color café obscuro a negro en los tallos. Cuando la enfermedad es severa, los brotes jóvenes se ennegrecen y se secan y los tallos son ceñidos por las lesiones. Las manchas café en las hojas son pequeñas, de forma irregular y se fusionan para formar áreas ennegrecidas. Las hojas infectadas se vuelven amarillas y se marchitan antes de caer. Manchas color café aparecen en las vainas bajo condiciones frescas de crecimiento.

DISTRIBUTION: Af., A., Aust., Eur., N.A., C.A., S.A.

150

Medicago sativa + Corynebacterium insidiosum (McCulloch) Jensen =

ENGLISH   Blight, Bacterial Wilt, Root Rot of Alfalfa
FRENCH   Flétrissure Bactérienne, Flétrissement, Jaunissement
SPANISH   Pudrición bacteriana del corazon, Tizón, Marchitez bacterial
PORTUGUESE
ITALIAN   Avvizzimento batteriaceo dell'erba medica
RUSSIAN
SCANDINAVIAN
GERMAN
DUTCH
HUNGARIAN
BULGARIAN
TURKISH
ARABIC
PERSIAN
BURMESE
THAI
NEPALI
HINDI
VIETNAMESE
CHINESE
JAPANESE   Ityð-saikin-byð   萎ちょう細菌病

The plants are reduced in vigor, the leaves yellow and blacken, and the plants die in late summer. The leaflets on infected plants are smaller and thicker prior to the loss of green color. The stems are smaller and are more numerous in the earlier stages of disease development. The main root shows a pale brown discoloration of the outer woody tissue. This is evident when the outer bark is peeled off. The wilted plants occur first in the lower portions of the fields, either as scattered plants or, more frequently, in groups of plants.

Le plantas perde lor vigor, le folios deveni jalne e nigre, e le plantas mori verso le fin del estate. Le foliettos in plantas infectate es plus minute e plus inspissate ante que illos perde lor color verde. Le pedunculos es plus minute e plus numerose durante le initial stadios del developpamento del morbo. Le radice principal monstra un discoloration brun clar del exterior texito lignose. Si on remove le exterior cortice, iste discoloration es evidente. Le plantas que suffre marcescentia se trova primo in le portiones plus basse del campos, o disperse, o, plus frequentemente, in gruppamentos de plantas.

Les plantes aiblissent, les feuilles jaunissent et noircissent et la mort survient vers la fin de l'été. Avant de perdre leur couleur verte, les folioles des plantes infectées sont plus petites et plus épaisses. Durant les premières phases de la maladie, les tiges sont plus petites et plus nombreuses. Les tissus ligneux externes de la racine principale se décolorent pour devenir brun pâle. On peut s'en rendre compte en soulevant l'écorce extérieure. C'est dans les parties basses des champs qu'on peut apercevoir les premières plantes flétries, soit dispersées, soit, plus fréquemment, en groupes.

Las plantas pierden vigor, las hojas se amarillan y ennegrecen y las plantas mueren al final del verano. Las hojillas' de las plantas infectadas son más pequeñas y gruesas antes de la pérdida del color verde. Los tallos son más pequeños y más numerosos en las etapas tempranas de desarrollo de la enfermedad. La raíz principal muestra una decoloración color café pálido del tejido leñoso externo. Esto es evidente al pelar la corteza externa. Las plantas marchitas aparecen primero en las partes más bajas de los campos, a veces plantas aisladas o más frecuentemente grupos de plantas.

DISTRIBUTION: Af., A., Aust., Eur., N.A.

Medicago spp. + Pseudopeziza medicaginis (Lib.) Sacc. =

ENGLISH	Leaf Spot
FRENCH	Taches des Feuilles de la Luzerne
SPANISH	Mancha negra de la hoja, Viruela, Peca, Mancha foliar, Mancha común
PORTUGUESE	Pinta preta
ITALIAN	Vaiolatura dell'erba medica
RUSSIAN	пятнистость бурая
SCANDINAVIAN	
GERMAN	Klappenschorf, Stengelbrand, Stengelbakteriose
DUTCH	Zwarte Bladvlekken
HUNGARIAN	Lucerna levélfoltossága
BULGARIAN	черни листни петна
TURKISH	Yonca yaprak benek lekesi
ARABIC	
PERSIAN	
BURMESE	
THAI	
NEPALI	
HINDI	
VIETNAMESE	
CHINESE	
JAPANESE	Kukigare-saikin-byô 茎枯細菌病

Circular, small brown spots occur on the leaflets. The spots are restricted in size, usually do not coalesce, and generally do not cause discoloration of the surrounding leaf tissue. The presence of small dark brown to black raised fruiting bodies of the fungus in the center of the mature spot is a distinguishing characteristic. Small, elliptical spots occur on the stems. Heavy infections cause defoliation, especially of the lower leaves.

Minute maculas circular brun se manifesta in le foliettos. Le maculas es limitate in dimension; illos usualmente non coalesce; e illos generalmente non causa coloration anormal del circumjacente texito foliar. Un characteristica distinctive es le minute structuras sporal brun obscur o nigre elevate in le centro del macula matur. Minute maculas elliptic se manifesta in le pedunculos. Infection sever causa defoliation, specialmente in le folios inferior.

Il se forme sur les folioles de petites taches rondes et brunes. Généralement, ces taches de dimensions restreintes ne se fusionnent pas et ne décolorent pas les tissus adjacents. La présence de petites fructifuciations brun foncé à noires, en relief, au centre des taches mûres, est un caractère distinctif du champignon. Sur les tiges, les taches sont petites et elliptiques. Les fortes infections causent la défoliation, surtout des feuilles inférieures.

En las hojitas aparecen manchas circulares, pequeñas, de color café. Las manchas están restringidas en tamaño, generalmente no se juntan, ni causan decoloración del tejido foliar que las rodea. Una característica distinctiva es que en el centro de la mancha madura hay cuerpos fructíferos levantados, pequeños y de color café a negro del hongo. En los tallos aparecen manchas pequeñas y elípticas. Las infecciones fuertes causan defoliación, especialmente de las hojas inferiores.

DISTRIBUTION: Af., A., Aust., Eur., N.A., C.A., S.A.

152

Medicago spp. + Pyrenopeziza medicaginis Fuckel =

ENGLISH	Yellow Leaf Blotch
FRENCH	Taches Brunes des Feuilles
SPANISH	Manchón amarillo de la hoja
PORTUGUESE	
ITALIAN	
RUSSIAN	
SCANDINAVIAN	
GERMAN	
DUTCH	Gele Bladvlekken
HUNGARIAN	
BULGARIAN	
TURKISH	
ARABIC	
PERSIAN	
BURMESE	
THAI	
NEPALI	
HINDI	
VIETNAMESE	
CHINESE	
JAPANESE	ôhan-byô 黄斑病

The young lesions appear as yellow stripes and blotches in elongated form parallel to the leaf veins. The lesions enlarge and the color changes to orange-yellow or brown, shading to yellow at the margins. Small orange to brown fruiting bodies of the fungus develop on the upper surface of the blotch. The stem lesions are elongated yellow blotches which soon turn dark brown.

Le juvene lesiones se manifesta como strias jalne e pustulas de forma elongate parallel al venas foliar. Le lesiones cresce e le color deveni jalne-orange o brun, tendente a jalne al margines. Minute structuras sporal del fungo orange o brun se forma in le superfacie superior del pustulas. Le lesiones del pedunculos es pustulas jalne elongate e illos rapidemente deveni brun obscur.

Les jeunes lésions forment des stries et des éclaboussures jaunes, allongées, parallèles aux nervures foliaires. Ces lésions s'agrandissent et tournent au jaune orangé ou au brun, avec des marges teintées de jaune. De petites fructifications du champignon orangées à brunes se forment à la face supérieure de l'éclaboussure. Sur la tige, les lésions se présentent sous forme d'éclaboussures jaunes, allongées, qui tournent rapidement au brun foncé.

Las lesiones jóvenes aparecen como rayas y manchones amarillos de forma alargada, paralelas a las venas de las hojas. Las lesiones crecen y el color cambia a anaranjado-amarillo o café, disminuyendo en amarillo en los márgenes. Pequeños cuerpos fructíferos del hongo, de color anaranjado a café se desarrollan en la superficie superior de manchón. Las lesiones del tallo son manchones amarillos, alargados, que pronto se vuelven de color café obscuro.

DISTRIBUTION: A., Aust., Eur., N.A., S.A.

Medicago spp. (and other hosts) + Uromyces striatus Schroet. =

ENGLISH	Rust
FRENCH	Rouille
SPANISH	Roya de las hojas, Polvillo
PORTUGUESE	Ferrugem
ITALIAN	Ruggine dell'erba medica
RUSSIAN	
SCANDINAVIAN	
GERMAN	
DUTCH	Roest
HUNGARIAN	
BULGARIAN	
TURKISH	
ARABIC	Sadaa صـدا
PERSIAN	Zȧngê yȯnjêh زنگ یونجه
BURMESE	
THAI	
NEPALI	
HINDI	
VIETNAMESE	
CHINESE	
JAPANESE	Sabi-byô さび病

The symptoms produced by this disease are typical of many of those caused by rust on foliage. Small, round, powdery, reddish-brown to dark brown fruiting bodies form mostly on the under-sides of the leaves. These may also occur on the stems. Seriously infected leaves may turn yellow and drop prematurely. Heavy infection results in greatly reduced production of forage and no seed if the crop is being grown for this purpose.

Le symptomas de iste maladia es typic pro (fer)rugine in foliage. Minute structuras sporal rotunde,pulverose, brun-rubiette, o brun obscur se forma principalmente in le superfacie inferior del folios. Illos pote occurrer anque in le pedunculos. Folios severmente infectate sovente deveni jalne e se separa e cade prematurmente. Infection sever causa un grande diminution de forrage e face le semines non formar se.(Le plantas es sovente seminate pro iste scopo.)

Cette maladie produit sur les feuilles plusieurs symptômes typiques dè la rouille. De petites fructifications rondes, poudreuses, brun rougeâtre à brun foncé se forment, surtout à la face inférieure des feuilles. Ces fructifications peuvent aussi apparaître sur les tiges. Les feuilles gravement infectées jaunissent et tombent prématurément. Une forte infection diminue de beau-coup le rendement en fourrage et supprime la production de graines.

Los síntomas producidos por esta enfermedad son típicos de aquellos causados por la roya en el follaje. Cuerpos fructíferos pequeños, redondos, polvorientos, café rojizo a café obscuro se forman generalmente en el envés de las hojas. Estos también pueden aparecer en los tallos. Las hojas afectadas severamente pueden volverse amarillas y caerse prematuramente. Las infecciones fuertes resultan en una producción grandemente reducida de forraje y nada de semilla si el cultivo se está haciendo con este propósito.

DISTRIBUTION: Af., A., Aust., Eur., N.A., C.A., S.A.

154

Medicago spp. + Urophlyctis alfalfae (Lagerh.) Magnus =

ENGLISH	Wart
FRENCH	Maladie des Tumeurs Marbrées de la Luzerne
SPANISH	Tumor del cuello de la raíz, Verrugas
PORTUGUESE	
ITALIAN	Mal del gozzo dell'erba medica
RUSSIAN	**бородавчатость корневой шейки**
SCANDINAVIAN	Lucernebrok
GERMAN	Wurzelkrebs
DUTCH	Knobbelvoet, Wortelknobbels
HUNGARIAN	
BULGARIAN	
TURKISH	
ARABIC	
PERSIAN	
BURMESE	
THAI	
NEPALI	
HINDI	
VIETNAMESE	
CHINESE	
JAPANESE	

The most characteristic symptom of the disease is the extensive production of galls which are composed of both stem and leaf tissue. These galls vary greatly in size and are white when young. They turn gray to brown as they decay and dry out in mid-season. The galls on the stems appear as swollen knots arising from the bud tissue, whereas the infected leaves and floral parts become thickened to form scale-like galls around the axis of the underdeveloped stem.

Le symptoma le plus characteristic de iste maladia es le apparition extensive de gallas le quales es composite de texito si ben de pedunculo como de folio. Iste gallas se varia grandemente in dimension e es blanc quando illos es juvene. Illos deveni gris o brun quando illos putresce e se desicca al medie-saison. Le gallas in le pedunculos se manifesta como nodos tumide emergente ab le texito de buttones. Folios e partes floral infectate deveni inspissate e forma gallas scaliose circum le axe del pedunculo subdeveloppate.

Le symptôme le plus caractéristique de cette maladie est la production abondante de galles formées des tissus de la tige et de la feuille. La grosseur des galles varie beaucoup. Les jeunes galles sont blanches. Elles vont du gris au brun à mesure qu'elles se décomposent et se dessèchent en mi-saison. Les galles des tiges prennent la forme de noeuds gonflés issus des tissus des bourgeons, tandis que les feuilles et les parties florales infectées s'épaississent pour former des galles écailleuses autour de l'axe de la tige sous-développée.

El síntoma más característico de la enfermedad es la gran producción de agallas, la cuáles están compuestas de tejido tanto del tallo como de la hoja. Estas agallas varían mucho en tamaño y son blancas cuando están jóvenes. Luego se vuelven de color gris a café a medida que se pudren o se secan a mediados de la estación. Las agallas en los tallos aparecen como nudos hinchados que se levantan del tejido del capullo, mientras que las hojas infectadas y partes florales se van engrosando para formar agallas parecidas a escamas alrededor del axis del tallo no desarrollado.

DISTRIBUTION: A., Aust., Eur., N.A., S.A.

Musa sapientum (Musa spp.) + Cordana musae (Zimm.) Höhnel (Scolecotrichum musae Zimm.)=

ENGLISH   Leaf Spot
FRENCH   Taches Foliaires dues à Cordana
SPANISH   Mancha de la hoja, Mancha parda de las hojas, Mancha cordana, Mancha foliar
PORTUGUESE   Manchas das fôlhas
ITALIAN
RUSSIAN   пяинистость корданозная (сколекотрихозная)
SCANDINAVIAN
GERMAN   Cordana-Blattfleckenkrankheit
DUTCH
HUNGARIAN
BULGARIAN
TURKISH
ARABIC
PERSIAN
BURMESE
THAI   Bi-Jut ใบจุด
NEPALI
HINDI
VIETNAMESE   Cháy lá
CHINESE   Shiang jaw yúan shing bìng   香蕉圓星病
JAPANESE

Small, brown spots enlarge to an oval or sometimes a diamond shape and develop a darker red-brown margin. Zonation becomes quite marked with age. A chlorotic halo is very noticeable on the lower surface. Dead tissue may occur as strips from edge to midrib, and a marginal necrosis with an uneven, zigzag, chlorotic edge separating healthy from diseased tissue may develop.

Minute maculas brun se developpa in un forma oval o a vices rhomboidal, e illos deveni brun-rubiette plus obscur in le margines. Zonation deveni assatis remarcabile post qualque tempore. Un halo chlorotic deveni multo remarcabile in le superfacie inferior. Texito morte sovente se manifesta como strias ab le margine al nervatura central. Un necrosis marginal con un bordo irregular, zigzag, e chlorotic, le qual devide le texito san e le texito morbide, pote formar se.

De petites taches brunes s'agrandissent en ovale ou parfois en losange et forment une bordure brun rouge plus foncée. La zonation devient bien marquée avec le temps. Un halo chlorotique est très voyant à la face inférieure. Les tissus morts peuvent apparaître en bandes depuis le bord des feuilles jusqu'à la nervure médiane. Il peut également se former une nécrose mar-ginale à bordure chlorotique, irrégulière et en zigzag, délimitant les tissus sains et les tissus affectés.

Manchas pequeñas, de color café crecen y adquieren forma ovalada o de diamante y desarrollan márgenes de color café-rojizo obscuro. Las zonas se vuelven muy marcadas con la edad. Un halo clorótico es bien notorio en el envés. El tejido muerto puede aparecer como rayas desde la orilla hasta la vena central, y puede desarrollarse una necrosis marginal con una orilla cloró-tica y desigual, en zig-zag, que separa el tejido sano del enfermo.

DISTRIBUTION: Af., A., Aust., N.A., C.A., S.A.

156

Musa sapientum (Musa spp.) + Fusarium oxysporum Schlecht ex Fr. f. sp. cubense (E. F. Sm.) Snyd. & Hansen =

ENGLISH	Wilt
FRENCH	Maladie de Panama, Flétrissement Fusarien, Trachéomycose
SPANISH	Mal de Panamá, Marchitez, Marchitamiento
PORTUGUESE	Mal do Panamá, Fusariose
ITALIAN	
RUSSIAN	увядание фузариозное, "панамская болезнь"
SCANDINAVIAN	
GERMAN	Fusarium-Welke, Panama-Krankheit
DUTCH	
HUNGARIAN	
BULGARIAN	
TURKISH	
ARABIC	
PERSIAN	
BURMESE	
THAI	Ty-Ply ทายทราย
NEPALI	
HINDI	
VIETNAMESE	Thối gốc và chết héo
CHINESE	
JAPANESE	

The first symptoms of this disease are faint yellow streaks in the stems of the oldest leaves. This may be followed by yellowing and wilting until the leaf collapses. Leaf collapse may also occur with little or no yellowing. The disease progresses from older to younger leaves and within four to six weeks only the dead trunk may remain. Discoloration of the leaf tissue accompanies infection. Streaks in these tissues may vary from light yellow to brown.

Le prime symptomas de iste maladia es le strias jalne pallide in le petiolos del folios le plus vetule. Jalnessa e marcescentia occurre usque a facer le folio suffrer collapso. Collapso foliar pote occurrer anque sin jalnesse. Le maladia se extende ab le folios plus vetule a illos plus juvene e a vices resta intra 4 o 6 septimanas solmente le trunco morte. Coloration anormal del texito foliar occurre con infection. Strias in iste texitos pote variar se ab jalne clar a brun.

Les premiers symptômes de cette maladie sont des stries jaune pâle sur les pétioles des plus vieilles feuilles. Cela peut être suivi du jaunisement et du flétrissement jusqu'à l'affaissement de la feuille. Il peut aussi y avoir affaissement sans jaunisement ou presque. La maladie se propage des vieilles feuilles aux jeunes, et peut en l'espace de quatre à six semaines, ne laisser qu'un tronc mort. L'infection s'accompagne d'une décoloration des tissus de la feuille. Dans ces tissus, les stries peuvent aller du jaune clair au brun.

Los primeros síntomas de esta enfermedad son rayas amarillo pálido en los pecíolos de las hojas más viejas. Esto puede ser seguido por un amarillamiento y marchitez hasta que la hoja sufre un colapso. El colapso de la hoja puede también ocurrir cuando hay poco o ningún amarillamiento. Esta enfermedad progresa de las hojas viejas hacia las jóvenes y en un lapso de cuatro a seis semanas puede quedar sólo el tronco muerto. La infeccion es acompañada por una decloración del tejido foliar. Las rayas en estos tejidos pueden variar de amarillo pálido a café.

DISTRIBUTION: Af., A., Aust., N.A., C.A., S.A.

<u>Musa</u> spp. + <u>Gloeosporium</u> <u>musarum</u> Cke. & Mass. =

ENGLISH	Spot, Rot, Black End
FRENCH	Anthracnose de la Banane, Macules sur Fruits
SPANISH	Moteado de la hoja, Mancha, Pudrición de la punta, Podredumbre, Punta negra
PORTUGUESE	Podridão dos frutos e cachos
ITALIAN	Antracnosi delle banane, Marciume apicole
RUSSIAN	**антракноз, пятнистость плодов глеоспориозная**
SCANDINAVIAN	
GERMAN	Anthracknose, Fruchtanthraknose
DUTCH	
HUNGARIAN	
BULGARIAN	
TURKISH	
ARABIC	
PERSIAN	
BURMESE	Thee-pok roga ၊ ဂလီယို ၀ ၇ ၇ ၀ ၆ ၂ ၎ ၀ ၇ ၆
THAI	Pol-Nao ผลเน่า
NEPALI	
HINDI	
VIETNAMESE	Đốm thâm trên cuống và trái
CHINESE	Shiang jaw tàn dzǔ bìng 香蕉炭疽病
JAPANESE	Tanso-byô 炭そ病

The pathogen associated with this disease attacks immature fruit in the field, also leaves and floral parts; however, the symptoms usually pass unnoticed in this stage of development of the plant. The presence of large black areas of the ripe and ripening fruit is the main characteristic symptom. Much damage occurs to the fruit following injuries of the large lesions in the form of scratches and other wounds during handling and transport.

Le pathogeno associate con iste maladia attacca fructos immatur in le campo e anque folios e partes floral. Le symptomas, autem, generalmente non es remarcate in iste stadio de developpamento del planta. Le symptoma characteristic le plus importante es le apparition de grande areas nigre in le fructo e matur e maturante se. Damno sever al fructo es le resultato de injurias del grande lesiones causate per grattamento e vulneres in tractamento e transportation.

Dans le champ, le pathogène associé à cette maladie attaque les fruits non mûrs, les feuilles et les parties florales. Cependant, les symptômes passent généralement inaperçus à ce stade de développement de la plante. Le principal symptôme caractéristique est la présence de grande plages noires sur le fruit mur ou en voie de mûrissement. Le fruit peut subir de graves avaries suite aux blessures de toutes sortes, égratignures, etc., que peuvent subir les grandes lésions lors des manipulations ou durant le transport.

El pategéno asociado con esta enfermedad ataca la fruta inmadura en el campo y también las hojas y partes florales; sin embargo, los síntomas generalmente pasan desapercibidos en este estado de desarrollo de la planta. El síntoma característico principal es la presencia de grandes áreas negras en la fruta madura y en maduración. Causa mucho daño a la fruta por las averías de las lesiones grandes como araños y otras heridas durante el manejo y transporte.

DISTRIBUTION: Af., A., Aust., Eur., N.A., C.A., S.A.

158

Musa sapientum (Musa spp.) + Helminthosporium torulosum (Sydow) Ashby (Deightoniella torulosa (Sydow) M. B. Ellis) =

ENGLISH   Black-tip, Black-end, Black-spot
FRENCH   Maladie du "Bout Noir", Helminthosporiose
SPANISH   Podredumbre de la vaina, Mancha de la hoja, Pudrición del fruto, Final negro
PORTUGUESE
ITALIAN
RUSSIAN   гельминтоспориоз
SCANDINAVIAN
GERMAN   Schwarze Blattfleckenkrankheit
DUTCH
HUNGARIAN
BULGARIAN
ARABIC
PERSIAN
BURMESE   Ah-net-kwet, Ah-net-pyauk roga   နေတို့ နဲ့ ဟယ လၢးတို့ ၃ ၀ လို ၀
THAI   Bi-Mhai   ใบไหม้
NEPALI
HINDI
VIETNAMESE   Cháy lá - Nám quả Thối ngang thân
CHINESE   Shiang jaw hei ban bing   香 蕉 黑 斑 病
JAPANESE

This disease causes a variety of symptoms on fruit and leaves. The skin of the fruit turns black and assumes a dry crumbling consistency. The symptom on the leaves consists of long and narrow spots which follow the veins. The spots are surrounded by a narrow, bright yellow peripheral band; the spots coalesce and large brown necrotic areas running the length of the leaf blade result.

Iste maladia causa symptomas varie in fructos e folios. Le pelle del fructo deveni nigre e monstra un character desiccate e pulverulente. Le symptoms in le folios es maculas elongate e stricte le quales occurre presso le venas. Le maculas es imbraciate per un stricte banda jalne brillante peripheric. Le maculas coalesce e le resultato es grande areas brun necrotic que se extende per tote le longitude del lamina foliar.

Cette maladie produit divers symptômes sur les fruits et les feuilles. La pelure du fruit devient noire et friable. Les symptômes sur les feuilles consistent en des taches longues et étroites le long des nervures. Les taches sont entourées d'une étroite bande périphérique jaune vif; les taches se fusionnent de façon à former de grandes plages nécrotiques brunes tout le long de la feuille.

Esta enfermedad causa una variedad de síntomas en frutos y hojas. La cáscara de la fruta se vuelve de color negro y adquiere na consistencia seca y friable. El síntoma en las hojas consiste en manchas largas y angostas que siguen a las venas. Las manchas están rodeadas por una banda angosta, periférica, de color amarillo brillante; las manchas se juntan y aparecen áreas necróticas grandes que están a lo largo de la hoja.

DISTRIBUTION: Af., A., Aust., N.A., C.A., S.A., W.I.

Musa spp. + Macrophoma musae (Cke.) Berl. & Vogl. (Phoma musae (Cke.) Sacc.) =

ENGLISH	Freckle, Black Spot of Fruit and Leaves
FRENCH	Moucheture Noire, Taches Noires des Fruits, Taches de Rousseur
SPANISH	Moteado de la hoja, Peca, Mancha negra de fruto y hojas
PORTUGUESE	
ITALIAN	Maculatura del banano
RUSSIAN	пятнистость макрофомозная плодов и листьев
SCANDINAVIAN	
GERMAN	Schwarze Frucht - und Blattfleckenkrankheit
DUTCH	
HUNGARIAN	
BULGARIAN	
TURKISH	
ARABIC	
PERSIAN	
BURMESE	Ah-net-set, Ah-net-pyauk roga ပင်္ဂ၍နိုဗာ:ဇရော:
THAI	
NEPALI	
HINDI	
VIETNAMESE	
CHINESE	Shiang jaw hei shing bìng 香蕉黑星病
JAPANESE	

This disease causes brown to black speckled stripes which extend through the midrib to the margin of the old leaves. Each individual spot has an indefinite border. The surface of the spots feels rough as a result of the fruiting bodies. The spots, usually heaviest on the upper surface, may coalesce to form blackened areas. The disease spreads from the leaves to fruits where similar spots occur and where the damage is greatest.

Iste maladia causa strias maculate brun o nigre, le quales se extende ab le nervatura central al margines de folios vetule. Cata macula monstra un bordo indefinite. Le superfacie del maculas es aspere a causa del structuras sporal. Le maculas es usualmente plus numerose in le superfacie superior. Illos pote coalescer usque a formar areas nigrate. Le maladia se extende ab le folios in le fructos ubi maculas similar se developpa. Le damno in le fructos es le plus sever.

Cette maladie produit des stries mouchetées brunes à noires qui progressent depuis la nervure centrale jusqu'aubord des vieilles feuilles. Chaque tache possède une bordure mal définie. Les fructifications donnent aux taches une surface granuleuse. Ces taches, généralement plus nombreuses à la face supérieure, peuvent confluer pour former des plages noircies. La maladie s'étend depuis les feuilles jusqu'aux fruits, où apparaissent des taches semblables et où les dégâts sont les plus importants.

Esta enfermedad causa rayas pecosas de color café a negro, las cuáles se extienden a través de la vena central hacia el margen de las hojas viejas. Cada mancha individual tiene un borde indefinido. La superficie de las manchas se siente rugosa como resultado de los cuerpos fructíferos. Las manchas, que generalmente son más numerosas en el haz, pueden juntarse para formar áreas ennegrecidas. La enfermedad se disemina de las hojas hacia los frutos, donde aparecen manchas similares y donde el daño es mayor.

DISTRIBUTION: A., Eur., N.A., S.A.

Musa sapientum (Musa spp.) + Mycosphaerella musicola Leach (Cercospora musae Zimm.) =

ENGLISH	Leaf Spot, Heart-leaf Rot, Sigatoka
FRENCH	Cercosporiose du Bananier, "Sigatoka Disease", Maladie de "Sigatoka"
SPANISH	Sigatoka, Chamusco, Mancha de las hojas, Podredumbre del corazón
PORTUGUESE	Mal de Sigatoka, Cercosporiose
ITALIAN	Cercosporiosi del banano, Malattia de Sigatoka
RUSSIAN	церкоспороз, "сигатока"
SCANDINAVIAN	
GERMAN	"Sigatoka" - Krankheit
DUTCH	
HUNGARIAN	
BULGARIAN	
TURKISH	
ARABIC	
PERSIAN	
BURMESE	
THAI	Bi-Jut ใบจุด
NEPALI	
HINDI	Parn dhabba परण घव्वा
VIETNAMESE	Đốm thâm kim trên lá
CHINESE	Shiang jaw ban yèh bìng 香蕉斑葉病
JAPANESE	Kurohosi-byð 黑星病

The first indication of infection of this disease is the presence of very small light yellow spots which are only visable if the leaf is held up to the light. Within a few days these specks enlarge to form loose dark brown spots. These spots are often surrounded by a light yellow zone. They unite to cover a large part of the leaf surface, giving it a burned appearance. Heavily infected plants will usually be under-sized and ripen prematurely.

Le prime symptomas de infection in iste maladia es multo minute maculas jalne clar, le quales es visibile solmente quando on examina le folio sub illumination brillante. Intra qualque dies illos cresce a maculas brun obscur laxe(non ligate). Frequentemente iste maculas es imbraciate per un zona jalne clar. Illos se uni usque a coperir un grande portion del superfacie foliar, causante un apparition de escaldatura. Plantas severmente infectate generalmente es sub-developpate e se matura ante le tempore ordinari.

La première indication de l'infection est la présence de très petites taches jaune clair, visibles seulement si la feuille est tenue sous la lumière. En quelques jours, ces tachetures s'agrandissent pour former des taches floues brun foncé. Ces taches sont souvent entourées d'une zone jaune pâle. Elles s'unissent pour couvrir une portion importante de la surface foliaire, lui donnant ainsi une apparence brûlée. Les plantes fortement infectées sont généralement sous développées et mûrissent prématurément.

La primera indicación de la infección de esta enfermedad es la presencia de pequeñas manchas amarillo claro, las cuáles son visibles si la hoja se sostiene contra la luz. Dentro de unos pocos días estas manchitas se agrandan hasta formar manchas sueltas café obscuro. Estas manchas están frecuentemente rodeadas por una zona amarillo claro. Estas se unen hasta cubrir una gran parte de la superficie de la hoja, dándole una apariencia de estar quemada. Las plantas severamente infectadas generalmente tienen menor tamaño y maduran prematuramente.

DISTRIBUTION: Af., A., Aust., N.A., C.A., S.A., W.I.

Nicotiana tabacum + Alternaria longipes (Ell. & Ev.) Mason =

ENGLISH	Brown Spot
FRENCH	Tache Brune, Alternariose
SPANISH	Alternaria, Mancha parda, Alternariosis, Mancha café
PORTUGUESE	Mancha de Alternaria, Mela, Requeima
ITALIAN	
RUSSIAN	**альтернариоз, пятнистость бурая сухая**
SCANDINAVIAN	
GERMAN	Blattbräune, Dürrfleckenkrankheit, Tabakschwamm
DUTCH	
HUNGARIAN	
BULGARIAN	
TURKISH	Kahverengi leke
ARABIC	
PERSIAN	Älternäreyä tōotōon آلترناریای نونون
BURMESE	
THAI	
NEPALI	
HINDI	
VIETNAMESE	
CHINESE	Yan tsǎw chǐr shing bǐng 煙草赤星病
JAPANESE	Akahosi-byð 赤星病

Lesions first appear on the lower and older leaves. The first indication of infection is the appearance of small water-soaked, circular areas which enlarge gradually. As the spots enlarge the centers die and become brown, leaving a sharp line of demarkation between diseased and healthy tissue. Sometimes there will be a halo of yellow tissue surrounding the brown lesions. The spots vary greatly in size with concentric markings which may fuse and render the leaf ragged. Sunken, elongated lesions appear on the stalks. They develop slowly but, if numerous, may girdle the plant.

Le lesiones se manifesta al initio in le folios inferior le plus vetule. Le prime symptoma de infection es minute areas aquose e circular, le quales cresce gradualmente. Quando le maculas cresce, le centros mori, deveni brun, e causa un demarcation multo evidente inter texito morbide e texito salubre. A vices un halo de texito jalne occurre circum le lesiones brun. Le maculas se varia grandemente in dimension e monstra marcationes concentric le quales pote unir se e facer le folio serrate. Depresse lesiones elongate se manifesta in le pedunculos. Illos cresce lentemente, ma si illos es numerose, illos pote imbraciar le planta.

Des lésions apparaissent d'abord sur les vieilles feuilles de la base. Le premier indice d'infection est l'apparition de petites plages circulaires détrempées qui s'agrandissent graduellement. A mesure que les taches s'agrandissent, leur centre meurt, devient brun, laissant une ligne de démarcation bien nette entre les tissus sains et les tissus malades. Parfois, un halo de tissus jaunes entoure les lésions brunes. Les taches ont des dimensions très variables et des marques concentriques qui peuvent se fondre et donner à la feuille un aspect déguenillé. Des lésions allongées et déprimées apparaissent sur les tiges. Ces lésions se forment lentement, mais peuvent, si elles sont nombreuses, encercler la plante.

Las lesiones aparecen primero en las hojas inferiores y más viejas. La primera indicación de infección es el aparecimiento de pequeñas áreas empapadas de agua, circulares, que se agrandan gradualmente. A medida que las manchas se agrandan, los centros mueren, se vuelven de color café, dejando una línea aguda de demarcación entre el tejido enfermo y el sano. Algunas veces habrá un halo de tejido amarillo que rodea las lesiones de color café. Las manchas varían grandemente en tamaño con marcas concéntricas que pueden fundirse y resulta en la hoja rasgada. En los tallos aparecen lesiones alargadas y hundidas en los tallos. Estas se desarrollan despacio, pero si son numerosas pueden rodear la planta.

DISTRIBUTION: Af., A., Aust., Eur., N.A., C.A., S.A.

162

Nicotiana tabacum + Cercospora nicotianae Ell. & Everh. =

ENGLISH	Frogeye Leaf Spot
FRENCH	Tache Ocellée, Cercosporiose, Taches Foliaires
SPANISH	Ojo de rana, Mancha circular, Cercosporiosis, Mancha parda, Ojo de sapo
PORTUGUESE	Cercosporiose, Mancha de Cercospora
ITALIAN	Cercosporiosi del tabacco
RUSSIAN	церкоспороз
SCANDINAVIAN	
GERMAN	Froschaugenkrankheit
DUTCH	
HUNGARIAN	
BULGARIAN	
TURKISH	
ARABIC	
PERSIAN	
BURMESE	Far-myet-si roga အာမျစ်ဝပ္ဖဲ့၊ ရ၁၄စိက္ဒိုူ ၄၄း
THAI	Ta-gop ทากบ
NEPALI	Surti Ko thople rog सुर्तिको थोप्ले रोग
HINDI	
VIETNAMESE	Đốm lá
CHINESE	Yan tsăw bór shing bìng 煙草白星病
JAPANESE	Sirahosi-byô 白星病

This disease may appear on seedlings, on plants in the field or on the harvested crop. The small lesions are brown, tan or dingy gray with parchment-like centers. Scattered through the centers are minute black dots. The spots usually occur on the lower leaves only and mature leaves are more susceptible than young leaves. Near harvest time the upper leaves may suddenly develop large necrotic spots that destroy the whole leaf.

Iste maladia pote manifestar se in plantulas, in plantas in le campo, o in le production post rendimento. Le minute lesiones es brun, bronzate, o gris con centros similar a pergamena. Minute punctos nigre es disperse per le centros. Le maculas usualmente occurre solmente in le folios inferior, e folios matur es plus susceptibile que folios juvene. Al tempore del rendimento se forma a vices, in le folios superior, grande areas necrotic le quales destrue tote le folio.

Cette maladie peut se montrer sur les semis, sur les plantes dans le champ ou sur la récolte. Les petites lésions sont brunes, couleur du tan ou gris terne et leur centre est parcheminé. De minuscules points noirs sont éparpillés dans le centre. C'est d'ordinaire sur les feuilles du bas seulement que les taches se forment, et les vieilles feuilles sont plus prédisposées que les jeunes. Vers le temps de la moisson, de grandes taches nécrotiques peuvent se former soudain à la partie supérieure et détruire toute la feuille.

Esta enfermedad puede aparecer en plántulas, en plantas en el campo o en la cosecha recolectada. Las pequeñas lesiones son de color café, crema o gris empañado, con centros que parecen de pergamino. Diseminadas a través de los centros hay manchitas negras pequeñísimas. Las manchas generalmente aparecen en las hojas inferiores y las hojas maduras son más susceptibles que las jóvenes. Cerca del tiempo de la cosecha las hojas superiores pueden repentinamente desarrollar grandes manchas necróticas que destruyen toda la hoja.

DISTRIBUTION: Af., A., Aust., Eur., N.A., C.A., S.A.

Nicotiana tabacum + Colletotrichum tabacum Böning (Colletotrichum nicotianae Averna-Sacca)=

ENGLISH   Anthracnose
FRENCH   Anthracnose
SPANISH   Antracnosis
PORTUGUESE   Antracnose
ITALIAN
RUSSIAN   антракноз
SCANDINAVIAN
GERMAN   Anthracknose, Brennfleckenkrankheit
DUTCH
HUNGARIAN
BULGARIAN
TURKISH
ARABIC
PERSIAN
BURMESE
THAI   Anthracnose   แอนแทรคโนส
NEPALI
HINDI
VIETNAMESE
CHINESE   Yan tsǎw tàn dzǔ bìng   煙草炭疽病
JAPANESE   Tanso-byô 炭そ病

On the leaves, young lesions are small, light green, water-soaked and considerably depressed. The spots soon enlarge to form large, circular areas. As the spots dry out they become papery, thin, gray-white and are surrounded by a raised water-soaked border which later becomes brownish. Larger spots may become zonate and have a dark brown center. Many small, lateral veins of the leaf are killed and turn brownish-black on the lower leaf surface. As a result, the affected leaf becomes wrinkled and distorted. On the stem, oblong, circular lesions appear which sometimes weaken the stem so much that it breaks. In the field the disease continues to develop by producing leaf spots and stem cankers.

In le folios, lesiones juvene es minute, verde clar, aquose, e assatis depresse. Le maculas rapidemente cresce in grande areas circular. Quando le maculas se desicca, illos deveni similar a papiro, tenue, blanc-gris, e imbraciate per un elevate bordo aquose, le qual plus tarde deveni quasi brun. Plus grande maculas pote devenir zonate e monstrar un centro brun obscur. Multo minute venas lateral del folio es mortificate e deveni nigre-brun in le inferior superfacie foliar. Le resultato es que le folio afflicte es crispate e distorte. In le pedunculos, lesiones oblonge o circular se forma, le quales a vices infirma le pedunculos de maniera que fissura occurre . In le campo le maladia continua su progresso, causante maculas foliar e canceres del pedunculos.

Sur les feuilles, les jeunes lésions sont petites, vert clair, délavées et très déprimées. Les taches ont tôt fait de s'agrandir pour former de grandes plages circulaires. A mesure qu'elles se dessèchent, les taches s'amincissent et prennent l'aspect d'un papier gris très pâle avec des marges relevées et délavées qui, par la suite, deviennent brunâtres. Les plus grandes taches peuvent se ceinturer en bandes concentriques autour d'un centre brun foncé. Plusieurs petites nervures latérales meurent et virent au brun très foncé à la face inférieure de la feuille. Il s'ensuit que la feuille atteinte se plisse et se tord. Sur la tige apparaissent des lésions oblongues ou circulaires qui, parfois, l'affaiblissent tellement qu'elle casse. Dans le champ, la maladie continue d'évoluer en produisant des taches sur les feuilles et des chancres sur les tiges.

Sobre las hojas, las lesiones jóvenes son pequeñas, verde pálido, empapadas de agua y considerablemente hundidas. Las manchas se agrandan pronto para formar áreas grandes, circulares A medida que las manchas se secan se vuelven como papel, delgadas, blanco-grisoso y están rodeadas por una orilla levantada y empapada de agua, que leugo se vuelve cafesosa. Las manchas más grandes pueden volverse zonadas y tienen un centro café obscuro. Muchas venas laterales pequeñas de la hoja mueren y se vuelven de color negro-cafesoso en el envés. Como resultado, la hoja afectada se vuelve arrugada y deforme. Lesiones alargadas, circulares aparecen en el tallo, las cuáles a veces debilitan el tallo tanto que se quiebra. En el campo la enfermedad continua desarrollándose al producir manchas foliares y cancros en los tallos.
DISTRIBUTION: A., Eur., N.A.

Nicotiana tabacum + Erwinia carotovora (L. R. Jones) Holland (Bacillus aroideae Townsend)=

ENGLISH	Hollow Stalk Rot, Black Leg
FRENCH	Jambe Noire, Tige Creuse, Bacteriose des Semis, Pourriture Bactérienne des tiges
SPANISH	Pudrición dell tallo, Podredumbre de tallo hueco, Pierna negra
PORTUGUESE	
ITALIAN	
RUSSIAN	гниль стеблей сердцевинная, пустостебельность
SCANDINAVIAN	
GERMAN	Sämlingsbakteriose, Bakterielle Stengelfäule
DUTCH	
HUNGARIAN	
BULGARIAN	
TURKISH	Yaş cürükügü
ARABIC	
PERSIAN	
BURMESE	
THAI	
NEPALI	
HINDI	
VIETNAMESE	
CHINESE	Yan tsǎw kung tung bìng 煙草空胴病
JAPANESE	Kūdō-byō 空洞病

In the seedling stage, this disease may occur during wet periods when the leaves touch the ground and become infected. The rot then spreads into the stems which may rot off or split open. The rotted areas turn black. In the field, the top leaves wilt and the infection spreads downward; the leaves droop and hang down or fall off, leaving the stalk bare. The stalk is usually hollow and the pith is brownish in color.

In plantulas, iste maladia occurre sub conditiones humide, quando le folios es in contacto con le terra e deveni infectate. Le putrefaction alora se extende in le pedunculos, le quales o cade o se finde. Le areas putrefacte deveni nigre. In le campo, le folios superior marcesce e le infection se avantia a basso. Le folios langue, se separa, e cade, lassante le pedunculo denudate. Le pedunculo usualmente es cave e le medulla es de color assatis brun.

Au stade plantule, cette maladie peut se montrer par temps humide, alors que les feuilles s'infectent au contact du sol. De là, la pourriture gagne les tiges qui peuvent pourrir complète-ment ou se fendre. Les aires pourries noircissent. Dans le champ, les feuilles du haut se flétrissent et l'infection se propage vers le bas; les feuilles se courbent et pendent ou tombent, laissant la tige dénudée. D'ordinaire, la tige est creuse et la moëlle brunâtre.

En las plántulas, la enfermedad peude aparecer durante períodos húmedos, cuando las hojas tocan el suelo y se infectan. La podredumbre entonces se disemina a los tallos, los cuáles pueden podrirse o abrirse por la mitad. Las áreas podridas se vuelven de color negro. En el campo, las hojas superiores se marchitan y la infección se disemina hacia abajo; las hojas se decaen y cuelgan o se caen, dejando el tronco desnudo. El tronco generalmente está hueco y la médula es de color cafesoso.

DISTRIBUTION: A., Eur., N.A.

Nicotiana tabacum + Peronospora tabacina Adam =

ENGLISH	Tobacco Blue Mold, Downy Mildew
FRENCH	Mildiou, Moisissure Bleue, Peronospora du Tabac
SPANISH	Mildiu, Moho, Mildiú velloso, Moho azul, Mildiú lanoso
PORTUGUESE	Mofo azul, Mildio
ITALIAN	Peronospora del tabacco
RUSSIAN	ложная мучнистая роса, пероноспороз
SCANDINAVIAN	
GERMAN	Blauschimmel, Falscher Mehltau
DUTCH	
HUNGARIAN	Dohányperonoszpóra
BULGARIAN	мана
TURKISH	Mavi küf
ARABIC	
PERSIAN	Sefedake dorõõghēye tõõtõõn سفیدک درونی توتون
BURMESE	
THAI	
NEPALI	
HINDI	
VIETNAMESE	
CHINESE	
JAPANESE	Beto-byõ べと病

The first evidence of infection is the appearance of round, yellow areas of diseased plants. These become cupped and have a gray or bluish mold growth on the lower surface. The upper surfaces of the leaves of infected plants remain normal in appearance for a day or two and then die and turn light brown in color. Diseased leaves sometimes become so twisted that the lower surfaces turn upward. Irregular lesions with necrotic central areas develop on the leaves, growth is retarded and the roots turn dark brown. Groups of yellow spots appear as circular blotches, then these spots coalesce to form necrotic areas.

Le prime symptoma de infection es rotunde areas jalne in la planta afflicte. Iste areas deveni cupuliforme e monstra un strato de mucor gris o quasi blau in le superfacie inferior. Le superfacie superior del folios de plantas infectate resta normal durante un o duo dies, alora mori e deveni brun clar. Folios afflicte es a vices distorte de maniera que le superfacies inferior es rotate in alto. Lesiones irregular con central areas necrotic se forma in le folios; le crescimento es impedite; e le radices deveni brun obscur. Gruppamentos de maculas jalne appare como pustulas circular. Alora iste maculas coalesce usque a formar areas necrotic.

Le premier indice d'infection est l'apparition de plages rondes et jaunes sur les plantes malades. Celles-ci prennent la forme d'une coupe et se couvrent, à la face inférieure des feuilles, d'une moisissure grise ou bleuâtre. La face supérieure des feuilles des plantes infectées garde une apparence normale durant un jour ou deux, et puis meurt et prend une couleur brun clair. Les feuilles malades deviennent parfois si tordues que leur face inférieure se retrousse. Des lésions irrégulières avec des aires centrales nécrotiques se forment sur les feuilles, la croissance est retardée et les racines deviennent brun foncé. Des groupes de taches jaunes apparaissent comme des éclaboussures circulaires, puis, ces taches se fusionnent pour former des aires nécrotiques.

La primera evidencia de infección es la aparición de áreas redondas, amarillas, en la planta enferma. Estas se vuelven acopadas y tienen un crecimiento mohoso gris o azuloso en el envés. El haz de las hojas de las plantas infectadas permanecen normales en apariencia por uno o dos días, leugo mueren y se vuelven de color café claro. Las hojas enfermas algunas veces se vuelven tan retorcidas que el envés se da vuelta hacia arriba. En las hojas se desarrollan lesiones irregulares con áreas necróticas centrales, el crecimiento se retarda y las raíces se vuelven de color café obscuro. Grupos de manchas amarillas aparecen como manchones irregulares, luego estas manchas se juntan para formar áreas necróticas.

DISTRIBUTION: Af., A., Aust., Eur., N.A., C.A., S.A.

Nicotiana tabacum + Phytophthora parasiticae var. nicotianae (v Breda de Haan) Tucker
(Phytophthora nicotianae v Breda de Haan) =

ENGLISH	Black Shank
FRENCH	Jambe Noire du Tabac, Phytophthorose, Fonte des Semis
SPANISH	Pierna negra, Pudrición negra del tallo, Pudrición descendente, Pata prieta
PORTUGUESE	Mildio
ITALIAN	Peronospora del tabacco
RUSSIAN	почернение стеблей, фитофтороз
SCANDINAVIAN	
GERMAN	Lanaskrankheit, Stammfäule, Schwarzbeinigkeit
DUTCH	
HUNGARIAN	
BULGARIAN	чернилка
TURKISH	
ARABIC	
PERSIAN	
BURMESE	
THAI	Khang-Dum แขงคำ
NEPALI	
HINDI	
VIETNAMESE	Đốm phấn - Héo lá
CHINESE	Yan tsǎw yǐh bǐng 煙草疫病
JAPANESE	Eki-byð 疫病

Young seedlings may rot off on the stem near the ground level and become dark brown or black. In the field, the disease is frequently found in low, wet areas. All or part of the root system may be infected and turn black, with the black stem lesions extending up the stalks several inches. Soon the leaves turn yellow to brown, shrivel, and in a few days the plants are dead.

Plantulas juvene pote putrescer al pedunculo presso le terra; illos deveni brun obscur o nigre. In le campo, le maladia occurre frequentemente in areas basse e humide. Tote le systema radical, o un portion, pote esser infectate e devenir nigre. Le lesiones nigre se extende in alto in le pedunculos usque a un decimetro o plus. Le folios rapidemente deveni jalne o brun, se crispa, e intra qualque dies le plantas es morte.

Les tiges des jeunes plants peuvent pourrir complètement au niveau du sol et devenir brun foncé ou noires. Dans le champ, la maladie se montre souvent sur des terrains bas et humides. Le système radiculaire peut être envahi en tout ou en partie, pendant que les lésions noires s'étendent sur plusieurs pouces vers le haut des tiges. Bientôt, les feuilles deviennent jaunes à brunes, se ratatinent, et les plantes meurent au bout de quelques jours.

Las plantitas jóvenes se pueden podrir en el tallo cerca de la superficie del suelo y volverse de color café obscuro o negro. En el campo, la enfermedad frecuentemente se encuentra en áreas bajas y húmedas. Todo o parte del sistema radicular puede infectarse y volverse negro, con las lesiones negras del tallo extendiéndose varias pulgadas hacia el tronco. Pronto las hojas se vuelven de color amarillo a café, se arrugan y en unos pocos días la planta muere.

DISTRIBUTION: Af., Aust., Eur., N.A., C.A., S.A.

Fig. 110. Barley-Covered Smut; R. A. Kilpatrick, ARS

Fig. 111. Barley-Loose Smut; J. A. Browning

Fig. 112. Sweet Potato-Leaf Spot; Charles Y. Yang, Taiwan, Republic of China

Fig. 113. Sweet Potato-Scab; Charles Y. Yang, Taiwan, Republic of China

Fig. 114. Sweet Potato-Fusarium Rot; ARS

Fig. 115. Sweet Potato-Scurf; Dept. of Pl. Path., Louisiana St. Univ.

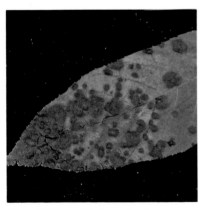

Fig. 116. Walnut-Anthracnose; Dept. of Pl. Path., Purdue Univ.

Fig. 117. Walnut-Bacterial Blight; ARS, Oregon St. Univ.

Fig. 118. Lettuce-Downy Mildew; ARS

PLATE 13

Fig. 119. Lettuce-Marginal Leaf Blight; Monterey Co. Dept. of Agric., Calif.

Fig. 120. Lettuce-Leaf Spot; Monterey Co. Dept. of Agric., Calif.

Fig. 121. Flax-Anthracnose

Fig. 122. Flax-Wilt; David E. Zimmer, N.D. St. Univ.

Fig. 123. Flax-Rust; W. H. Bragonier

Fig. 124. Flax-Pasmo

Fig. 125. Flax-Browning

Fig. 126. Tomato-Nailhead Spot; ARS

Fig. 127. Tomato-Leaf Mold; Arden F. Sherf, Cornell Univ.

PLATE 14

Fig. 128. Tomato-Anthracnose; ARS

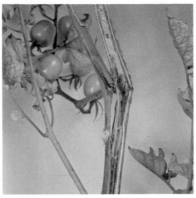

Fig. 129. Tomato-Bacterial Canker; John L. Maas

Fig. 130. Tomato-Stem Rot

Fig. 131. Tomato-Fusarium Wilt; Dept. of Pl. Path., Clemson Univ.

Fig. 132. Tomato-Ring Rot of Fruit

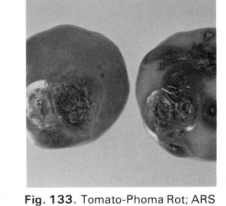

Fig. 133. Tomato-Phoma Rot; ARS

Fig. 134. Tomato-Leaf Spot; ARS

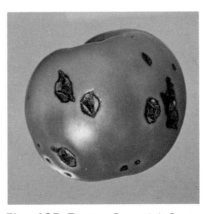

Fig. 135. Tomato-Bacterial Spot; ARS

Fig. 136. Apple-Mushroom Root Rot; Charles R. Drake, Dept. of Pl. Path., V.P.I.

PLATE 15

Fig. 137. Apple-Crown Gall

Fig. 138. Apple-Botryosphaeria Rot; ARS

Fig. 139. Apple-Fire Blight; ARS

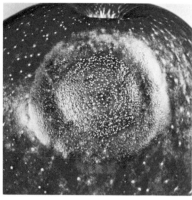

Fig. 140. Apple-Bitter Rot; ARS

Fig. 141. Apple-Leaf Spot

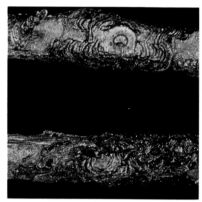

Fig. 142. Apple-Canker; Dept. of Pl. Path., Wash. St. Univ.

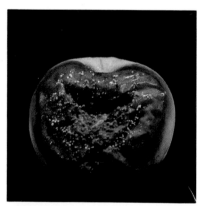

Fig. 143. Apple-Blue Mold; Charles R. Drake, Dept. of Pl. Path., V.P.I.

Fig. 144. Apple-Collar Rot; Dept. of Botany & Pl. Path., U. of Md.

Fig. 145. Apple-Powdery Mildew; ARS

PLATE 16

Fig. 146. Apple-Brown Rot; Dept. of Pl. Path., Wash. St. Univ.

Fig. 147. Apple-Black Rot; ARS

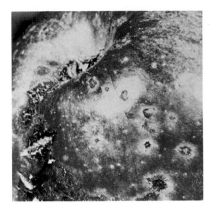

Fig. 148. Apple-Scab; ARS

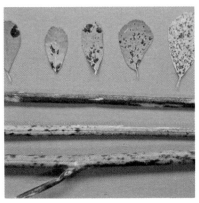

Fig. 149. Alfalfa-Black Stem; ARS

Fig. 150. Alfalfa-Bacterial Wilt; ARS

Fig. 151. Alfalfa-Leaf Spot; ARS

Fig. 152. Alfalfa-Yellow Leaf Blotch; ARS

Fig. 153. Alfalfa-Rust; ARS

Fig. 154. Alfalfa-Wart; ARS

PLATE 17

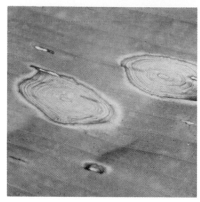

Fig. 155. Banana-Leaf Spot; Rohm and Haas

Fig. 156. Banana-Wilt; R. H. Stover, Div. of Trop. Res., Honduras

Fig. 157. Banana-Rot; ARS

Fig. 158. Banana-Black Tip; Rohm and Haas

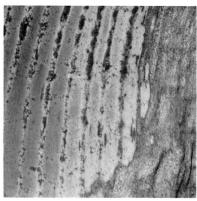

Fig. 159. Banana-Freckle; R. H. Stover, Div. of Trop. Res., Honduras

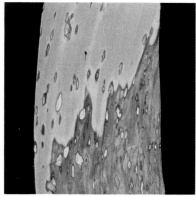

Fig. 160. Banana-Sigatoka; R. H. Stover, Div. of Trop. Res., Honduras

Fig. 161. Tobacco-Brown Spot; Tobacco Laboratory, ARS

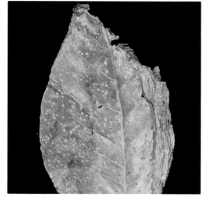

Fig. 162. Tobacco-Frogeye Leaf Spot; Tobacco Laboratory, ARS

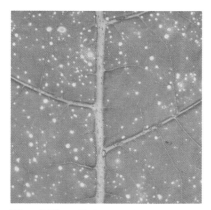

Fig. 163. Tobacco-Anthracnose; ARS, Oxford, N.C.

PLATE 18

Fig. 164. Tobacco-Hollow Stalk Rot; ARS

Fig. 165. Tobacco-Blue Mold; Tobacco Laboratory, ARS

Fig. 166. Tobacco-Black Shank; Tobacco Laboratory, ARS

Fig. 167. Tobacco-Blackfire; ARS, Oxford, N.C.

Fig. 168. Tobacco-Wildfire; ARS, Oxford, N.C.

Fig. 169. Tobacco-Seedling Blight

Fig. 170. Tobacco-Black Root Rot; Tobacco Laboratory, ARS

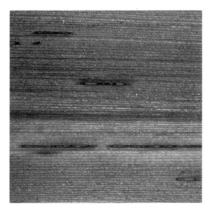

Fig. 171. Rice-Narrow Brown Spot; M. C. Rush, La. St. Univ.

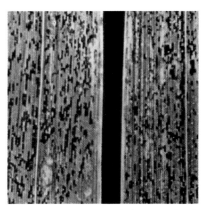

Fig. 172. Rice-Leaf Smut; M. C. Rush, La. St. Univ.

PLATE 19

Fig. 173. Rice-Brown Spot; M. A. Marchetti, ARS, Beaumont, Texas

Fig. 174. Rice-Stem Rot; International Rice Research Institute

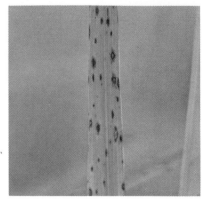

Fig. 175. Rice-Blast; M. A. Marchetti, ARS, Beaumont, Texas

Fig. 176. Rice-Sheath Blight-M. A. Marchetti, ARS, Beaumont, Texas

Fig. 177. Rice-False Smut; M. C. Rush, La. St. Univ.

Fig. 178. Millet-Head Smut

Fig. 179. Pearl Millet-Downy Mildew

Fig. 180. Bean-Leaf Spot; Marvin Williams, N.C. St. Univ.

Fig. 181. Bean-Leaf Spot

PLATE 20

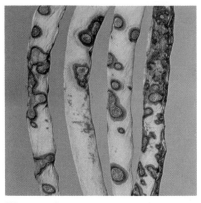

Fig. 182. Bean-Anthracnose; Arden F. Sherf, Cornell Univ.

Fig. 183. Bean-Bacterial Wilt

Fig. 184. Bean-Angular Leaf Spot; Arden F. Sherf, Cornell Univ.

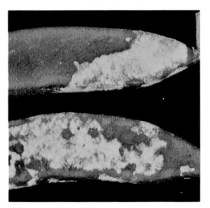

Fig. 185. Lima Bean-Downy Mildew; ARS

Fig. 186. Bean-Halo Blight; ARS

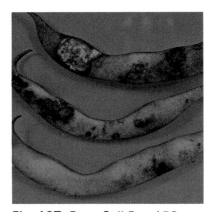

Fig. 187. Bean-Soil Rot; ARS

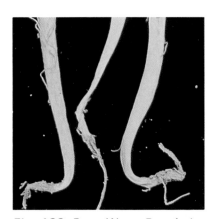

Fig. 188. Bean-Water Rot; Arden F. Sherf, Cornell Univ.

Fig. 189. Bean-Rust; ARS

Fig. 190. Bean-Bacterial Blight; Arden F. Sherf, Cornell Univ.

PLATE 21

Fig. 191. Pine-Rust; Marvin Williams, N.C. St. Univ.

Fig. 192. Pine-Blister Rust; Pl. Path. Dept., U. of Wisc.

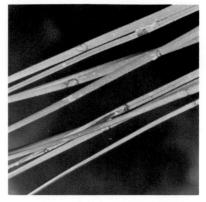

Fig. 193. Pine-Needle Blight; Dept. of Pl. Path., Clemson Univ.

Fig. 194. Pine-Needle Cast; Marvin Williams, N.C. St. Univ.

Fig. 195. Pine-Butt Rot; Marvin Williams, N.C. St. Univ.

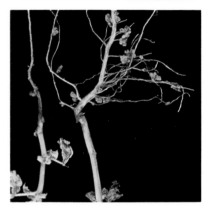

Fig. 196. Pea-Root Rot; Donald J. Hagedorn, Dept. of Pl. Path., Univ. of Wisc.

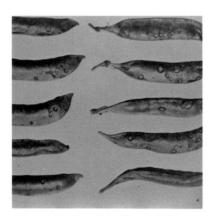

Fig. 197. Pea-Pod Spot; ARS

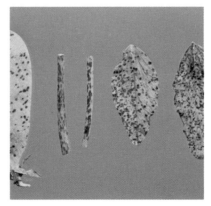

Fig. 198. Pea-Blight; Donald J. Hagedorn, Dept. of Pl. Path., Univ. of Wisc.

Fig. 199. Pea-Bacterial Blight; ARS

PLATE 22

Fig. 200. Pea-Leaf Blotch; Donald J. Hagedorn, Dept. of Pl. Path., Univ. of Wisc.

Fig. 201. Pea-Rust

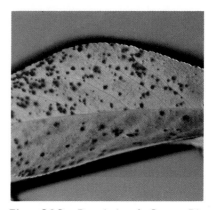

Fig. 202. Peach-Leaf Spot; Pl. Path. Div., Penn. St. Univ.

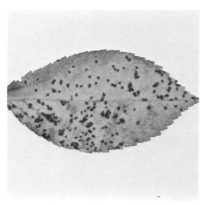

Fig. 203. Cherry-Leaf Blight; ARS

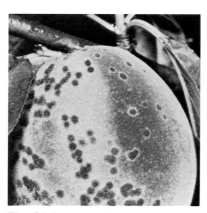

Fig. 204. Peach-Scab; ARS

Fig. 205. Peach-Scab; Charles R. Drake, Dept. of Pl. Path., V.P.I.

Fig. 206. Cherry-Brown Rot; Charles R. Drake, Dept. of Pl. Path., V.P.I.

Fig. 207. Peach-Powdery Mildew; John L. Maas

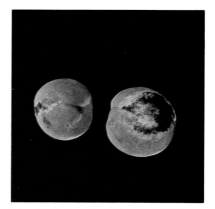

Fig. 208. Peach-Rhizopus Soft Rot; Charles R. Drake, Dept. of Pl. Path., V.P.I.

PLATE 23

Fig. 209. Peach-Spur Blight; ARS

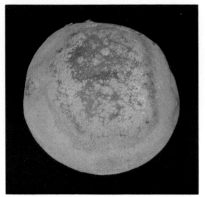

Fig. 210. Peach-Powdery Mildew; ARS

Fig. 211. Peach-Sliver Leaf; J. M. Ogawa, Univ. of Calif.

Fig. 212. Peach-Witches'-broom; J. M. Ogawa, Univ. of Calif.

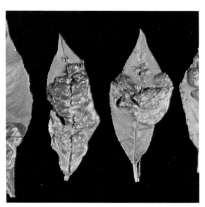

Fig. 213. Peach-Leaf Curl; Charles R. Drake, Dept. of Pl. Path., V.P.I.

Fig. 214. Peach-Plum Pockets; Paul C. Pecknold, Calif. St. Dept. of Agric.

Fig. 215. Peach-Rust; ARS

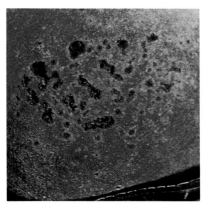

Fig. 216. Peach-Bacterial Spot; Charles R. Drake, Dept. of Pl. Path., V.P.I.

Fig. 217. Pear-Rust; ARS

PLATE 24

Nicotiana tabacum + Pseudomonas angulata (Fromme & Murray) Holland =

ENGLISH	Blackfire, Angular Leaf Spot
FRENCH	Tache Angulaire, Bactériose du Tabac
SPANISH	Fuego negro, Mancha foliar angular
PORTUGUESE	
ITALIAN	Macchie angolari del tabacco, Maculatura angolare
RUSSIAN	пятнистость бактериальная угловатая
SCANDINAVIAN	
GERMAN	Eckige Blattfleckenkrankheit, Schwarzer Bakterienbrand
DUTCH	
HUNGARIAN	
BULGARIAN	
TURKISH	Köşeli yaprak lekesi hast.
ARABIC	
PERSIAN	
BURMESE	
THAI	
NEPALI	
HINDI	
VIETNAMESE	Đốm gốc làm cháy lá
CHINESE	Yan tsǎw jiáo ban bìng 煙草角斑病
JAPANESE	Kakuhan-byð 角斑病

The lesions of this disease in the seedbed are black, commonly angular and bounded by veins. In the field, the lesions may involve large areas of the leaf with the color of the spots varying from tan to black. The lesions are irregular and on rapidly growing tobacco the leaves become puckered and torn; the centers of the lesions fall out and in severe infections little besides the veins remains.

Le lesiones de iste maladia del seminario es nigre, usualmente angular, e limitate per venas. In le campo, le lesiones pote coperir grande areas del folio, le maculas se variante in color ab bronzate a nigre. Le lesiones es irregular, e, in tabaco que cresce rapidemente, le folios deveni corrugate e fisse. Le centros del lesiones se separa e cade; in infection sever quasi solmente le venas resta.

Les lésions que cette maladie produit dans les couches de semis sont noires, ordinairement angulaires et limitées par les nervures. Dans le champ, les lésions peuvent couvrir de grandes aires de la feuille; la couleur des taches peut varier du beige au noir. Les lésions sont irrégulières, et, sur le tabac dont la croissance est rapide, les feuilles deviennent plissées et déchirées. Le centre des lésions tombe, et, dans les cas graves, il ne reste guère plus que les nervures.

Las lesiones de est enfermedad en el semillero son negras, generalmente angulares y rodeadas por venas. En el campo las lesiones peuden cubrir grandes áreas de la hoja, variando el color de las manchas de crema a negro. Las lesiones son irregulares y en el tabaco que está creciendo rápidamente las hojas se vuelven fruncidas y desgarradas; los centros de las lesiones se caen y en infecciones severas queda poco, aparte de las venas.

DISTRIBUTION: Af., A., Eur., N.A., S.A.

Nicotiana tabacum + Pseudomonas tabaci (Wolf & Foster) Stev. =

ENGLISH	Wildfire
FRENCH	Feu Sauvage, Bactériose du Tabac, Maladie du Feu Sauvage
SPANISH	Fuego salvaje
PORTUGUESE	Queima bacteriana
ITALIAN	Fuoco selvatico, Brucciatura, Batteriosi del tabacco
RUSSIAN	рябуха бактериальная
SCANDINAVIAN	
GERMAN	Wildfeuer
DUTCH	
HUNGARIAN	
BULGARIAN	див огън
TURKISH	Tütün vahşi ateşi
ARABIC	
PERSIAN	Ätáshåkê tōotōon آتشک توتون
BURMESE	
THAI	
NEPALI	
HINDI	
VIETNAMESE	
CHINESE	Yan tsǎw yǐe hwǒh bìng 煙草野火病
JAPANESE	Nobi-byô 野火病

This disease occurs in both the plant bed and in the field. The most conspicuous symptoms are localized chlorotic halos which surround a brown necrotic spot that is usually very small in size. The first symptoms of the disease are circular, yellowish-green areas. Within a day, brown dead specks appear in the center of the spots and the yellow-green halo becomes more prominent. Within a few more days the dead center and the halo increase greatly in size and become surrounded by a water-soaked border around the margin of the dead brown center.

Iste maladia occurre e in le seminario e in le campo. Le symptomas le plus remarcabile es halos chlorotic localisate le quales imbracia un macula necrotic brun, usualmente de dimension multo minute. Le prime symptomas del morbo es areas circular verde-jalnette. Intra un sol die, brun maculas morte se manifesta in le centros e le halo verde-jalne deveni plus visibile. Intra qualque dies le centro morte e le halo cresce grandemente in dimension e deveni imbraciate per un bordo aquose circum le margine del brun centro morte.

Cette maladie se présente dans les couches de semis et dans le champ. Les symptômes les plus apparents sont des halos chlorotiques localisés qui entourent une tache nécrotique brune, généralement très petite. Les premiers symptômes de la maladie sont des plages circulaires vert jaunâtre. En moins d'un jour, des points morts de couleur brune apparaissent au centre des taches, et le halo vert jaune devient plus proéminent. Au bout de quelques jours, le centre mort et le halo deviennent très grands et s'entourent d'une bordure détrempée à la marge du centre mort brun.

Esta enfermedad aparece tanto en el semillero como en el campo. Los síntomas más conspicuos son halos cloróticos localizados, los cuáles rodean una mancha necrótica café que generalmente es de tamaño muy pequeño. Los primeros síntomas de esta enfermedad son áreas circulares, verde-amarillosas. En el lapso de un día, aparecen manchitas muertas de color café en el centro de las otras manchas y el halo verde-amarilloso se vuelve más prominente. En unos pocos días más el centro muerto y el halo aumentan grandemente en tamaño y se rodean de una orilla empapada de agua alrededor del margen del centro café muerto.

DISTRIBUTION: Af., A., Eur., N.A., S.A.

<u>Nicotiana</u> <u>tabacum</u> + <u>Sclerotinia</u> <u>sclerotiorum</u> (Lib.) d By. =

ENGLISH	Seedling blight, Bed Rot, Sclerotinia Disease
FRENCH	Pourriture Blanche des Tiges, Sclérotiniose
SPANISH	Tizón de la plántula, Podredumbre del semillero, Enfermedad esclerotinia
PORTUGUESE	
ITALIAN	
RUSSIAN	**гниль стеблей склеротиниозная, склеротиниоз**
SCANDINAVIAN	
GERMAN	Sklerotinienkrankheit, Stengelfäule
DUTCH	
HUNGARIAN	
BULGARIAN	
TURKISH	Sap çürüklüğü
ARABIC	
PERSIAN	
BURMESE	
THAI	
NEPALI	
HINDI	
VIETNAMESE	
CHINESE	Yan tsǎw jìunn hór bìng 煙草菌核病
JAPANESE	Kinkaku-byð 菌核病

Seedlings may be attacked by this fungus disease at any time up to transplant size with infection usually occurring at or just beneath the groundline. Affected stems develop a brown soft rot which spreads into the leaves and rapid destruction of the seedlings occurs. Frequently there can be seen a white cottony fungus growth growing over the rotted plants. Under favorable weather conditions for disease development this fungus may infect the flowers which fall on the leaves, causing them to rot and hang limply.

Plantulas es attaccate per iste maladia fungal usque al transportation, le infection usualmente occurrente a o sub le terreno. Pedunculos afflicte monstra un brun putrefaction molle, que se extende in le folios; e le plantulas es rapidemente destructe. Frequentemente on vide un strato blanc similar a coton super le plantas putrefacte. Sub favorabile conditiones de tempore, iste fungo pote infectar le flores, le quales alora cade super le folios e causa que iste putresce e langue.

Cette maladie fongueuse peut attaquer les jeunes plants à n'importe quel moment jusqu'à ce qu'ils soient prêts à transplanter; l'infection se produit généralement au niveau du sol ou juste un peu plus bas. Les tiges atteintes produisent une pourriture brune molle qui se répand dans les feuilles, et les plantules ne tardent pas à succomber. On peut fréquemment observer sur les plants pourris, le mycélium blanc cotonneux du pathogène. Dans des conditions climatiques favorables à la maladie, ce champignon peut infecter les fleurs; celles-ci tombent sur les feuilles, qui pourrissent à leur tour et pendent mollement.

Las plántulas peuden ser atacadas por esta enfermedad fungosa en caulquier momento hasta que tengan tamaño para trasplantarlas, apareciendo la infección generalmente sobre o un poquito abajo de la superficie del suelo. Los tallos afectados desarrollan una podredumbre café sauve que se disemina hacia las hojas y destruye rápidamente las plántulas. Frecuentemente puede verse un crecimiento algodonoso blanco que crece sobre las plantas podridas. Bajo condiciones ambientales para el desarrollo de la enfermedad, este hongo peude infectar las flores que caen sobre las hojas, causando que éstas se pudran y cuelguen flácidamente.

DISTRIBUTION: A., Eur., N.A.

Nicotiana tabacum (and others) + Thielaviopsis basicola (Berk. & Br.) Ferraris (Thielavia basicola Zopf) =

ENGLISH	Black Root Rot
FRENCH	Pourriture Noire des Racines, Pourridié noir
SPANISH	Podredumbre negra de la raíz
PORTUGUESE	Podridão radicular
ITALIAN	Marciume radicale delle piantine di tabacco
RUSSIAN	гниль корней черная
SCANDINAVIAN	Rodbrand
GERMAN	Wurzelbraüne, Wurzelfäule
DUTCH	Kiemplantdoder, Voetziekte, Wortelrot
HUNGARIAN	
BULGARIAN	
TURKISH	
ARABIC	
PERSIAN	
BURMESE	
THAI	
NEPALI	
HINDI	
VIETNAMESE	Thối dầu rể
CHINESE	Yan tsǎw hei sèh gen fǔ bìng 煙草黑色根腐病
JAPANESE	Kurone-byõ 黒根病

This disease attacks the roots whether in the plant bed or the field, rotting them and eventually turning them black. In severe infection, the disease is recognized by slow growth of the plants and yellowing of the leaves. The yellow stunted plants tend to wilt and flower prematurely.

Iste maladia attacca le radices o in le seminario o in le campo, causante putrefaction e al fin nigration. In infection sever, le maladia es recognoscibile per crescimento tardive del plantas e per jalnessa del folios. Le jalne plantas cuje crescimento es impedite propende a marcescer e florar prematurmente.

Cette maladie attaque les racines, aussi bien dans la couche de semis que dans le champ; elle fait pourrir les plantes et les fait noircir éventuellement. Dans les cas graves, la maladie se reconnaît à la croissance lente des plantes et au jaunessement du feuillage. Les plantes jaunes et rabougries tendent à se flétrir et à fleurir prématurément.

Esta enfermedad ataca las raíces ya sea en el semillero o en el campo, pudriéndolas y volviéndolas de color negro. Cuando la infección es severa, la enfermedad se reconoce por el lento crecimiento de las plantas y el amarillamiento de las hojas. Las plantas achaparradas y amarillas tienden a marchitarse y a florecer prematuramente.

DISTRIBUTION: Af., A., Aust., Eur., N.A., C.A., S.A.

Oryza sativa + Cercospora oryzae Miyake =

ENGLISH	Narrow Brown Spot, Glume Spot
FRENCH	Macules Foliaires du Riz, Cercosporose
SPANISH	Mancha café angosta, Mancha de la hoja y la gluma, Mancha castaña
PORTUGUESE	Cercosporiose, Mancha estreita de fôlha
ITALIAN	Cercosporiosi del riso
RUSSIAN	церкоспороз
SCANDINAVIAN	
GERMAN	Cercospora-Blattfleckenkrankheit
DUTCH	
HUNGARIAN	
BULGARIAN	
TURKISH	
ARABIC	
PERSIAN	
BURMESE	
THAI	Bi-Kheed-Si-Numtan ใบขีดสีน้ำตาล
NEPALI	
HINDI	Sankeirn Bhura संकिनि भूरा
VIETNAMESE	Gạch nâu
CHINESE	Shoěi dàw tiáu yèh ku bǐng 水稻條葉枯病
JAPANESE	Suzihagare-byô すじ葉枯病

This disease is characterized by the narrow linear spots which are reddish-brown to dark reddish-brown, depending upon the crop variety. On susceptible varieties, the lesions fade to a lighter brown along the margins and to light gray-brown in the older center. On the resistant varieties, the lesions are smaller and uniform in color. The lesions are generally more abundant on the leaves, although spots on the stems and floral parts are present in heavy infections.

Iste maladia se characterisa per le stricte maculas linear brun-rubiette clar o obscur secundo le varietate. In le varietates susceptibile, le color del lesiones se cambia in un brun plus clar in le margines e in brun-gris clar in le centro plus vetule. In le varietates resistente, le lesiones es plus minute e de color uniforme. Le lesiones es usualmente plus numerose in le folios, ma maculas se manifesta in le pedunculos e in le partes floral in infection sever.

Cette maladie est caractérisée par d'étroites taches linéaires, dont la couleur peut être brun rougeâtre à brun rougeâtre foncé, selon la variété. Chez les variétiés prédisposées, les lésions se déteignent pour devenir brun clair à la marge et gris brun clair au centre, avec le temps. Chez les variétés résistantes, les lésions sont plus petites et de couleur uniforme. Les lésions sont généralement plus abondantes sur les feuilles, bien qu'il y ait, lors d'infections considér-ables, des taches sur les tiges et les parties florales.

Esta enfermedad se caracteriza por las manchas lineales angostas que son café-rojizas a café obscuro rojizo, dependiendo de la variedad atacada. En variedades susceptibles, las lesiones se destiñen a un café más claro a lo largo de los márgenes y a un gris-café claro en el centro más viejo. En las variedades resistentes, las lesiones son más pequeñas y uniformes en color. Las lesions son generalmente más abundantes en las hojas, aunque las manchas en los tallos y partes florales están presentes en las infecciones severas.

DISTRIBUTION: Af., A., Aust., N.A., C.A., S.A.

172

Oryza sativa + Entyloma oryzae H. & P. Syd. =

ENGLISH	Leaf Smut
FRENCH	Charbon des Feuilles
SPANISH	Otras manchas foliares, Carbón de la hoja, Tizón de la hoja
PORTUGUESE	
ITALIAN	
RUSSIAN	головня листьев
SCANDINAVIAN	
GERMAN	Blätterbrand
DUTCH	
HUNGARIAN	
BULGARIAN	
TURKISH	
ARABIC	
PERSIAN	
BURMESE	
THAI	
NEPALI	
HINDI	Parn Kand परण कंद
VIETNAMESE	
CHINESE	Shŏěi dǎw hei jŏng bìng 水稻黑腫病
JAPANESE	Kokusyu-byô 黒しゅ病

This disease appears on both upper and lower surfaces of leaves as linear, rectangular or angular black spots not usually united. The spots are covered by the epidermis which ruptures where wet. Where infection is heavy the leaves turn chlorotic and split.

Iste maladia se manifesta in le superfacies e superior e inferior del folios como maculas nigre, linear, rectangular, o angular, usualmente non unite. Le maculas es coperte per le epidermis le qual se finde quando le tempore es humide. Si le infection es sever, le folios deveni chlorotic e se finde.

Cette maladie se manifeste à la face supérieure et a là face inférieure des feuilles sous forme de taches noires linéaires, rectangulaires ou angulaires, et qui, d'ordinaire, ne s'unissent pas. Les taches sont recouvertes de l'épiderme qui se brise aux points humides. Lorsque l'infection est abondante, les feuilles deviennent chlorotiques et se déchirent.

Esta enfermedad aparece tanto en el haz como en el envés de las hojas, como manchas lineales, rectangulares o angulares de color negro, generalmente sin unirse. Las manchas están cubiertas por la epidermis la cuál se rompe al mojarse. Cuando la infección es severa, las hojas se vuelven cloróticas y se rajan.

DISTRIBUTION: Af., A ., Aust., Eur., N.A., C.A., S.A.

Oryza sativa (Oryza spp.) + Helminthosporium oryzae (Cochliobolus miyabeanus (Ito & Kur.) Drechsler ex Dastur) =

ENGLISH   Brown Spot
FRENCH   Helminthosporiose du Riz, Brûlure
SPANISH   Mancha carmelita, Tizón de la hoja, Mancha café del arroz, Mancha parda
PORTUGUESE   Helminthosporiose, Mancha parda
ITALIAN   Elmintosporiosi del riso
RUSSIAN   пятнистостьглазковая, гельминтоспориоз бурый
SCANDINAVIAN
GERMAN   Helminthosporium-Fusskrankheit, Helminthosporiose
DUTCH
HUNGARIAN
BULGARIAN
TURKISH   Celtik kahverengi yaprak lekesi
ARABIC   Takhtit تخطيط
PERSIAN   Låkêhê ghåhvêēyê bêrênj كله قهوه‌ای برنج
BURMESE
THAI   Bi-Jut-Si-Numtan ใบจุดสีน้ำตาล
NEPALI   Dhan Ko Khairo-thople pat धानको खैरो थाम्प्ले पात, Dhan Ka khaira rog धानका खैरा रोग
HINDI   Helminthosporium Angmaari हेलमिनथोसपोरयम अंगमारी
VIETNAMESE   Đốm nâu
CHINESE   Shoĕi dàw hú má yĕh ku bĭng 水稻胡麻葉枯病
JAPANESE   Gomahagare-byô ごま葉枯病

Seedling blight occurs before or after emergence. Brown leaf spots develop on the less severe-ly infected seedlings; circular to elongate brown leaf spots are first small without marked watersoaking and later spread with reddish-brown margins to gray centers. On severely in-fected plants, the leaves dry out before the plants are mature. Lesions also occur on the stems. These are brown necrotic areas which result in the production of shriveled grain.

Necrosis del plantulas occurre o ante o post emergentia. Maculas foliar brun se forma in le plantulas minus severmente infectate. Maculas foliar, o circular o elongate, es al initio minute sin esser aquose; plus tarde illos se extende con margines brun-rubiette e centros gris. In plantas severmente infectate, le folios se desicca ante que le plantas es matur. Lesiones occurre anque in le pedunculos. Istos es areas necrotic brun le quales face le granos crispar se.

La brûlure des semis survient avant ou après l'émergence. Des taches brunes se forment sur les feuilles des plantules les moins gravement infectées; au début, ces taches brunes, cir-culaires à allongées, sont petites et ne paraissent guère détrempeés, mais, plus tard, elles s'agrandissent, ont des marges brun rougeâtre et un centre gris. Sur les plantes gravement attaquées, les feuilles se dessèchent avant la maturité des plantes. Il se produit aussi des lésions sur les tiges. Ces lésions forment des aires nécrotiques brunes qui ne permettent d'obtenir qu'un grain ratatiné.

El tizón de las plántulas aparece antes o después de la emergencia. Las manchas foliares de color café se desarrollan en las plántulas menos severamente infectadas; las manchas foliares circulares o alargadas, de color café, son pequeñas al principio sin empapamiento de agua marcado y luego crecen con márgenes de color café-rojizo a centros grises. En las plantas severamente infectadas, las hojas se secan antes de que las plantas maduren. Las lesiones aparecen también en los tallos. Estas son áreas de color café, necróticas, las cuáles dan como resultado granos arrugados.

DISTRIBUTION: Af., A., Aust., Eur., N.A., C.A., S.A.

174

Oryza sativa + Leptosphaeria salvanii Catt. (Sclerotium oryzae Catt.) =

ENGLISH	Stem Rot
FRENCH	Maladie à Sclérotes du Riz, Sclérotiose
SPANISH	Podredumbre del tallo, Pudrición del tallo
PORTUGUESE	
ITALIAN	Mal dello sclerozio del riso, Gentiluomo del riso
RUSSIAN	гниль склероциальная, "кустистость"
SCANDINAVIAN	
GERMAN	Sklerotienkrankheit, Leptosphaeria-Blattfleckenkrankheit
DUTCH	
HUNGARIAN	
BULGARIAN	
TURKISH	Çeltik sap çürüklüğü
ARABIC	Affan El-Sak غن الساق
PERSIAN	
BURMESE	Pin-see-pok, Gwa-bo roga ပင်လ ပ င်္ချို ဂ ုပ် ရိုဝ ၆ဆ၆ ၆ ၌ဂ ်ဒဝ ၆ ' .
THAI	Lam-Ton-Nao ลำต้นเน่า
NEPALI	Dhan Ko Danth-kuhine rog धानको दाँठ कुहिने रोग, Danth sarra rog दाँठ सराँ रोग
HINDI	Stambh Viglan स्तम्भ विगलान
VIETNAMESE	Gây bệnh hạch khuẩn dưới nhiều hình thức
CHINESE	Shoĕi dăw shĭao chio jĭunn hér bìng 水稻小球菌核病
JAPANESE	Syôkyûkinkaku-byô 小球菌核病

The symptoms of this disease are usually seen at the later growth stages. Necrotic lesions begin on the outer leaves and then spread to the inner ones and the stem base. At about maturity the fruiting bodies of the fungus may occur within the plant tissue. The disease causes the plant to fall over at the groundline and to bear lightweight grain. With severe infection, it causes partial grain sterility.

Le symptomas de iste maladia es usualmente remarcate in le ultime stadios de crescimento. Lesiones necrotic comencia al folios exterior e se extende al folios inferior e al pedé del pedunculo. Circa le tempore de maturation, le structuras sporal del fungo pote formar se intra le texito del planta. Le maladia attacca le planta al pede e face lo cader al terra e producer granos legier. In infection sever, le maladia causa sterilitate de un parte del granos.

C'est généralement vers la fin de la saison de végétation qu'on peut observer les symptômes de cette maladie. Des lésions nécrotiques se montrent d'abord sur les feuilles de l'extérieur, pour se propager ensuite aux feuilles de l'intérieur , pour se propager ensuite aux feuilles de l'intérieur et à la base de la tige. Vers la maturité, les fructifications du champignon peuvent se former à l'intérieur des tissus de la plante. La maladie cause le renversement de la plante à fleur de sol et fait que les grains ont peu de poids. Dans les cas graves, l'infection provoque la stérilité partielle des grains.

Los síntomas de esta enfermedad se ven generalmente en los estados posteriores de crecimiento. Las lesions necróticas comienzan en las hojas exteriores y luego se disemina hacia las interiores y la base del tallo. Cerca de la maduración los cuerpos fructíferos del hongo pueden aparecer dentro del tejido de la planta. La enfermedad hace que la planta se doble desde el suelo y produzca grano de poco peso. Cuando la infección es severa, causa una esterilidad parcial del grano.

DISTRIBUTION: Af., A., Aust., Eur., N.A., C.A., S.A.

Oryza sativa (Gramineae) + Pyricularia oryzae Cav. =

ENGLISH Blast, Rotten Neck, Seedling Blight
FRENCH Brunissure du Riz, Piriculariose, Nieille du Riz
SPANISH Quemado, Tizón de los nudos y de las hojas, Pudrición basal de la espiga, Brusone
PORTUGUESE Brusone
ITALIAN Brusone del riso, Bruseggio
RUSSIAN **запал, гниль узла метелки, "брузоне", пирикуляриоз**
SCANDINAVIAN
GERMAN Piricularia-Fleckenkrankheit, "Brusone"-Krankheit
DUTCH
HUNGARIAN
BULGARIAN **чалгън**
TURKISH Kurt boğazı
ARABIC Lafha لفحة
PERSIAN Lákēhē ābēyē bērēi کلهٔ آبی برنج
BURMESE Gok-kyo roga ပိုင ငိုကြူလေး ငိုယား အောဒ် ငိုက်ဝေး
THAI Bi-Mhai ใบไหม้
NEPALI Dhan Ko neel kanthi धानको निलकंण्ठी , Marra kal मरा काल
HINDI Blast ब्लास्ट
VIETNAMESE Đốm úa
CHINESE Shoěi dǎw dǎw ruǒh bìng 水稻稻熱病
JAPANESE Imoti-byô いもち病

This disease occurs on the leaves, stems and floral structures. The most conspicuous symptom is the appearance of lesions on the neck of the branches, near the base. These lesions prevent the grain from filling. These dark brown lesions are indicative of severe necrosis of the tissue, and they prevent the grain from filling. Brown lesions at the crown and nodes also occur. The leaf spots on young leaves are linear, and on older leaves they are small and circular. Leaf lesions occur on seedlings and as the plant approaches maturity, small circular brown lesions occur on the grain.

Iste maladia occurre in le folios, le pedunculos, e le partes floral. Le symptomas le plus remarcabile es lesiones in le collos del ramos, presso le pede. Iste lesiones brun obscur indica necrosis sever del texito, e illos causa que le le granos non es complete. Lesiones brun obscur occurre anque in le nodos e in le juncturas de pedunculo e radices. Le maculas in folios juvene es linear e in folios plus vetule illos es minute e circular. Lesiones foliar occurre in plantulas e, quando le planta se approxima a maturation, minute lesiones circular brun occurre in le granos.

Cette maladie se présente sur les fueilles, les tiges et les parties florales. Le symptôme le plus manifeste est l'apparition de lésions sur le collet, près du pied des chaumes. Ces lésions brun foncé indiquent une grave nécrose des tissus et empêchent le grain de se remplir. Il peut y avoir aussi des lésions brunes à la couronne et aux noeuds. Sur les jeunes feuilles, les taches sont linéaires, alors qu'elles sont petites et circulaires sur les vieilles feuilles. Il peut y avoir des lésions sur les feuilles des plantules. Lorsque la plante arrive à maturité, il se forme sur le grain de petites lésions circulaires brunes.

Esta enfermedad aparece en las hojas, tallos y estructuras florales. El síntoma más conspicuo es la aparición de lesiones en el cuello de las ramas, cerca de la base. Estas lesiones impiden que el grano se llene. Estas lesiones de color café obscuro son indicativas de una necrosis severa del tejido e impiden que el grano se llene. También aparecen lesiones de color café en la corona y nudos. Las manchas en las hojas jóvenes son lineales y en las hojas maduras son pequeñas y circulares. Las lesiones foliares aparecen en las plántulas y a medida que la planta se acerca a la madurez, aparecen pequeñas manchas circulares, de color café, en el grano.
DISTRIBUTION: Af., A., Aust., Eur., N.A., C.A., S.A.

Oryza sativa (Glycine max) + Rhizoctonia solani Kuehn (Corticium sasakii (Shirai) T. Matsu.) =

ENGLISH   Sheath Blight, Stem and Crown Blight
FRENCH   Fonte des Semis
SPANISH   Pudrición del talluelo, Pudrición de la vaina, Tizón de la vaina
PORTUGUESE
ITALIAN
RUSSIAN   "белая ножка", войлочная болезнь, ризоктониоз
SCANDINAVIAN
GERMAN   Weisshosigkeit, Umfallen
DUTCH
HUNGARIAN
BULGARIAN
TURKISH
ARABIC
PERSIAN
BURMESE
THAI   Khap-Bi-Hang   กาบใบแห้ง
NEPALI
HINDI   Acchad Angmaari   अछद अंगमारी
VIETNAMESE
CHINESE   Shoěi dàw wén ku bìng   水稻紋枯病
JAPANESE   Mongare-byŏ   紋枯病

Large brownish-red areas of the leaves just above the water level are killed during warm, moist weather. The disease is observed in thick stands of the crop. The lesions are large, irregular, elliptical with reddish-brown margins, straw-color, and light yellow or greenish-yellow centers. White or light tan strands of the fungus develop under moist conditions. Seedlings and mature plants are blighted under conditions for disease development.

Grande areas rubie-brunette del folios es mortifacte un pauc super le plano de aqua durante tempore calide e humide. Le maladia se manifesta plantationes dense del plantas. Le lesiones es grande, irregular, elliptic, de color como palea, con margines brun-rubiette e centros jalne clar o jalne-verde. Hyphas blanc o bronzate clar se forma sub conditiones humide. Plantulas e plantas maturate suffre necrosis sub conditiones favorabile pro le extension del maladia.

Par temps humide et chaud, de grandes aires rouge brunâtre des feuilles sont tuées juste au-dessus du niveau de l'eau. La maladie se voit dans des peuplements denses. Les lésions, de couleur paille, sont grandes, irrégulières, elliptiques, à marges brun rougeâtre et à centre jaune clair ou jaune verdâtre. Des cordons blancs ou chamois clair du champignon se forment à l'humidité. Dans des conditions favorables à la maladie, les semis et les plantes adultes sont détruites.

Durante épocas cálidas y húmedas, aparecen grandes áreas rojo-cafesosas en las hojas, un poco arriba del nivel del agua. La enfermedad se observa en poblaciones densas del cultivo. Las lesiones son grandes, irregulares, elípticas con márgenes café-rojizos, de color paja y centros amarillo claro o amarillo verdoso. Hileras blancas o café claro del hongo se desarrollan bajo condiciones húmedas. Las plántulas y plantas adultas aparecen tiznadas bajo condiciones para el desarrollo de la enfermedad.

DISTRIBUTION: A., Eur., N.A., C.A.

Oryza sativa (Gramineae) + Ustilaginoidea virens (Cooke) Tak. =

ENGLISH	False Smut
FRENCH	Faux Charbon, Charbon Vert du Riz
SPANISH	Falso carbón, Tizón Falso
PORTUGUESE	Carvão Verde, Falso carvão
ITALIAN	Carbone verde
RUSSIAN	головня ложная (в завязях), головня зеленая
SCANDINAVIAN	
GERMAN	Falscher Reisbrand
DUTCH	
HUNGARIAN	
BULGARIAN	
TURKISH	
ARABIC	
PERSIAN	
BURMESE	
THAI	Dog-Kra-Tin ดอกกระถิน
NEPALI	Dhan Ko hariyo poke धानको हरियो पोके , Hara kal हरा काल
HINDI	Abhasi Kand अभासी कंद
VIETNAMESE	Bông trái
CHINESE	Shoěi dàw dàw chíu bìng 水稻稻麴病
JAPANESE	Inakŏji-byŏ 稲こおじ病

This disease appears as olive-green, velvety, globose masses in the floral parts. The fruiting bodies of the fungus develop and replace the flowers. Usually only a few flowers are infected per plant. The large, black fungus fruiting bodies can usually be removed during the cleaning and threshing of the crop.

Iste maladia se manifesta como massas in le partes floral, verde como olivas, similar a vil- luto, e globose. Le structuras sporal del fungo se forma e reimplacia le flores. Usualmente solmente alicunes del flores es infectate in cata planta. On pote usualmente remover le grande structuras sporal nigre del fungo durante le purification e le disgranamento.

Cette maladie se présente sous forme de masses globuleuses, vert olive, veloutées, dans les parties florales. Les fructifications du champignon prennent la place des fleurs. Il n'y a généralement que quelque fleurs infectées par plante. Il est d'ordinaire possible d'éliminer, lors du nettoyage et du battage de la récolte, les grosses fructifications noires du champignon.

Esta enfermedad aparece como masas de color verde olivo, aterciopeladas, globosas en las partes florales. Los cuerpos fructíferos del hongo se desarrollan y sustituyen a las flores. Generalmente sólo unas pocas flores por planta se infectan. Los cuerpos fructíferos negros, grandes, generalmente pueden ser quitados durante la limpieza y trillado de la cosecha.

DISTRIBUTION: Af., A., Aust., Eur., N.A., C.A., S.A.

178

Panicum spp. + _Spacelotheca destruens_ (Schlecht.) Stevenson & A. G. Johnson =

ENGLISH	Head Smut
FRENCH	Charbon du Millet, Charbon des Panicules
SPANISH	Tizón de la cabeza
PORTUGUESE	
ITALIAN	Carbone del miglio
RUSSIAN	головня метелок
SCANDINAVIAN	
GERMAN	Sphacelotheca-Rispenbrand
DUTCH	
HUNGARIAN	
BULGARIAN	
TURKISH	
ARABIC	
PERSIAN	
BURMESE	
THAI	
NEPALI	
HINDI	Sheersh Kand शीरश कंद
VIETNAMESE	
CHINESE	Shǐao mǐ swāi hei swāi bīng 小米穗黑穗病
JAPANESE	Kuroho-byô 黒穂病

This fungus disease infects the seedlings and later causes the normal grain to be replaced by an oval gray sac. The tough membraned sac is full of black dust consisting of the fungus spores. The color of the affected grain may vary from bright green to brown or black.

Iste maladia fungal infecta le plantulas e plus tarde causa que le grano normal es reimplaciate per un sacco oval gris. Iste sacco membranacee dur es plen del pulvere nigre del sporas fungal. Le color del grano afflicte pote variar se ab verde brillante usque a brun o nigre.

Cette maladie fongueuse infecte les semis et par la suite fait remplacer le grain normal par un sachet ovale gris. Ce petit sac membraneux et résistant est rempli d'une poussière noire qui consiste en spores du champignon. La couleur du grain malade peut aller du vert éclatant au brun ou au noir.

Esta enfermedad fungosa infecta las plántulas y luego ocasiona que el grano normal sea sustituido por un saco gris ovalado. El duro saco membranado está lleno de polvo negro formado por las esporas del hongo. El color del grano afectado puede variar de verde brillante a café o negro.

DISTRIBUTION: Af., A., Aust., Eur., N.A., S.A.

Pennisetum typhoides (Setaria itatica and other Gramineae) + Sclerospora graminicola (Sacc.) Schroet. =

ENGLISH	Downy Mildew	
FRENCH	Mildiou des Céréales	
SPANISH	Mildiú lanoso	
PORTUGUESE		
ITALIAN		
RUSSIAN		
SCANDINAVIAN		
GERMAN		
DUTCH		
HUNGARIAN		
BULGARIAN		
TURKISH		
ARABIC		
PERSIAN		
BURMESE		
THAI		
NEPALI		
HINDI	Harit Baal Rog	हरित बाल रोग
VIETNAMESE		
CHINESE		
JAPANESE		

The principal symptoms of this disease are the dwarfed plants and the excessive development of branches. The development of leaf-like malformations of the floral parts is also characteristic. This condition prevents grain development. Leaf necrosis and browning are followed by splitting and shredding of invaded tissue, especially as the plant approaches maturity.

Le symptomas principal de iste maladia es le plantas nanos e le superdeveloppamento del ramos. Le developpamento de malformationes similar a folios in le partes floral es anque characteristic. Iste condition impedi le developpamento de granos. Necrosis e brunimento foliar es secute per fissura e fragmentation del texito invase, specialmente quando le plantas se approxima a maturation.

Les symptômes principaux de cette maladie sont le nanisme des plantes et la prolifération excessive des chaumes. L'apparition de malformations foliacées des parties florales est une autre caractéristique. Ce désordre empêche le grain de se former. Après la nécrose et le brunissement des feuilles, les tissus envahis éclatent et s'effilochent, surtout à l'approche de la maturité des plants.

Los síntomas principales de esta enfermedad son las plantas achaparradas y el desarrollo excesivo de ramas. El desarrollo de malformaciones en forma de hoja de las partes florales es también característico. Esta condición impide el desarrollo del grano. La necrosis y color café que toma la hoja son seguidas por que el tejido invadido se raja y se hace jirones, especialmente cuando la planta se acerca a la madurez.

DISTRIBUTION: Af., A., Aust., N.A., S.A.

180

Phaseolus spp. (Vigna spp.) + Ascochyta phaseolorum Sacc. =

ENGLISH	Leaf Spot
FRENCH	Ascochytose, Macules Foliaires
SPANISH	Mancha en anillos de la hoja, Mancha foliar
PORTUGUESE	Mancha folhar de Ascochyta
ITALIAN	
RUSSIAN	аскохитоз
SCANDINAVIAN	
GERMAN	Ascochyta-Blattfleckenkrankheit
DUTCH	Bladvlekken
HUNGARIAN	
BULGARIAN	
TURKISH	
ARABIC	
PERSIAN	
BURMESE	
THAI	
NEPALI	
HINDI	Sem Ki Angmaari सेम की अंगमारी
VIETNAMESE	
CHINESE	
JAPANESE	Kappan-byô 褐斑病

The fungus causes spots on leaves, pods and stems from the seedling stage until maturity. Spots are circular, brown and studded with minute black fruiting bodies which are large enough to be seen with the unaided eye. The spots usually are marked with concentric zones and dark margins. In severe attacks the leaves die, beginning with the lowest and progressing upward. The lesions are elongate on pods, stems and petioles and have fewer fruiting bodies. Lesions appearing at the soil level may extend downward into the roots or girdle the stem, killing the entire plant.

Le fungo produce maculas in folios, siliquas, e pedunculos ab le stadio del plantulas usque a maturation. Le maculas es circular, brun, e marcate per minute structuras sporal nigre le quales es tanto grande que on los vide sin instrumento. Le maculas es usualmente marcate con zonas concentric e margines obscur. In attaccos sever le folios mori, al initio le inferiores alora le superiores. Le lesiones es elongate in siliquas, pedunculos, e petiolos e monstra minus de structuras sporal. Lesiones que se manifesta al plano del terra pote extender se a basso in le radices o imbraciar le pedunculo, causante que tote le planta mori.

Depuis le semis jusqu'à la maturité, ce champignon produit des taches sur les feuilles, les gousses et les tiges. Les taches sont circulaires, brunes et parsemées de minuscules fructifications noires visibles à l'oeil nu. Les taches sont en général dotées de ceintures concentriques et de marges sombres. Dans les cas graves, les feuilles meurent, depuis celles du bas vers le haut. Sur les gousses, les tiges et les pétioles, les lésions sont allongées et portent moins de fructifications. Les lésions à fleur de sol peuvent descendre aux racines ou encercler la tige, de façon à detruire toute la plante.

El hongo produce manchas en las hojas, vainas y tallos desde el estado de plántula hasta la madurez. Las manchas son circulares, de color café, que tienen cuerpos fructíferos diminutos, pero que pueden ser vistos a simple vista. Las manchas generalmente están marcadas con zonas concéntricas y márgenes obscuros. En ataques severos las hojas mueren, comenzando con las inferiores y progresando hacia arriba. Las lesiones son alargadas en las vainas, pedúnculos y pecíolos y tienen pocos cuerpos fructíferos. Las lesiones que aparecen a nivel del suelo pueden extenderse hacia abajo a las raíces o rodear el tallo, matando toda la planta.

DISTRIBUTION: Af., A., Aust., Eur., N.A., S.A.

Phaseolus vulgaris + Cercospora phaseoli Dearn. & Barth. (Cercospora cruenta Sacc.) =

ENGLISH	Leaf Spot
FRENCH	Cercosporose, Taches Foliaires
SPANISH	Cercosporiosis, Mancha foliar
PORTUGUESE	
ITALIAN	
RUSSIAN	**пятнистость церкоспорозная, церкоспороз**
SCANDINAVIAN	
GERMAN	Cercospora-Fleckenkrankheit
DUTCH	
HUNGARIAN	
BULGARIAN	
TURKISH	
ARABIC	
PERSIAN	
BURMESE	Ywet-pyauk roga ဣဒၵု အာကိုစပိုးရာခရွယင်း က
THAI	Bi-Jut ใบจุด
NEPALI	
HINDI	Parn dhabba पर्ण धब्बा
VIETNAMESE	
CHINESE	
JAPANESE	Susumon-byô すす斑病

This fungus attacks mature leaves and produces brown, rust-colored patches of various sizes and shapes. The spread of the spots, which are mostly angular, is restricted by the veins. Usually the spots are surrounded by a dark-colored border. The center of these lesions frequently dry, crack and drop out, leaving ragged holes. When the disease spreads to various areas of the plant at one time, the result is premature dropping of leaves.

Iste fungo attacca folios matur e produce areas brun, del color de ferrugine, de dimensiones e contornos diverse. Le extension del maculas, le quales es generalmente angular, es restricte per le venas. Usualmente le maculas es imbraciate per un bordo de color obscur. Le centros de iste lesiones frequentemente se desicca, se finde, e cade, lassante perforationes zigzag. Si le morbo se extende a areas diverse del planta al mesme tempore, defoliation prematur occurre.

Ce champignon attaque les feuilles adultes et y produit des plaques brun rouille de diverses formes et dimensions. L'agrandissement des taches, qui sont le plus souvent angulaires, est restreint par les nervures. D'ordinaire, les taches sont entourées d'une bordure de couleur foncée. Souvent, le centre de ces lésions se dessèche, se fissure et tombe, laissant des trous à bords déguenillés. Quand la maladie s'étend à diverses aires de la plante en même temps, il s'ensuit une chute prématurée des feuilles.

Este hongo ataca las hojas maduras y produce parches de color café o color herrumbroso de varios tamaños y formas. La diseminación de las manchas, que son más que nada angulares, está restringida por las venas. Generalmente las manchas están rodeadas por un borde de color obscuo. El centro de estas lesiones frecuentemente se seca, se raja, se caen y dejan agujeros. Cuando la enfermedad se disemina a varias partes de la planta en una sola vez, el resultado es la caída prematura de las hojas.

DISTRIBUTION: A., N.A.

182

Phaseolus vulgaris + Colletotrichum lindemuthianum (Sacc. & Magn.) Bri. & Cav. =

ENGLISH	Anthracnose
FRENCH	Anthracnose
SPANISH	Antracnosis
PORTUGUESE	Antracnose
ITALIAN	Antracnosi
RUSSIAN	антракноз
SCANDINAVIAN	Bønnesyge, Bönfläcksjuka
GERMAN	Brennfleckenkrankheit
DUTCH	Vlekkenziekte, Vlekziekte
HUNGARIAN	Kolletotrihumos levél-, szár-és termésfoltos-ság
BULGARIAN	антракноза
TURKISH	Fasulye antraknozu
ARABIC	
PERSIAN	
BURMESE	Twet-pyauk roga ၈က ၁လကတုၐြ၌၄ ၁ ၆လ၄ ၃ ၄ံ� ၄ ၆
THAI	Anthracnose แอนแทรคโนส
NEPALI	
HINDI	Anthracnose ऐनथ्रेक्नोज
VIETNAMESE	Đén lá và trái
CHINESE	
JAPANESE	Tanso-byð 炭そ病

Yellowish, brown or black sunken cankers on the pods, stems and primary leaves are the most characteristic symptoms. The center of these lesions contain salmon-colored ooze. The most noticeable symptoms on the foliage are the blackened dead portions of the veins on the leaf underside. Stems may rot off just as they emerge from the ground or the primary leaves may be destroyed, killing the entire plant. Later infections harm mostly the pods and their seeds which may or may not be discolored.

Le symptomas le plus characteristic es depresse canceres jalnette, brun, o nigre in le siliquas, pedunculos, e folios primari. Le centros de iste lesiones monstra un exsudation de color de salmon. Le symptomas le plus remarcabile in le foliage es le morte portiones nigrate del venas in le superfacie inferior del folios. Pedunculos a vices putresce se tosto que illos emerge ab le terra, o le folios primari es a vices destructe, facente tote le planta morir. Infectiones plus tarde attacca principalmente le tote le planta morir. Infectiones plus tarde attacca principalmente le siliquas e lor granos, le quales pote monstrar coloration anormal o non.

Les symptómes les plus caractéristiques sont des chancres déprimés, jaunâtres, bruns ou noirs sur les gousses, les tiges et les feuilles primaires. Le centre de ces lésions contient un exsudat couleur saumon. Sur le feuillage, les symptômes les plus remarquables sont des portions mortes noircies des nervures, à la face inférieure de la feuille. Les tiges peuvent pourrir dès l'émergence, ou bien les feuilles primaires peuvent être détruites, ce qui entraîne la mort de la plante. Les infections plus tardives endommagent surtout les gousses et les graines, qui peuvent être décolorées ou non.

Los síntomas más característicos son cancros amarillosos, café o negros, hundidos, que aparecen en las vainas, tallos y hojas primarias. El centro de estas lesiones contiene exudaciones de color salmón. El síntoma más notorio en el follaje son las prociones muertas y ennegrecidas de las venas en el envés de la hoja. Los tallos pueden podrirse en cuanto brotan del suelo o las hojas primarias pueden ser destruídas, matando toda la planta. Las infecciones posteriores deñan principalmente las vainas y sus semillas, las cuáles pueden o no estar decoloradas.

DISTRIBUTION: Af., A., Aust., Eur., N.A., C.A., S.A..

Phaseolus vulgaris (P. spp., Glycine max) +Corynebacterium flaccumfaciens (Hedges)Dowson=

ENGLISH	Bacterial Wilt
FRENCH	Flétrissure Bactérienne
SPANISH	Marchitez bacteriana
PORTUGUESE	
ITALIAN	Avvizzimento batteriaceo del fagiolo
RUSSIAN	увядание бактериальное
SCANDINAVIAN	
GERMAN	Bakterielle Welke
DUTCH	
HUNGARIAN	
BULGARIAN	бактериално увяхване
TURKISH	Fasulye solgunluğu
ARABIC	
PERSIAN	
BURMESE	
THAI	
NEPALI	
HINDI	
VIETNAMESE	
CHINESE	
JAPANESE	

This seedborne disease attacks the very young seedlings. Young plants two to three inches tall are usually killed when attacked. If they do survive, they may live throughout the season. The leaves may become flaccid and hang limply on the stems and branches, especially during the warmest part of the day. If the vascular bundles become clogged and cut off the water supply, the leaves will brown and drop off. Sometimes when these typical symptoms are not apparent, golden-yellow necrotic leaf lesions can be seen. If the disease appears on the pods, water-soaked spots occur. Wilt is more conspicuous on ripe pods where it produces an olive-green color.

Iste maladia, transportate per le semines, attacca le plantulas ancora juvene. Si attaccate, plantas juvene con un altitude de 3 o 4 cm usualmente mori. Si illos supervive le attacco, illos pote restar vivente per tote le saison. Le folios a vices langue sin vigor ab le pedunculos e ramos, specialmente durante le parte le plus calide del die. Si le fasces vascular se occlude e le approvisionamento de aqua es impedite, le folios deveni brun e cade. A vices, si iste symptomas typic non es evidente, on vide necrotic lesiones foliar, aurate. Si le maladia se manifesta in le siliquas, maculas aquose es visibile. In siliquas maturate, marcescentia es multo evidente, producente un color verde como olivas.

Cette maladie transmise par les graines attaque les très jeunes plantules. Si elles subissent l'attaque alors qu'elles n'ont que deux ou trois pouces de haut, elles succombent généralement. Certaines d'entre elles peuvent pourtant survivre durant toute la saison. Les feuilles peuvent devenir flasques et pendre mollement sur les tiges et les rameaux, surtout durant la grande chaleur du jour. Si les faisceaux vasculaires sont bloqués et coupent l'approvisonnement d'eau, les feuilles brunissent et tombent. Parfois, lorsqu'on ne peut observer ces symptômes typiques, il est possible de voir des lésions nécrotiques jaune doré sur les feuilles. Si la maladie se présente sur les gousses, il s'y produit des taches délavées. La maladie est plus apparente sur les gousses mûres, qu'elle colore en vert olive.

Esta enfermedad que se origina en la semilla ataca las plántulas más jóvenes. Las plántulas de dos a tres pulgadas de alto generalmente mueren al ser atacadas. Las hojas pueden volverse flácidas y colgar débilmente en los tallos y ramas, especialmente durante la parte más cálida del día. Si los bultos vasculares se tapan y cortan el aprovisionamiento de agua, las hojas se volverán de color café y se caerán. Algunas veces, cuando estos síntomas típicos no son aparentes, pueden verse lesiones foliares necróticas de color amarillo-dorado. Si esta enfermedad aparece en las vainas, surgen manchas empapadas en agua. La marchitez es más conspicua én las vainas maduras, donde produce un color verde-olivo.
DISTRIBUTION: Aust., Eur., N.A.

184

Phaseolus spp. + Phaeoisariopsis griseola (Sacc.) Ferr. (Isariopsis griseola Sacc.) =

ENGLISH	Angular Leaf Spot
FRENCH	Maladie des Taches Anguleuses
SPANISH	Mancha angular de la hoja, Mancha foliar angular
PORTUGUESE	Mancha angular
ITALIAN	
RUSSIAN	пятнистость угловатая (грибная), плесень серая
SCANDINAVIAN	
GERMAN	Isariopsis-Blattbräune
DUTCH	
HUNGARIAN	
BULGARIAN	
TURKISH	
ARABIC	
PERSIAN	
BURMESE	
THAI	
NEPALI	
HINDI	
VIETMANESE	
CHINESE	Tsày dòu jǎw ban bìng 菜豆角斑病
JAPANESE	Kakuhan-byð 角斑病

Spots, which originate on the underside of the leaf, are delimited by the veins and veinlets. The lesions are gray at first and brown later with no colored borders. This absence of color in the leaf and the angular shape of the spots are characteristic. On the pods, the spots have sharply defined, nearly black borders and reddish-brown centers. They vary in size, ultimately becoming so crowded they cover the width of the pod.

Maculas, que comencia in le superfacie inferior del folio, es delimitate per le venas e le venulas. Le lesiones, al initio gris e plus tarde brun, ha bordos sin color. Iste absentia de color e le contorno angular del maculas es characteristic. In le siliquas, le maculas monstra ben definite, quasi nigre bordos e centros brun-rubiette. Illos se varia in dimension, ma, al fin, illos deveni numerose usque a coperir tote le latitude del siliqua.

Les taches, qui prennent naissance à la face inferieure de la feuille, sont délimitées par les nervures, grandes ou petites. Les lésions sont grises au début, et brunes par la suite, sans bordures colorées. Cette absence de couleur dans la feuille et la forme angulaire des taches sont des points caractéristiques. Sur les gousses, les taches ont des bords presque noirs et nettement définis et un centre brun rougeâtre. Leurs dimensions varient, mais, à la fin, les taches sont si tassées qu'elles recouvrent toute la gousse.

Las manchas que se originan en el envés de las hojas, son delimitadas por las venas y venillas. Las lesiones son grises primero y luego café, con bordes no coloreados. Esta ausencia de color en la hoja y la forma angular de las manchas son características. En las vainas, las manchas tienen orillas bien definidas, casi negras y centros café-rojizos. Estas varían en tamaño, ultimadamente volviéndos tan amontonadas que cubren el ancho de la vaina.

DISTRIBUTION: Af., A., Aust., Eur., N.A., C.A., S.A.

Phaseolus limensis (and others) + Phytophthora phaseoli Thaxter =

ENGLISH   Downy Mildew
FRENCH   Mildiou, Phytophthorose, Pourriture des Gousses
SPANISH   Mildiú lanoso
PORTUGUESE
ITALIAN   Peronospora dei fagioli di Lima
RUSSIAN   **гниль бобов фитофторозная, фитофтороз**
SCANDINAVIAN
GERMAN   Phytophthora-Hülsenfäule
DUTCH
HUNGARIAN
BULGARIAN
TURKISH   Fasulye mildiyösü
ARABIC
PERSIAN
BURMESE
THAI
NEPALI
HINDI
VIETNAMESE
CHINESE
JAPANESE

Symptoms of this disease can be seen on all aboveground parts of the plant. Lesions are irregularly shaped and of purplish color. At high humidities they spread over the pods **and** become
covered with a dense coating of a white fungus growth which produces spores. There usually
are reddish bands around the infected areas.

Le symptomas de iste maladia es visibile in tote le partes epigee del planta. Le lesiones es
de contorno irregular e de color purpuree. Sub conditiones de humiditate illos se extende super
le siliquas e es coperte de un dense strato fungal blanc le qual produce sporas. Usualmente
bandas rubiette occurre circum le areas infectate.

Les symptômes de cette maladie peuvent apparaître sur toutes les parties aériennes de la
plante. Les lésions sont de forme irrégulière et de couleur pourprée. A la faveur d'une grande
humidité, elles envahissent les gousses et se couvrent d'une épaisse couche d'un mycélium
blanc porteur des spores du champignon. Des bandes rougeâtres entourent généralement l'aire
infectée.

Los síntomas de esta enfermedad pueden verse en todas las partes aéreas de la planta. Las
lesiones son de forma irregular y de color purpúreo. Bajo humedades altas se diseminan sobre
las vainas y se cubren con una densa capa de un crecimiento fungoso blanco, el cuál produce
esporas. Generalmente hay bandas de color rojizo alrededor del área infectada.

DISTRIBUTION: Af., A., Eur., N.A., C.A.

Phaseolus vulgaris + Pseudomonas phaseolicola (Burkh.) Dowson (Pseudomonas medicaginis f. sp. phaseolicola (Burkh.) Dowson) =

ENGLISH   Halo Blight, Grease Spot, Bacterial Blight
FRENCH   Maladie de la Graisse
SPANISH   Tizón de halo, Quemazón bacteriana, Hielo amarillo, Mancha grasosa
PORTUGUESE
ITALIAN   Batteriosi, Grassume del fagiolo
RUSSIAN   бактериоз угловатый, пятнистость бактериальная угловатая
SCANDINAVIAN   Fedtpletsyge
GERMAN   Fettfleckenkrankheit, Eckige Fleckigkeit, Bakterille Fettfleckenkrankheit
DUTCH   Vetvlekkenziekte
HUNGARIAN   Baktériumos paszulyvész
BULGARIAN   пригор
TURKISH
ARABIC
PERSIAN
BURMESE
THAI
NEPALI
HINDI
VIETNAMESE
CHINESE
JAPANESE   Basagare-byô   かさ枯病

Small, water-soaked spots with halo-like zones of greenish-yellow tissue enlarge and coalesce on the underside of the leaflets and leaves. These areas become flaccid, turn brown, become necrotic and may cause extensive defoliation. Light cream-or silver-colored exudate is sometimes noted in the lesions. If the vascular tissue of the leaf is affected, the initial infection usually progresses from the veinlets to large veins and to the midrib. Reddish discolorations with watersoaked tissues adjacent to the veins are produced. If the leaf infection starts from the petiole, the main vein and its branches appear watersoaked at first and then a brick-red color.

Minute maculas aquose con zonas similar a halos, de texito jalne-verdette, cresce e coalesce in le superfacie inferior del foliettos e del folios. Le areas infectate perde turgor, deveni brun e necrotic, e pote causar defoliation extensive. Exsudate, de color de crema clar o argentee, es a vices visibile in le lesiones. Si le texito vascular del folio es afflicte, le infection usualmente progrede ab le venulas a grande venas e usque al nervatura central del folio. Discolorationes rubiette con texito aquose adjacente al venas es le resultato. Si le infection foliar comencia ab le petiolo, le vena principal e su ramos pare aquose al initio e plus tarde rubie como briccas.

De petites taches délavées et ceinturées par des tissus jaune verdâtre qui forment une sorte de halo s'agrandissent et se fusionnent à la face inférieure des folioles et des feuilles. Les aires infectées deviennent flasques, brunes, nécrotiques, et peuvent causer beaucoup de défoliation. Un exsudat crème clair ou argenté s'observe parfois dans les lésions. Si les tissus vasculaires de la feuille sont atteints, l'infection se propage généralement des petites aux grosses nervures, ainsi qu'à la nervure médiane. Des tissus délavés adjacents aux nervures prennent une couleur rougeâtre. Si l'infection de la feuille vient du pétiole, la nervure principale et ses ramifications paraissent d'abord délavées, puis se colorent en rouge brique.

Manchas pequeñas, empapadas de agua, con zonas parecidas a halos de tejido amarillo-verdoso se agrandan y se juntan en el envés de las hojas y hojillas. Las áreas infectadas se vuelven flácidas, adquieren color café, se vuelven necróticas y pueden causar una extensa defoliación. Una exudación de color crema claro o de color plateado se nota generalmente en las lesiones. Si el tejido vascular de la hoja se ve afectado, la infección generalmente progresa desde las venillas hacia las venas grandes y la vena del centro. Aparecen decoloraciones rojizas con tejidos empapados de agua, adyacentes a las venas. Si la infección de la hoja comienza del pecíolo, la vena principal y sus ramas aparecen primero empapadas de agua y luego toman un color rojo ladrillo.
DISTRIBUTION: Af., Aust., Eur., N.A., C.A., S.A.

<u>Phaseolus</u> <u>vulgaris</u> (many others) + <u>Pythium</u> <u>aphanidermatum</u> (Edson) Fitzp. =

ENGLISH	Wilt, Soil Rot
FRENCH	
SPANISH	Marchitez, Podredumbre del suelo, Mal del talluelo
PORTUGUESE	
ITALIAN	
RUSSIAN	
SCANDINAVIAN	
GERMAN	
DUTCH	
HUNGARIAN	
BULGARIAN	сечене
TURKISH	
ARABIC	
PERSIAN	
BURMESE	
THAI	
NEPALI	
HINDI	
VIETNAMESE	
CHINESE	
JAPANESE	Watagusare-byô 綿腐病

The different species of pythia vary in their temperature requirements. This particular fungus is associated with high temperatures and characteristic wilt symptoms, but only the growth of the organism in cultures can identify the specific species. In general, <u>Phythium</u> causes wet rot and damping-off of young plants. In somes cases the pith of the stem is destroyed; a semi-soft, colorless to dark-brown rot often spreads up and down the stem, sometimes as far as the lateral leaves. If the weather is moist a cotton-like growth is visible.

Le differente species del genere <u>pythium</u> se varia in lor exigentias temperatura. Iste fungo se associa con temperaturas elevate e con symptomas que es characteristic de marcescentia. Ma on pote indentificar le specie precise solmente per cultivation del organismo in medios artificial. Generalmente, <u>pythium</u> causa putrefaction humide al pede de plantulas. A vices le medulla del pedunculo es destructe; un putrefaction semi-molle, sin color o brun obscur, frequentemente se extende in alto e in basso per le pedunculo, a vices usque al folios lateral. Si le tempore es humide, un strato similar a coton es visibile.

Les différentes espèces de <u>Pythium</u> varient quant à leurs exigences de température. Le champignon dont il s'agit se complaît à des températures élevées et s'accompagne de symptómes typiques de flétrissure, mais seule l'étude de sa crossance en culture permet de déterminer l'espèce à laquelle cet organisme appartient. En général, le <u>Pythium</u> cause un pourriture aquese et la fonte des semis. Dans certains cas, la moëlle de la tige est détruite; une pourriture mi-molle, incolore à brun foncé se propage souvent vers le haut et le bas de la tige, parfois jusqu'aux feuilles latérales. Si le temps est humide, on peut voir une végétation cotonneuse.

Las diferentes especies de pythia varían en sus requerimientos de temperatura. Este hongo en particular está asociado con altas temperaturas y síntomas de marchitez característicos, pero sólo el crecimiento del organismo bajo cultivo peude identificar las especies específicas. En general, <u>Phythium</u> causa podredumbre húmeda y mal del talluelo de plantas jóvenes. En algunos casos la médula del tallo es destruída; una podredumbre semi-suave, incolora a café obscuro frecuentemente se disemina arriba y abajo del tallo, algunas veces llegando hasta las hojas laterales. Si el tiempo es húmedo, se puede ver un crecimiento algodonoso.

DISTRIBUTION: Af., A., Aust., Eur., N.A., C.A., S.A.

188

Phaseolus spp., (Gossypium spp., Solanum tuberosum) + Pythium debaryanum Hesse =

ENGLISH	Water Rot, Leak, Damping-off
FRENCH	Fonte des Semis et des Boutures
SPANISH	Mal del talluelo, Podredumbre de agua, Goteo
PORTUGUESE	
ITALIAN	
RUSSIAN	
SCANDINAVIAN	
GERMAN	
DUTCH	
HUNGARIAN	
BULGARIAN	сечене
TURKISH	Çökerten
ARABIC	
PERSIAN	
BURMESE	
THAI	
NEPALI	
HINDI	
VIETNAMESE	Thối rễ, hại cây con
CHINESE	
JAPANESE	

The different species of pythia vary in their temperature requirements. This particular fungus is associated with cool temperatures. Only the growth of the organism in cultures can identify the specific species however. In general, Pythium causes wet rot and damping-off of young plants. In some cases the pith of the stem is destroyed; a semi-soft, colorless to dark brown rot often spreads up and down the stem, sometimes reaching as far as the lateral leaves. If the weather is moist a cotton-like growth is visible.

Le differente species del genere pythium se varia in lor exigentias de temperatura. Iste fungo se associa con temperaturas fresc. Nonobstante, on pote identificar le specie precise solmente per cultivation del organismo in medios artificial. Generalmente, pythium produce putrefaction humide al pede del plantulas. A vices le medulla del pedunculo es destructe; un putrefaction semi-molle, sin color e brun obscur, frequentemente se extende in alto e in basso per le pedunculo, a vices usque al folios lateral. Si le tempore es humide, un strato similar a coton es visibile.

Les différentes espèces de Pythium varient quant à leurs exigences de température. Le champignon dont il s'agit se complaît à des températures fraîches. Cependant, seule l'étude de sa crossance en culture permet de déterminer l'espèce à laquelle cet organisme appartient. En général, le Pythium cause une pourriture aqueuse et la fonte des semis. Dans certains cas, la moëlle de la tige est détruite; une pourriture mi-molle, incolore à brun foncé, se propage souvent vers le haut et le bas de la tige, parfois jusqu'aux feuilles latérales. Si le temps est humide, on peut voir une végétation cotonneuse.

Las diferentes especies de pythia varían en sus requerimientos de temperatura. Este hongo en particular está asociado con temperaturas frías. Sólo el crecimiento de este organismo bajo cultivo puede identificar las especies específicas. En general Pythium causa podredumbre húmeda y mal del talluelo de las plantitas. En algunos casos la médula del tallo es destruída; una podredumbre semi-suave, incolora a café obscuro frecuentemente se disemina arriba y abajo del tallo, algunas veces llegando hasta las hojas laterales. Si el tiempo es húmedo, se puede ver un crecimiento algodonoso.

DISTRIBUTION: Af., A., Aust., Eur., N.A., C.A., S.A.

Phaseolus spp., (Vigna spp. and others) + Uromyces phaseoli (Pers.) Wint. (Uromyces appendiculatus (Pers.) Unger) =

ENGLISH	Rust, Brown Rust
FRENCH	Rouille du Haricot
SPANISH	Roya, Chahuixtle, Polvillo, Roya café
PORTUGUESE	Ferrugem
ITALIAN	Ruggine del fagiolo
RUSSIAN	ржавчина
SCANDINAVIAN	Bønnerust, Bønrost
GERMAN	Bohnenrost, Rost
DUTCH	Roest, Bonenroest
HUNGARIAN	
BULGARIAN	оидиум
TURKISH	Fasulye pası
ARABIC	Sadaa اد صـ
PERSIAN	Zángē lōōbēyā زنگ لوبیا
BURMESE	Than-chi roga ဤ ၌ ၎ကိ၍ ၆ အ ၈ ၃ကု လ ၆ ၆
THAI	Ra-Snim-Leck ราสนิมเหล็ก
NEPALI	
HINDI	Ratua रतुवा
VIETNAMESE	Rỉ lá và trái
CHINESE	Tsày dòu shǐou bìng 菜豆銹病
JAPANESE	Sabi-byð さび病

Bean rust attacks the leaves, pods and, rarely, the tender parts of the stem and branches. It is most abundant on the leaves however. The symptoms first appear on the lower surface of the leaves in the form of minute, white, slightly raised spots. These soon enlarge, turn reddish and then dark brown. The leaves turn yellow, then brown and dry.

(Fer)rugine in phaseolo attacca le folios, siliquas, e, rarmente, le partes tenere del pedunculos e del ramos. Illo es le plus abundante in le folios. Le symptomas se manifesta al initio in le superfacie inferior del folios como minute maculas blanc e un pauc elevate. Illos cresce rapidemente, deveni rubiette e alora brun obscur. Le folios deveni jalne, alora brun e desiccate.

La rouille du haricot attaque les feuilles, les gousses et, plus rarement, les parties tendres de la tige et des rameaux. C'est sur les feuilles, toutefois, qu'elle abonde le plus. Les symptômes apparaissent d'abord à la face inférieure des feuilles sous forme de minuscules taches blanches légèrement surélevées. Ces taches s'agrandissent bientôt, deviennent rougeâtres, et ensuite brun foncé. Les feuilles jaunissent, puis brunissent et se dessèchent.

La roya del frijol ataca las hojas, vainas y raramente, las partes tiernas del tallo y ramas. Es más abundante, sinembargo, en las hojas. Los síntomas aparecen primero en el envés de las hojas en forma de manchas miniatura, blancas, algo levantadas. Estas crecen pronto, se vuelven de color rojizo y leugo café obscuro. Las hojas se vuelven amarillas, luego café y secas.

DISTRIBUTION: Af., A., Aust., Eur., N.A., C.A., S.A.

190

Phaseolus vulgaris (Phaseolus spp., Lablab niger) + Xanthomonas phaseoli (E. F. Smith)
Dowson =

ENGLISH   Bacterial Blight
FRENCH   Brûlure Bactérienne, Bactériose
SPANISH   Tizón común, Mancha bacterial, Bacteriosis, Quemazón bacterial común
PORTUGUESE   Crestamento comun
ITALIAN   Batteriosi del fagiolo, Nebbia batterica
RUSSIAN   пятнистость бактериальная бурая, ожог стеблей
SCANDINAVIAN
GERMAN   Fettfleckenkrankheit, Bakterielle Brandfleckigkeit
DUTCH
HUNGARIAN
BULGARIAN   бактериален пригор
TURKISH   Fasulye adi yaprak yanıklığı
ARABIC
PERSIAN
BURMESE
THAI
NEPALI
HINDI   Sem Ki Jeewanik Angmaari   सेम की जीवनिक अंगमारी.
VIETNAMESE
CHINESE
JAPANESE   Hayake-byð   葉焼病

The first symptoms of blight are small, water-soaked areas on the leaves as well as the pods.
Areas between the lesions may turn yellow and die. Often the spots coalesce to form large,
brown lesions of irregular size and shape with lemon-yellow margins. On the petioles and
stems the infection may appear as cankers.

Le prime symptomas de bacteriosis es minute areas aquose in le folios e in le siliquas. Areas
inter le lesiones a vices deveni jalne e moriSovente le maculas coalesce usque a formar grande
lesiones brun de dimension e contorno irregular con margines jalne como limon. In le petiolos
e in le pedunculos le infection pote manifestar se como canceres.

Les premiers symptômes de brulûre sont de petites plages délavées sur les feuilles et sur
les gousses. Les aires situées entre les lésions peuvent jaunir et mourir. Souvent, les taches
se fusionnent pour former de grandes lésions brunes de grosseur et de forme variables, avec
des marges jaune citron. Sur les pétioles et les tiges, l'infection peut apparaître sous forme
de chancres.

Los primeros síntomas del tizón son pequeñas áreas empápadas de agua sobre las hojas, así
como sobre las vainas. Las áreas entre las lesiones pueden volverse amarillas y morir. Frecuentemente las manchas se unen para formar lesiones grandes, de color café, de tamaño y
forma irregular con márgenes de color amarillo limón. En los pecíolos y tallos la infección
peude aparecer como cancros.

DISTRIBUTION: Af., A., Aust., Eur., N.A., C.A., S.A.

Pinus spp. (Ipomoea batatas and others) + Coleosporium ipomoeae (Schw.) Burrill =

ENGLISH   Rust, Needle Rust
FRENCH   Rouille des Aiguilles du Pin
SPANISH   Roya de las hojas del pino, Roya de la acícula
PORTUGUESE
ITALIAN   Ruggine delle foglie del pino
RUSSIAN
SCANDINAVIAN
GERMAN   Kiefernnadelblasenrost
DUTCH
HUNGARIAN
BULGARIAN
TURKISH
ARABIC
PERSIAN
BURMESE
THAI
NEPALI
HINDI
VIETNAMESE
CHINESE
JAPANESE

This disease develops in the spring as cream- to orange-colored, bag-like pustules on the needles. Needle rust may cause defoliation and stunt young pine trees.

Iste maladia se forma in le primavera como pustulas similar a saccos, del color de crema o de orange, in le agulias. (Fer)rugine del agulias pote causar defoliation e impedir le crescimento del juvene pinos.

Cette maladie se manifeste sur les aiguilles au printemps, sous forme de pustules de couleur crème à orange, en forme de sachets. La rouille des aiguilles peut causer la défoliation et le rabougrissement des jeunes pins.

Esta enfermedad se desarrolla en la primavera como pústulas en forma de bolsa, de color crema a naranja, en las acículas del pino. La roya de las acículas puede causar defoliación y achaparramiento de los árboles jóvenes de pino.

DISTRIBUTION: Af., A., N.A., C.A., S.A.

192

Pinus strobus (Ribes grossularia) + Cronartium ribicola J. C. Fisch. =

ENGLISH   Blister Rust
FRENCH   Rouille Vésiculeuse, Rouille du Pin Weymouth
SPANISH   Roya de ampolla
PORTUGUESE
ITALIAN   Ruggine del semprevivi
RUSSIAN
SCANDINAVIAN   Weymouthsfyrrens blærerust, Weymouthtallens törskaterost
GERMAN   Weymouthskiefernblasenrost
DUTCH   Roest van Ribessoorten
HUNGARIAN
BULGARIAN
TURKISH
ARABIC
PERSIAN
BURMESE
THAI
NEPALI
HINDI
VIETNAMESE
CHINESE
JAPANESE   Hassinsabi-byð   発疹さび病

This is a stem disease found on white pine. It grows from the needles into the bark of twigs and branches where it produces swollen, oval cankers. These cankers enlarge, girdle and finally kill infected stems. In late spring, orange to yellow fruiting bodies appear on breaks in the diseased bark. Small orange to yellow pustules are produced on the leaves of the shrubs during the summer.

Isto es un maladia del trunco del pino de Weymouth. Illos se extende ab le agulias in le cortice del ramettos e del ramos, ubi illo produce tumide canceres oval. Iste canceres cresce usque a imbraciar e destruer truncos infectate. Tarde in le primavera, structuras sporal orange o jalne se manifesta in vulneres del cortice morbide. In le estate, minute pustulas orange o jalne se forma in le folios del arbustos.

C'est une maladie de tronc qui s'attaque au pin blanc. La rouille vésiculeuse croît depuis les aiguilles jusqu'à l'écorce des ramilles et des branches où elle produit des chancres boursouflés, de forme ovale. Ces chancres s'agrandissent et encerclent les tiges pour finalement les tuer. Tard au printemps, des fructifications orangées a jaunes apparaissent dans des fis sures de l'ecorce malade. De petites pustules orangées à jaunes se forment sur les feuilles des arbustes durant l'été.

Esta es una enfermedad del tallo que se encuentra en el pino blanco. Crece de las acículas hacia la corteza de ramillas y ramas, donde produce cancros ovalados e hinchados. Estos cancros crecen, rodean los tallos infectados y finalmente los matan. Tarde en la primavera aparecen cuerpos fructíferos de color anaranjado a amarillo, en rajaduras de la corteza enferma. En las hojas de los arbustos, durante el verano, aparecen pequeñas pústulas anaranjadas.

DISTRIBUTION: Af., Eur., N.A.

Pinus spp. + <u>Dothistroma</u> <u>pini</u> Hulbary (<u>Scirrhia</u> <u>pini</u> Funk & Parker) =

ENGLISH   Needle Blight, Brown Needle Spot
FRENCH
SPANISH   Tizón de la acícula; Mancha café de la acícula
PORTUGUESE   Queima das acículas
ITALIAN
RUSSIAN
SCANDINAVIAN
GERMAN
DUTCH
HUNGARIAN
BULGARIAN
TURKISH
ARABIC
PERSIAN
BURMESE
THAI
NEPALI
HINDI
VIETNAMESE
CHINESE
JAPANESE

This leaf disease appears in late summer as slightly swollen, dark spots or bands on 1-year-old needles. The part of a diseased needle from the swollen area to the tip turns light brown and dies. The swollen area remains a constant size throughout the fall and winter months and begins to enlarge in late winter. By the springtime they appear as dark brown to black, raised fruiting bodies. Affected trees show sparse foliage, since diseased needles drop prematurely.

Iste maladia foliar se manifesta tarde in le estate como legiermente tumide maculas o bandas obscur in le agulias que ha un anno. Le portion de un agulia morbide ab le area tumide usque al extremitate deveni brun clar e mori. Le area tumide resta le mesme durante le autumno e hiberno, ma tarde in le hiberno comencia crescer. In le primavera se manifesta elevate structuras sporal brun obscur o nigre. Arbores infectate ha sparse foliage perque le agulias morbide cade prematurmente.

Cette maladie des feuilles apparaît à la fin de l'été sous forme de taches ou bandes sombres légèrement renflées sur les aiguilles d'un an. La partie d'une aiguille malade qui s'étend de la plage renflée jusqu'au bout vire au brun clair et meurt. La partie renflée ne change pas de grosseur en automne ni en hiver, mais elle commence à grossir à la fin de l'hiver. Au printemps, on y voit apparaître des fructifications surélevées, brun foncé à noires. Comme les aiguilles malades tombent prématurément, les arbres atteints présentent un feuillage épars.

Esta enfermedad foliar aparece a finales del verano como manchas obscuras, un poco hinchadas o bandas en acículas de un año de edad. La parte de una acícula enferma desde el área hinchada hasta la punta se vuelve de color café claro y muere. El área hinchada permanece del mismo tamaño durante el otoño y el invierno y comienza a agrandarse a finales del invierno. Para la primavera, aparecen como cuerpos fructíferos de color café obscuro a negro y levantados. Los árboles atacados muestran follaje escaso, ya que por la enfermedad las acículas se caen prematuramente.

DISTRIBUTION: Af., A., Aust., Eur., N.A., S.A.

194

Pinus spp. + Lophodermium pinastri (Schrad. ex Fr.) Chev. =

ENGLISH   Needle Cast
FRENCH   Rouge, Maladie Rouge du Pin
SPANISH   Caída de las acículas
PORTUGUESE
ITALIAN   Arrossamento, Caduta delle foglie di pino
RUSSIAN
SCANDINAVIAN   Fyrrens spraekkesvamp, Tallskytte
GERMAN   Kiefernritzenschorf
DUTCH   Dennenschotziekte
HUNGARIAN
BULGARIAN
TURKISH
ARABIC
PERSIAN
BURMESE
THAI
NEPALI
HINDI
VIETNAMESE
CHINESE
JAPANESE   Hahurui-byô   葉ふるい病

This leaf disease produces reddish-brown spots or elongated areas on needles. Small, black, oval fruiting bodies appear on the affected needles. Affected needles may or may not fall off.

Iste maladia foliar produce maculas brun-rubiette o areas elongate in agulias. Minute structuras sporal nigre e oval se manifesta in le agulias afflicte. Agulias afflicte pote separar se e cader o non.

Cette maladie des feuilles produit des taches ou des plages allongées brun rougeâtre sur les aiguilles. De petites fructifications noires, ovales, apparaissent sur les aiguilles atteintes. Les aiguilles malades peuvent tomber ou non.

Esta enfermedad foliar produce manchas café-rojizas o áreas alargadas en las acículas. Pequeños cuerpos fructíferos de color negro, ovalados aparecen sobre las acículas infectadas. Las acículas afectadas pueden o no caerse.

DISTRIBUTION: Af., A., Aust., Eur., N.A., C.A.

<u>Pinus sylvestris</u> + <u>Polyporus schweinitzii</u> Fr. =

ENGLISH	Butt Rot
FRENCH	Carie Brune Cubique
SPANISH	Pudrición del tronco
PORTUGUESE	
ITALIAN	
RUSSIAN	
SCANDINAVIAN	
GERMAN	Kiefernstockfäule
DUTCH	
HUNGARIAN	
BULGARIAN	
TURKISH	
ARABIC	
PERSIAN	
BURMESE	
THAI	
NEPALI	
HINDI	
VIETNAMESE	
CHINESE	
JAPANESE	Husin-byð 腐心病

This fungus causes butt and root rot, weakening trees and making them easily blown down in high winds. Wood decayed by the fungus is yellowish to red-brown and tends to be brittle and easily crumbled. The fruiting bodies, when viewed from above, appear as velvety caps, reddish-brown in color with yellow-brown margins. The undersurface is green-brown and has large irregular pores.

Iste fungo produce putrefaction brun de radices e del pede del trunco, lassante arbores infirme e facilemente jectate al terra per ventos tempestuose. Ligno putrefacte per iste fungo es jalnette o brun-rubiette, e illos esfragile e facilemente pulverisate. Le structuras sporal, quando on los examina de supra, se manifesta como cappellos de villuto, de color brun-rubiette con margines brun-jalne. Le superfacie inferior es brun-verde e monstra grande poros irregular.

Ce champignon cause une pourriture du pied et de la racine, qui affaiblit l'arbre et l'expose à se faire renverser plus facilement par les grands vents. Le bois que le champignon a carié est jaunâtre à brun rouge, et devient plus fragile et facile à émietter. Les fructifications, vues de haut, apparaissent comme des bonnets de velours, brun rougeâtre, avec des marges brun jaune. Le dessous est brun vert et a de gros pores irréguliers.

Este hongo causa pudrición del tronco y de la raíz, debilitando los árboles y haciéndolos susceptibiles a que el viento fuerte los doble. La madera podrida por el hongo es de color amarilloso a café rojizo y tiende a ser quebradiza y fácilmente se hace migajas. Los cuerpos fructíferos, cuando son vistos desde arriba, aparecen como cápsulas aterciopeladas, de color café rojizo, con márgenes café-amarillosos. El envés es de color café-verdoso y tiene grandes poros irregulares.

DISTRIBUTION: A., Aust., Eur., N.A., C.A., S.A.

Pisum sativum + Aphanomyces euteiches Drechsler =

ENGLISH	Root Rot
FRENCH	Pourridié, Pourriture des Racines et de la Base de la Tige du Pois
SPANISH	Pudrición radicular
PORTUGUESE	
ITALIAN	Marciume radicale del pisello
RUSSIAN	гниль корней афаномицетная
SCANDINAVIAN	Visnesyke
GERMAN	Wurzelfäule der Erbse
DUTCH	
HUNGARIAN	
BULGARIAN	
TURKISH	
ARABIC	
PERSIAN	
BURMESE	
THAI	Rak-Nao รากเน่า
NEPALI	Kerau Ko jaro-kuhine rog केराउको जरो कुहिने रोग
HINDI	
VIETNAMESE	
CHINESE	Wuan dòu gen fǔ bìng 豌豆根腐病
JAPANESE	Negusare-byŏ 根腐病

The first symptoms of this disease may be water-soaked lesions which turn grayish-brown or black; however, some plants may bear reddish streaks on the main tap root or on the stem near the groundline. Similar symptoms may be seen on the smaller rootlets, progressing throughout and destroying the entire root system. In some instances the infection may extend a short distance above the soil line. Above-ground symptoms are stunting, yellowing or wilting of foliage.

Le prime symptomas de iste maladia a vices es lesiones aquose le quales deveni brun-gris o nigre; nonobstante, alicun plantas a vices monstra strias rubiette in le radice principal vertical o in le pedunculo presso le terra. Symptomas similar se manifesta in le radiculas plus minute, extendente se in omne partes e destruente tote le systema radical. A vices le infection pote extender se un pauc super le terra. Symptomas epigee es obstruction de crescimento, jalnessa, e marcescentia de foliage.

Les premiers symptômes de cette maladie peuvent être des lésions délavées qui deviennent brun grisâtre ou noires; certaines plantes peuvent toutefois porter des stries rougeâtres sur la racine principale (pivotante) ou sur la tige au niveau du sol. On peut observer sur les petites radicelles des symptômes semblables, qui accusent des progrès rapides jusqu'à la destruction complète du système radicularie. Dans certains cas, l'infection peut parcourir une courte distance au-dessus du niveau du sol. Les symptômes au-dessus du sol sont le rabougrissement, le jaunissement ou le flétrissement du feuillage.

Los primeros síntomas de esta enfermedad pueden ser lesiones empapadas de agua, las cuáles se vuelven de color café-grisáceo o negro; sin embargo, algunas plantas pueden tener rayas rojizas en la raíz principal o en el tallo cerca de la superficie del suelo. Síntomas similares pueden ser vistos en las raicillas más pequeñas, progresando hacia dentro y destruyendo todo el sistema radicular. En algunos casos la infección puede extenderse a corta distancia arriba de la superficie del suelo. Los síntomas aéreos son achaparramiento, amarillamiento o marchitez del follaje.
DISTRIBUTION: Aust., Eur., N.A.

Pisum sativum (Vicia faba, Medicago sativa and others) + Ascochyta pisi Lib. =

ENGLISH	Lead Spot, Pod Spot
FRENCH	Ascochytose, Anthracnose du pois
SPANISH	Anthracnosis, Ascochyta, Mancha de las hojas y de las vaines, Quemazón
PORTUGUESE	
ITALIAN	Antracnosi del pisello, Seccume del pisello
RUSSIAN	аскохитоз
SCANDINAVIAN	Aertesyge, Ärtfläcksjuka
GERMAN	Brennflecken, Keimlings-Fusskrankheit,Ascochyta-Brennfleckenkrankheit
DUTCH	Vlekkenziekte, Lichte Vlekkenziekte op Erwt
HUNGARIAN	Aszkohitás levél-, szár-és hüvelyfoltosság
BULGARIAN	аскохитоза
TURKISH	
ARABIC	
PERSIAN	
BURMESE	Ywet-hnit-thee-pyauk roga အက်စကိုဒိုက်တ ၁ ပိုက်ဆိုင်
THAI	
NEPALI	
HINDI	
VIETNAMESE	
CHINESE	Wuan dòu hor ban bìng 豌豆褐斑病
JAPANESE	Kappan-byô 褐斑病

The symptoms of this disease can be found on all aboveground parts of the plant. On the pods and stems, the diseased areas are sunken and tan with dark margins. Lesions on the leaves are similar but are not sunken. The lesions often contain minute black fruiting bodies.

Le symptomas de iste maladia se manifesta in tote le partes epigee del planta. In le siliquas e in le pedunculos le areas morbide es depresse e bronzate con margines obscur. Lesiones in le folios es similar ma non depresse. Le lesiones sovente contine minute structuras sporal nigre.

Les symptômes de cette maladie apparaissent sur toutes les parties aériennes de la plante. Sur les gousses et les tiges, les parties malades sont déprimées et couleur du tan avec des marges sombres. Les lésions contiennent souvent de minuscules fructifications noires.

Los síntomas de esta enfermedad pueden encontrarse en todas las partes aéreas de la planta. En las vainas y tallos las áreas enfermas están hundidas, de color crema con márgenes obscuros. Las lesiones en las hojas son similares pero no están hundidas. Las lesiones frecuentemente contienen cuerpos fructíferos diminutos.

DISTRIBUTION: Af., A., Aust., Eur., N.A., C.A., S.A.

P isum sativum (other Leguminosae) + Mycosphaerclla pinodes (Berk. & Blox.) Vestergr. =

ENGLISH	Blight, Foot Rot
FRENCH	Brûlure Ascochytique, Taches Foliaires, Mycosphaerellose
SPANISH	Tizón, Podredumbre de pié
PORTUGUESE	
ITALIAN	Antracnosi, Seccume, Nebbia del pisello
RUSSIAN	**пятнистость микосфереллезная, микосфереллез**
SCANDINAVIAN	
GERMAN	Brennfleckenkrankheit
DUTCH	Vlekkenziekte, Voetziekte
HUNGARIAN	Aszkohitás levél-, szár-és hüvelyfoltosság
BULGARIAN	
TURKISH	
ARABIC	
PERSIAN	
BURMESE	
THAI	
NEPALI	
HINDI	
VIETNAMESE	
CHINESE	
JAPANESE	Katumon-byô 褐紋病

Symptoms appear on all above-ground parts of the plant and on the root system. Lesions on the above-ground parts begin as small, purplish spots which enlarge, become more or less zonate and dark brown without a definite margin; they may be circular or irregular in shape with dark centers. Infection spreads from the petiole to the stem, causing girdling lesions; flowers become spotted and pods are poorly filled. Post- and pre-emergence damping-off, death or dwarfing of older plants and discoloration and shrinkage of seed may occur.

Symptomas se manifesta in tote le partes epigee del planta e in le systema radical. Lesiones in le partes epigee comencia como minute maculas purpuree le quales cresce, deveni plus o minus zonate e brun obscur sin un margine definite. Illos pote esser de contorno circular o irregular con centros obscur. Infection se extende ab le petiolo al pedunculo, producente lesiones imbraciante. Flores es maculate e siliquas non es complete. Putrefaction o ante o post emergentia, morte o obstruction de crescimento de plantas plus vetule, e discoloration e contraction del granos a vices occurre.

Les symptômes apparaissent sur toutes les parties aériennes et sur le système radiculaire de la plante. Les lésions sur les parties aériennes sont d'abord de petites taches pourprées, qui s'agrandissent, deviennent plus ou moins ceinturées et brun foncé sans marge définie; elles peuvent être circulaires ou de forme irrégulière avec un centre sombre. L'infection se propage du pétiole à la tige, ce qui produit des lésions encerclantes; les fleurs deviennent tachées et les gousses sont mal remplies. Il peut y avoir fonte de préémergence ou de post-émergence, mort ou nanisme des plantes adultes, altération de la couleur et ratatinement des graines.

Los síntomas aparecen en todas las partes aéreas de la planta y en el sistema radicular. Las lesiones en las partes aéreas comienzan como manchas pequeñas, purpúreas, las cuáles crecen y se vuelven más o menos zonadas y de color café obscuro, sin un margen definido; pueden ser circulares o irregulares en tamaño, con centros obscuros. La infección se disemina del pecíolo hacia el tallo, causando infecciones que rodean; las flores se manchan y las vainas se llenan pobremente. Puede ocurrir mal del talluelo post y pre-emergente, muerte o achaparramiento de las plantas mayores y decoloración y encogimiento de la semilla.

DISTRIBUTION: Af., A., Aust., Eur., N.A., C.A., S.A.

Pisum sativum (other Leguminosae) + Pseudomonas pisi Sackett =

ENGLISH	Bacterial Blight
FRENCH	Brûlure Bactérienne, Bactériose du pois
SPANISH	Tizón bacteriano, Quemazón bacterial
PORTUGUESE	
ITALIAN	Avvizzimento dei piselli
RUSSIAN	ожог бактериальный стеблой и бобов, пятнистость бактериальная
SCANDINAVIAN	
GERMAN	Stengelbrand der Erbse, Erbsenstengelbrand, Hülsenflecken
DUTCH	
HUNGARIAN	
BULGARIAN	бактериоза
TURKISH	
ARABIC	
PERSIAN	
BURMESE	
THAI	
NEPALI	
HINDI	
VIETNAMESE	
CHINESE	
JAPANESE	Turugare-saikin-byô つる枯細菌病

The first symptoms are water-soaked lesions on the pods, stems or leaves. These often enlarge and a creamy-colored slimy ooze may be seen on the surface of the lesions. The infected stem and pod tissue become an olive-green to olive-brown color. The diseased areas of the leaves turn brown and become papery in texture.

Le prime symptomas es lesiones aquose in le siliquas, pedunculos, o folios. Istos sovente cresce e on vide un exsudato viscose del color de crema al superfacie del lesiones. Le texito infectate del pedunculo e del siliquas deveni verde o brun como olivas. Le areas morbide del folios deveni brun e de textura papyracee.

Les premiers symptômes sont des lésiones délavées sur les gousses, les tiges et les feuilles. Ces lésions s'agrandissent souvent, et l'on peut voir, à leur surface, un exsudat visqueux de couleur crème. Les tissus infectés de la tige et des gousses se colorent en vert olive ou brun olive. Les parties malades des feuilles brunissent et prennent la consistance du papier.

Los primeros síntomas aparecen como lesiones empapadas de agua en las vainas, tallos u hojas. Estas frecuentemente se agrandan y una exudación de color cremoso peude verse en la superficie de las lesiones. El tejido infectado de tallos y vainas se vuelve de color verde olivo a verde cafesoso. Las áreas enfermas de las hojas se vuelven de color café y de textura como papel.

DISTRIBUTION: Af., A., Aust., Eur., N.A., C.A., S.A.

200

Pisum sativum + Septoria pisi West =

ENGLISH	Leaf Blotch
FRENCH	Tache Septorienne, Taches Foliaires, Septoriose
SPANISH	Mancha foliar, Mancha de la hoja
PORTUGUESE	
ITALIAN	
RUSSIAN	септориоз
SCANDINAVIAN	
GERMAN	Septoria-Blattfleckenkrankheit
DUTCH	
HUNGARIAN	
BULGARIAN	
TURKISH	
ARABIC	
PERSIAN	
BURMESE	
THAI	
NEPALI	
HINDI	
VIETNAMESE	
CHINESE	
JAPANESE	Kokuhan-byô　黒斑病

This disease causes large, effuse, pale tan spots marked by minute black fruiting bodies. The infection usually begins at the edge of the leaf as a yellowish discoloration that spreads until the entire leaflet is affected. From the leaflet the infection passes to the nodes of the stem which become discolored and shrunken. A few fruiting bodies may be seen near the base of the stem. Young plants are often killed.

Iste maladia produce grande maculas diffuse bronzate clar, marcate per minute structuras sporal nigre. Le infection usualmente comencia al margine del folio como un discoloration jalnette le qual se extende usque a afficer tote le folietto. Ab le folio le infection se extende al nodos del pedunculo, le quales suffre discoloration e contraction. Alicun structuras sporal es visibile presso le pede del pedunculo. Iste maladia frequentemente destrue plantas juvene.

Cette maladie cause de grandes taches débordantes, chamois pâle, parsemées de minuscules fructifications noires. L'infection débute généralement au bord de la feuille sous forme d'une décoloration jaunâtre qui s'étend jusqu'à ce que toute la foliole soit affectée. De la foliole, l'infection va aux noeuds de la tige, qui perdent leur couleur et se ratatinent. On peut voir quelques fructifications près de la base de la tige. Les jeunes plantes sont souvent détruites.

Esta enfermedad causa manchas grandes, esparcidas, de color café claro, marcadas por cuerpos fructíferos diminutos. La infección generalmente comienza a la orilla de la hoja como una decoloración amarillosa, que se disemina hasta que toda la hoja se ve afectada. De la hojilla la infección pasa hacia los nudos del tallo los cuáles se decoloran y se encogen. Pueden verse algunos pocos cuerpos fructíferos cerca de la base del tallo. La plantas jóvenes general-mente mueren.

DISTRIBUTION: Aust., Eur., N.A.

Pisum sativum (other Leguminosae) + Uromyces pisi (DC.) Wint. (Uromyces pisi-sativi (Pers.)
Liro) =

ENGLISH	Rust	
FRENCH	Rouille du Pois	
SPANISH	Roya	
PORTUGUESE		
ITALIAN	Ruggine del pisello	
RUSSIAN	ржавчина	
SCANDINAVIAN		
GERMAN	Erbsenrost	
DUTCH	Roest, Erwtenroest	
HUNGARIAN		
BULGARIAN	ръжда	
TURKISH	Bezelye pası	
ARABIC	Sadaa	سدأ
PERSIAN		
BURMESE		
THAI		
NEPALI	Kerau Ko sindure rog	केराउको सिन्दुरे रोग
HINDI	Ratua	रतुवा
VIETNAMESE		
CHINESE	Wuan dòu shìou bìng	豌豆銹病
JAPANESE	Sabi-byô	さび病

Rust pustules appear on all aboveground parts of the plant, being most numerous on the underside of the leaves. Infection is first evident as minute, almost white, slightly raised pustules
which later become a distinct, reddish-brown color. When leaves become thoroughly infected,
they shrivel and fall off. When infection is severe, stems become defoliated.

Pustulas de (fer)rugine se manifesta in tote le partes epigee del planta, ma illos es le plus
numerose in le superfaacie inferior del folios. Infection se manifesta al initio como minute pustulas quasi blanc e un pauc elevate, le quales plus tarde ha un color distinctive brun-rubiette.
Quando le folios es totalmente infectate, illos se crispa e cade. Si le infection es sever, defoliation del pedunculos occurre.

Des pustules de rouille apparaissent sur toutes les parties aériennes de la plante, mais en plus
grand nombre sur la face inférieure des feuilles. L'infection apparaît d'abord sous forme de
pustules minuscules, presque blanches, légèrement proéminentes, qui deviennent plus tard
d'un brun rougeâtre distinctif. Les feuilles tout à fait infectées se ratatinent et tombent.
Dans les cas graves, toutes les feuilles tombent.

Pústulas de roya aparecen en todas las partes aéreas de la planta, siendo más numerosas
en el envés de las hojas. La infección se ve primero como pústulas diminutas, casi blancas,
un poco levantadas, que luego se vuelven de un distinctivo color café rojizo. Cuando las hojas
están bastante infectadas, se arrugan y caen. Cuando la infección es severa, los tallos se de-
folian.

DISTRIBUTION: Af., A., Eur., S.A.

Prunus persica + Cercosporella persica Sacc. (Mycosphaerella pruni-persicae Deighton) =

ENGLISH   Leaf Spot

FRENCH   Taches Foliaires, Cercosporellose

SPANISH   Mancha foliar

PORTUGUESE

ITALIAN

RUSSIAN   пятнистость церкоспореллезная, церкоспореллез

SCANDINAVIAN

GERMAN   Cercosporella-Blattfleckenkrankheit

DUTCH

HUNGARIAN

BULGARIAN

TURKISH

ARABIC

PERSIAN

BURMESE

THAI

NEPALI

HINDI

VIETNAMESE

CHINESE

JAPANESE

This disease usually occurs in shaded localities, especially on trees of dense foliage. The first symptoms are pale yellowish leaf spots. These show on the underside, a delicate frostlike appearance due to the growth of white fungus. With severe infection the leaves drop prematurely.

Iste maladia occurre usualmente in locos umbrose, specialmente in arbores con foliage dense. Le prime symptomas es maculas foliar pallide e jalnette. In le superfacie inferior se manifesta un strato delicate como pruina a causa del crescimento de un fungo blanc. In infection sever le folios cade prematurmente.

Cette maladie se manifeste ordinairement dan les endroits ombrageux, particulièrement sur les arbres à feuillage dense. Les premiers symptômes sont des taches jaunâtre pâle. A l'envers des feuilles, ces taches donnent l'impression d'un léger frimas produit par le mycélium blanc du pathogène. Dans les cas graves, les feuilles tombent prématurément.

Esta enfermedad generalmente aparece en lugares sombreados, especialmente en árboles de follaje denso. Los primeros síntomas son manchas amàrillo pálido en las hojas. Estas mues-tran en el envés de las hojas una apariencia de escarcha debido al crecimiento del hongo blanco. Con infección severa, las hojas se caen prematuramente.

DISTRIBUTION: A., Eur., N.A., S.A.

Prunus cerasus + Coccomyces hiemalis Higgins =

ENGLISH   Leaf Blight, Shot Hole
FRENCH   Criblure
SPANISH   Tizón foliar, Agujero de bala
PORTUGUESE
ITALIAN
RUSSIAN
SCANDINAVIAN
GERMAN
DUTCH
HUNGARIAN
BULGARIAN
TURKISH
ARABIC
PERSIAN
BURMESE
THAI
NEPALI
HINDI
VIETNAMESE
CHINESE
JAPANESE

The disease lesions may be formed on leaves and fruit. These appear as minute purple spots on the upper surface of the leaves, especially on young seedlings. These spots enlarge slightly, but remain small when infection is heavy and the spots numerous. The outline of the lesions is usually not well-defined in its early stages. Later the area in the center of the spot may die and fall out, giving a shot-hole effect. Many of the leaves turn yellow before falling. Complete defoliation frequently occurs in nursery stock or young trees in the orchard.

Le lesiones de iste maladia a vices se forma in folios e in fructos. Illos se manifesta como minute maculas purpuree in le superfacie superior del folios, specialmente in plantulas juvene. Iste maculas cresce un pauc, ma resta minute si le infection es sever e le maculas es numerose. Le contorno del lesion non es usualmente ben definite in le prime stadios. Plus tarde, le area in le centro del macula pote morir e cader, lassante perforationes. Multe folios deveni jalne ante que illos cade. Defoliation total occurre frequentemente in le seminario o in arbores juvene in le verdiero.

Les lésiones peuvent se former sur les feuilles et le fruit. Elles apparaisent sous forme de minuscules taches pourprées sur la face supérieure des feuilles, particulièrement sur les semis. Ces taches s'accroissent un peu, mais restent petites si l'infection est grave et si leur nombre est élevé. Au début de l'infection, les contours de la lésion sont généralement mal définis. Plus tard, le centre de la tache peut mourir et tomber en laissant comme un trou de balle. Beaucoup de feuilles jaunissent avant de tomber. La défoliation totale des jeunes arbres se produit souvent dans les pépinières et dans les vergers.

Las lesiones de esta enfermedad peuden formarse en hojas y frutas. Estas aparecen como manchas diminutas de color púrpura en el haz de las hojas, especialmente en plántulas pequeñas. Estas manchas crecen un poco, pero permanecen pequeñas cuando la infección es fuerte y las manchas son numerosas. La forma de la lesión generalmente no se ve bien definida al principio. Luego, el área en el centro de la mancha puede morir y caerse, dando un efecto de agujero de bala. Muchas de las hojas se vuelven de color amarillo antes de caer. Frecuentemente ocurre una defoliación completa en el almaciguero o en árboles jóvenes en el campo.

DISTRIBUTION: Af., A. Aust., Eur., N.A.

Prunus spp. + Coryneum carpophilum (Lév.) Jauch (Cladosporium carpophilum Theum.) =

ENGLISH   Scab
FRENCH   Tavelure
SPANISH   Viruela, Viruela holandesa, Mal de munición, Sarna, Gomosis, Roña del fruto
PORTUGUESE   Sarna, Pinta preta
ITALIAN   Scabbia, Vaiolo del pesco
RUSSIAN   парша
SCANDINAVIAN
GERMAN   Pfirsichschorf
DUTCH   Perzikschurft
HUNGARIAN
BULGARIAN   струпясване, сачмянка
TURKISH   Şeftali karalekesi
ARABIC
PERSIAN
BURMESE
THAI
NEPALI
HINDI
VIETNAMESE
CHINESE
JAPANESE   Kurohosi-byô   黒星病

On the fruit, the first lesions to appear are small, circular, greenish-colored patches. Soon the lesions become yellow and then brown and have a corky appearance. With rapid growth of the fruit, the lesions cause these to crack open. The lesions on the twigs are similar to those on the fruit. The symptoms on the leaves are not conspicuous. They are greenish-brown and usually occur on the midribs or veins on the lower surface.

In le fructo, le prime lesiones que se manifesta es minute, circular areas quasi verde. Le lesiones rapidemente deveni jalne e alora brun e similar a corco. Si le fructos cresce rapidemente, le lesiones los finde. Le lesiones in le ramettos es similar a illos in le fructos. Le symptomas in le folios non es ramarcabile. Illos es brun-verde e usualmente occurre al nervatura central o al venas del superfacie inferior.

Les premièrs lésions sur le fruit sont de petites plaques rondes, verdâtres. Les lésions ne tardent pas à jaunir, puis à brunir, et ressemblent à du liège. A la faveur d'une croissance rapide, les lésions font éclater le fruit. Les lésions sur les ramilles sont semblables. Sur les feuilles, les symptômes sont peu apparents. C'est généralement à la face inférieure, sur la nervure médiane ou sur des nervures secondaires, qu'on peut déceler des lésions brun verdâtre.

En la fruta, las primeras lesiones aparecen como parches pequeños, circulares, de color verdoso. Pronto las lesiones se vuelven de color amarillo y luego café y tienen apariencia de corcho. Con un crecimiento rápido de la fruta, las lesiones hacen que ésta se parta. Las lesiones en las ramillas son similares a las de la fruta. Los síntomas en las hojas no son conspicuos. Estos son café-verdoso y generalmente aparecen en las venas centrales o en las venas del envés.

DISTRIBUTION: Af., A., Aust., Eur., N.A., S.A.

Prunus cerasus + Fusicladium cerasi (Rab.) Sacc. (Venturia cerasi Aderh.) =

ENGLISH	Scab
FRENCH	Tavelure du Cerisier
SPANISH	Roña
PORTUGUESE	
ITALIAN	Ticchiolatura delle ciliegie
RUSSIAN	
SCANDINAVIAN	Kirsebærskurv, Körsbärsskorv
GERMAN	Kirschenschorf, Pfirsichschorf, Schorf
DUTCH	Kersenschurft
HUNGARIAN	
BULGARIAN	
TURKISH	Kiraz kara lekesi
ARABIC	
PERSIAN	
BURMESE	
THAI	
NEPALI	
HINDI	
VIETNAMESE	
CHINESE	
JAPANESE	

Severe infection of this disease results in misshapen fruit which frequently become cracked open. The fungus produces numerous small, oval, brown, slightly raised lesions on twigs that may retard their growth. Leaf symptoms are inconspicuous.

In iste maladia infection sever produce fructos deforme le quales frequentemente se finde. Le fungo produce numerose lesiones minute, oval, brun, legiermente elevate in ramettos e pote impedir lor crescimento. Symptomas in le folios non es remarcabile.

Une infection grave peut déformer les fruits et souvent les faire éclater. Le champignon produit de nombreuses petites lésions ovales, brunes, légèrement surélevées sur les ramilles dont la croissance peut être ralentie. Les symptômes sur les feuilles sont peu apparents.

La infección severa de esta enfermedad resulta en frutos mal formados, los cuáles frecuente-mente se rajan. El hongo produce numerosas lesiones pequeñas, ovaladas, café, un poco levan-tadas en las ramillas que pueden retardar su crecimiento. Los síntomas foliares son incon-spicuos.

DISTRIBUTION: A., Eur., N.A., S.A.

Prunus spp. + Monilinia fructicola (Wint.) Honey =

ENGLISH	Brown Rot, Twig Canker
FRENCH	Pourriture Brune des Fruits, Rot Brun, Moniliose
SPANISH	Podredumbre morena, Pudrición café, Cancro de la ramilla, Podredumbre café
PORTUGUESE	Podridão parda
ITALIAN	
RUSSIAN	гниль плодов монилиальная ("концентрическая"), монилиоз
SCANDINAVIAN	
GERMAN	Monilia-Fruchtfäule, Polsterschimmel
DUTCH	
HUNGARIAN	
BULGARIAN	
TURKISH	Meyve monilyasi
ARABIC	
PERSIAN	
BURMESE	
THAI	
NEPALI	
HINDI	
VIETNAMESE	
CHINESE	
JAPANESE	

This fungus attacks the host at any stage of growth. It causes blossom blight, leaf and twig blight or a brown rot of the fruit. The disease is most serious on the fruit where it first appears as small, circular, brown spots. During moist weather the spots may enlarge rapidly. The diseased, mummified fruit may drop or remain hanging. On twigs and branches, the fungus causes cankers. Occasionally, the twigs are killed beyond the point of infection.

Iste fungo attacca su planta-hospite per tote le stadios de crescimento. Illo produce necrosis del flores, folios, e ramettos o un putrefaction brun del fructo. Le morbo es le plus damnose in le fructo ubi illo se manifesta al initio como minute maculas circular brun. In tempore humide le maculas pote crescer rapidemente. Le morbide fructo mumificate pote cader o restar al arbore. In ramettos e ramos le fungo produce canceres. A vices, le ramettos mori ultra le extension del infection initial.

Ce champignon attaque son hôte à n'importe quel moment de sa croissance. Il cause la brûlure des fleurs, des feuilles et des ramilles, et une pourriture brune du fruit. La maladie est surtout grave sur le fruit où elle se montre d'abord sous forme de petites taches brunes, circulaires. Par temps humide, les taches peuvent s'accroître rapidement. Le fruit malade momifié peut tomber ou rester pendant. Sur les branches et les ramilles, le champignon cause des chancres. Parfois, les ramilles meurent au-dessus du point d'infection.

Este hongo ataca su hospedera en cualquier época de crecimiento. Causa tizón de las flores, tizón de la hoja y ramilla o una podredumbre café del fruto. Esta enfermedad e más seria en la fruta donde aparece primero como manchas pequeñas, circulares, de color café. Durante tiempo húmedo las manchas crecen rápidamente. La fruta enferma momificada puede caerse o quedar colgada. El hongo causa cancros en las ramillas y ramas. Ocasionalmente, las ramillas mueren más allá del punto de infección.

DISTRIBUTION: Aust., N.A., C.A., S.A.

Prunus cerasus (Prunus domestica) + Podosphaera oxycanthae (DC.) De Bary (Podosphaera tridactyla (Wallroth) de Bary) =

ENGLISH	Powdery Mildew
FRENCH	Oïdium
SPANISH	Oidium, Mildiú polvoriento
PORTUGUESE	
ITALIAN	Nebbia del biancospino
RUSSIAN	мучнистая роса
SCANDINAVIAN	
GERMAN	Mehltau
DUTCH	
HUNGARIAN	
BULGARIAN	
TURKISH	Erik küllemesi
ARABIC	Bayad Dakiky بياض دقيقى
PERSIAN	
BURMESE	
THAI	
NEPALI	
HINDI	
VIETNAMESE	
CHINESE	
JAPANESE	Udonko-byô うどんこ病

This disease produces symptoms on the leaves, young shoots and fruit. The young leaves may become entirely coated with a thick layer of the fungus, causing them to curl and become narrow as they expand. The white fungus powdery layer may extend over the entire terminal portion of the growing shoot. On the fruit, the disease first appears in the form of white round spots which increase in size until the whole surface of the fruit is involved. The fruit takes on a pinkish color, which later turns to a dark brown. Sometimes fruit cracking occurs to render it worthless.

Iste maladia produce symptomas in le folios, ramos juvene, e fructos. Le folios juvene pote devenir totalmente coperte de un strato dense del fungo, facente los crispar se e devenir stricte quando illos cresce. Le strato pulverose del fungo blanc pote extender se super tote le portion terminal del ramo crescente. In le fructos le morbo se manifesta al initio como maculas rotunde blanc, le quales cresce in dimension usque a involver tote le superfacie del fructo. Le fructo deveni rosee e plus tarde brun obscur. A vices le fructo se finde e perde su valor economic.

Les symptômes de cette maladie se montrent sur les feuilles, les jeunes pousses et le fruit. Les jeunes feuilles peuvent être couvertes entièrement d'une épaisse couche du champignon qui les fait s'enrouler et se rétrécir à mesure qu'elles croissent. La couche poudreuse de champignon blanc peut gagner toute le partie supérieure de la pousse. Sur le fruit, la maladie paraît d'abord sous forme de taches blanches, circulaires, qui s'agrandissent jusqu'à recouvrir tout le fruit. Le fruit devient rosâtre, puis brun foncé. Parfois le fruit se fendille et perd toute valeur.

Esta enfermedad produce síntomas en las hojas, brotes jóvenes y frutos. Las hojas jóvenes peuden volverse enteramente cubiertas con una gruesa capa del hongo, causando que éstas se encrespen y se vuelvan delgadas a medida que se expanden. La capa polvorienta y blanca del hongo peude extenderse sobre toda la porción terminal del brote en crecimiento. En el fruto la enfermedad aparece primero en forma de manchas blancas y redondas, las cuáles aumentan de tamaño hasta que toda la superficie del fruto está cubierta. El fruto adquiere un color rosadoso, el cuál más tarde se vuelve café obscuro. Algunas veces el fruto se raja, inutilizándolo.

DISTRIBUTION: Af., A., N.A., S.A.

<u>Prunus</u> <u>persica</u> (<u>Fragaria</u> <u>chiloensis</u>, <u>Ipomoea</u> <u>batatas</u>) + <u>Rhizopus</u> <u>nigricans</u> Ehrenb. ex Fr. =

ENGLISH	Fruit Rot, Rhizopus Soft Rot
FRENCH	Moisissure Chevelue, Pourriture des Fruits
SPANISH	Podredumbre del fruto, Podredumbre blanda de Rhizopus
PORTUGUESE	
ITALIAN	
RUSSIAN	гниль плодов ризопусная
SCANDINAVIAN	
GERMAN	Rhizopus-Fruchtfäule
DUTCH	
HUNGARIAN	
BULGARIAN	протичане
TURKISH	Kara küf
ARABIC	
PERSIAN	
BURMESE	
THAI	
NEPALI	
HINDI	
VIETNAMESE	
CHINESE	
JAPANESE	Kurokabi-byô 黒かび病

This disease occurs in the field, but is especially serious on the fruit during transit and storage. The fungus spreads rapidly through the containers and often infection on a single fruit may spread to all within a few days. The most characteristic symptom is the black fungus which produces strands on the fruit resembling whiskers.

Iste maladia occurre in le campo ma es specialmente damnose al fructo durante transporation e immagasinage. Le fungo se extende rapidemente per le receptaculos, e sovente le infection de un sol fructo pote extender se a tote le fructos intra alicun dies. Le symptoma le plus characteristic es le fungo nigre le qual produce in le fructo cordas similar a barbas.

Cette maladie se déclare dans le champ, mais devient particulièrement grave pendant le transport et l'entreposage du fruit. Le champignon se propage rapidement dans les contenants, et, souvent, un suel fruit malade peut contaminer tous les autres en quelques jours. Le symptôme le plus caractéristique est le champignon noir qui produit sur le fruit des touffes de filaments qui évoquent des moustaches de chat.

Esta enfermedad aparece en el campo, pero es especialmente seria mientras la fruta está en tránsito y almacenada. El hongo se disemina rápidamente a través del empaqué y frecuentemente uns sola fruta infectada puede contaminar a las otras en unos pocos días. El síntoma más característico es el hongo negro que produce hebras en el fruto, semejando bigotes.

DISTRIBUTION: Af., A., Aust., Eur., N.A., C.A., S.A.

Prunus persica (Prunus spp., Pyrus communis and others) + Sclerotinia laxa Aderh. & Ruhl. (Sclerotinia fructicola (Wint.) Rehm.) =

ENGLISH   Blossom Wilt, Spur Blight, Wither Tip
FRENCH   Rot-brun des Arbres Fruitiers, Pourriture des Fruits, Moniliose, Brûlure
SPANISH   Podredumbre morena, Podredumbre del fruto, Marchitez del extremo, Tizón
PORTUGUESE
ITALIAN   Muffa delle albicocche
RUSSIAN   гниль плодов монилиальная ("рассеянная"), монилиоз
SCANDINAVIAN   Grå monilia
GERMAN   Sclerotinia-Braunfäule
DUTCH
HUNGARIAN
BULGARIAN
TURKISH   Çiçek monilyası
ARABIC
PERSIAN
BURMESE
THAI
NEPALI
HINDI
VIETNAMESE
CHINESE
JAPANESE   Haibosi-byô   灰星病

This disease originates in the orchard, but may cause damage at any time during the marketing process. In the orchard, it occurs on the leaves, blossoms, fruits, twigs and limbs. The disease on the fruit appears at first as small, circular, light brown spots, which under favorable conditions for infection may bring about complete decay in 24 hours. The spots at no stage are sunken and the flesh remains firm. In late stages the skin turns dark brown or black.

Iste maladia comencia in le verdiero ma pote affliger le fructo usque al mercato. In le verdiero illo occurre in le folios, flores, fructos, ramettos, e ramos. In le fructo, le morbo se manifesta al initio como minute maculas circular brun clar, le quales, sub conditiones favorabile pro infection, pote causar putrefaction total intra un sol die. Le maculas non es jammais depresse, e le pulpa resta firme. In stadios tarde le pelle deveni brun obscur o nigre.

Cette maladie commence dans le verger, mais peut avoir des effets nocifs à n'importe quel moment de la mise en vente. Dans le verger, ce sont les feuilles, les fleurs, les fruits, les ramilles et les branches qui sont atteints. La maladie sur le fruit apparaît d'abord sous forme de petites taches circulaires, brun clair, qui peuvent provoquer, dans des conditions favorables à l'infection, la pourriture complète en 24 heures. Les taches ne sont jamais déprimées, et la chair reste ferme. Vers les dernières phases, la peau devient brun foncé ou noire.

Esta enfermedad se origina en el huerto, pero puede causar daño en cualquier momento durante el proceso de mercadeo. En el huerto ataca las hojas, flores, frutos, ramillas y ramas. La enfermedad en el fruto aparece primero como pequeñas manchas circulares de color café, las que bajo condiciones favorables para la infección pueden causar una podredumbre completa en 24 horas. Las manchas nunca están hundidas y la carne permanece firme. En estado tardío la piel se vuelve de color café obscuro o negro.

DISTRIBUTION: Af., A., Eur., N.A., S.A.

210

Prunus persica + Sphaerotheca pannosa Wallr. ex Lév. =

ENGLISH	Powdery Mildew
FRENCH	Blanc, Oidium du Pêcher
SPANISH	Mildiú polvoriento, Oidium, Oidio, Blanco
PORTUGUESE	Oidio
ITALIAN	Nebbia oidio, Mal bianco del pesco
RUSSIAN	
SCANDINAVIAN	Rosenmeldug Ferskenmeldug, Rosmjöldagg
GERMAN	Echter Mehltau, Pfirsichmehltau
DUTCH	Witziekte van de Perzik, Meeldauw van Perzik
HUNGARIAN	
BULGARIAN	брашнеста мана
TURKISH	Külleme hastalığı
ARABIC	Bayad Dakiky بياض دقيقى
PERSIAN	Sēfēdàkê hàghēghēyê hōolōo سفیدک حقیقی هلو
BURMESE	
THAI	
NEPALI	
HINDI	Phaphundi फाफुन्दी
VIETNAMESE	
CHINESE	
JAPANESE	

This fungus attacks leaves, twigs and fruit, producing on all of them the characteristic irregular white patches of mold. Underneath these patches of the fruit, the skin takes on a brown or dead color, the underlying flesh is rather hard, water-soaked in appearance and usually extends at least half the distance to the center.

Iste fungo attacca folios, ramettos, e fructos, producente in illos le areas irregular e blanc characteristic de mucor. Sub iste areas in le fructo le pelle es brun o morte, le pulpa subjacente es assatis dur, aquose, e usualmente iste symptoma se extende al minus in le medie distantia verso le centro del fructo.

Ce champignon s'attaque aux feuilles, aux ramilles et aux fruits, produisant des plaques caractéristiques, irrégulières de moisissure blanche. Sous les plaques qui couvrent le fruit, la peau brunit comme à la mort, la chair reste plutôt ferme et prend un aspect délavé qui s'étend d'habitude au moins jusqu'à mi-chemin du centre.

El hongo ataca las hojas, ramillas y frutos, produciendo en todos ellos parches característicos de moho blanco. Bajo estos parches en el fruto, la piel adquiere un color café o muerto, la piel que está abajo es algo dura, de apariencia empapada de agua y que generalmente se extiende por lo menos la mitad de la distancia hacia el centro.

DISTRIBUTION: Af., A., Eur., N.A., S.A.

Prunus domestica (Malus pumila = Malus sylvestris and many hardwoods) + Stereum purpureum (Pers. ex Fr.) Fr. =

ENGLISH	Silver Leaf
FRENCH	Plomb, Maladie du Plomb des Arbres Fruitiers et d'ornement
SPANISH	Plateado, Roja de plata
PORTUGUESE	
ITALIAN	Mal del piomo del pesco
RUSSIAN	"млечный блеск"
SCANDINAVIAN	Sølvglans, Silverglans
GERMAN	Milchglanz, Bleiglanz der Olstbäume
DUTCH	Loodglans, Loodglansziekte
HUNGARIAN	
BULGARIAN	сребърен лист
TURKISH	Gümus hastalıgı
ARABIC	
PERSIAN	
BURMESE	
THAI	
NEPALI	
HINDI	
VIETNAMESE	
CHINESE	
JAPANESE	Ginyô-byô 銀葉病

The first evidence of this disease is on the foliage which shows a metallic luster in marked contrast to the normal green of healthy trees. A cross section of the limb of a diseased tree will show a dark discoloration of the heartwood where the fungus is growing. Death of the infected branches or of the entire tree results within a year or two and stunting of growth is often apparent before the tree dies. The fungus fruiting bodies may appear on the surface of cracks in the bark or be entirely absent.

Le prime signos de iste maladia occurre in le foliage, le qual ha un lustro metallic in contradistinction al verde normal de arbores salubre. In section transversal, le ramo de un arbore morbide va monstrar un anormal coloration obscur del duramen ubi se trova le fungo. Le morte del ramos infectate o de tote le arbore occurre intra un o duo annos, e impedimento de crescimento es sovente evidente ante que le arbore mori. Le structuras sporal del fungo pote manifestar se in le superfaice de vulneres in le cortice o non esser visibile.

Le premier indice de cette maladie paraît sur le feuillage qui présente un lustre métallique formant un contraste marqué avec le vert normal des arbres sains. Une coupe transversale d'une branche malade révèle une coloration foncée du coeur du bois ou pousse le champignon. La mort des branches infectées ou de l'arbre tout entier s'ensuit en un ou deux ans, et l'arrêt de croissance est souvent manifeste avant la mort de l'arbre. Les fructifications du champignon peuvent apparaître à la surface des crevasses dans l'écorce, ou faire complètement défaut.

La primera evidencia de esta enfermedad se ve en el follaje, el cuál muestra un brillo metálico en marcado contraste con el verde normal de los árboles sanos. Una sección cruzada de una rama de un árbol enfermo mostrará una decoloración obscura de la madera del corazón donde está creciendo el hongo. La muerte de las ramas infectadas o de todo el árbol resulta en uno o dos años y el achaparramiento es frecuentemente aparente antes de que el árbol muera. Los cuerpos fructíferos del hongo pueden aparecer en la superficie de las rajaduras en la corteza o peuden estar totalmente ausentes.

DISTRIBUTION: Af., A., Aust., Eur., N.A., S.A.

212

Prunus cerasus (Prunus spp.) + Taphrina cerasi (Fuckel) Sadeb. =

ENGLISH   Witches'-broom
FRENCH   Balai de Sorcière du Cerisier
SPANISH   Escoba de bruja
PORTUGUESE
ITALIAN   Scopazzi del ciliegio
RUSSIAN
SCANDINAVIAN   Heksefost paa Kirsebær
GERMAN   Hexenbesen, Kirschenhexenbesen
DUTCH   Heksenbezem der Kerseboom
HUNGARIAN
BULGARIAN   самодивски метли
TURKISH   Cadi süpürgesi
ARABIC
PERSIAN
BURMESE
THAI
NEPALI
HINDI
VIETNAMESE
CHINESE
JAPANESE   Tengusu-byô   てんぐす病

This disease causes a typical witches'-broom effect. The excessive branching and leaf production may affect a single branch or the whole tree. If the infection is severe, small fruit
results and if the disease occurs consecutively for a number of years, the trees die prematurely.

Iste maladia produce un effecto typic de 'scopa del maga'. Le excessive ramification e production de folios pote afficer un sol ramo o tote le arbore. Si le infection es sever, le resultato es fructos minute. Si le maladia infecta le arbore durante alicun annos consecutive, illo
mori prematurmente.

Cette maladie produit un balai de sorcière typique. La production excessive de ramilles et de
feuilles peut affecter une seule branche ou tout l'arbre. Si infection est grave, le fruit reste
petit, et si la maladie survient plusieurs années de suite, l'arbre meurt prématurément.

Esta enfermedad causa un típico efecto de escoba de bruja. La producción excesiva de ramas
y hojas peude afectar una sola rama o todo el árbol. Si la infección es severa, resultan frutos
pequeños y si la enfermedad aparece consecutivamente por varios años, los árboles mueren
prematuramente.

DISTRIBUTION: Af., A., Aust., Eur., N.A., S.A.

Prunus persica (P. amygdalus) + Taphrina deformans (Berk.) Tul. =

ENGLISH   Peach Leaf Curl, Leaf Curl, Leaf Blister
FRENCH   Cloque
SPANISH   Hoja crespa, Cloca, Torque, Enrulamiento de las hojas, Enroscamiento
PORTUGUESE   Crespeira
ITALIAN   Accartocciamento, Bolla delle foglie
RUSSIAN   курчавость листьев
SCANDINAVIAN   Ferskenblæresyge, Persikkrussjuka, Blærsyge
GERMAN   Kräuselkrankheit der Pfirsiche
DUTCH   Krul, Krulziekte der Perzik
HUNGARIAN
BULGARIAN   къдравост
TURKISH   Şeftali yaprak kıvırcıklığı
ARABIC   Tagaood Awrak   تجعد اورات
PERSIAN   Lâb shôtôryê hōolōo   لب شتری هلو
BURMESE
THAI
NEPALI   Arhu ko pat kopra   आरुको पात कोप्रा
HINDI
VIETNAMESE
CHINESE
JAPANESE   Syukuyô-byô   縮葉病

Curling, puckering, swelling and discoloration of affected leaves are typical symptoms of this disease. The color ranges from red to purple in very young leaves to yellow in older foliage. The symptoms are noticed as soon as the leaves emerge. After they become chlorotic, the leaves die and fall. Diseased trees are conspicuous for their discoloration and early defoliation. Terminal twigs and sometimes blossoms and young fruit are also affected. Older fruit will show discolored, irregular, wrinkled areas of the skin.

Inrolamento, crispation, inflation, e coloration anormal del folios afflicte es symptomas typic de iste maladia. Le color se varia ab rubie usque a purpuree in le folios multo juvene e es jalne in foliage plus vetule. Le symptomas es remarcate si tosto que le folios emerge. Post que illos deveni chlorotic, le folios mori e cade. Arbores morbide es conspicue a causa de lor coloration anormal e defoliation prematur. Ramettos terminal e a vices flores e fructos juvene es anque afflicte. Fructos plus vetule monstra areas discolorate, irregular, e corrugate del pelle.

L'enroulement, le gaufrage, le gonflement et la décoloration des feuilles atteintes sont les symptômes typiques de cette maladie. La couleur va du rouge au pourpre chez les feuilles très jeunes, au jaune chez les plus vieilles feuilles. Les symptômes apparaissent en même temps que les feuilles. Celles-ci devenues chlorotiques meurent et tombent. Les arbres malades se reconnaissent facilement à la décoloration et à la chute prématurée de leurs feuilles. Les ramilles terminales et parfois les fleurs et les jeunes fruits sont également atteints. La pelure des fruits plus avancés présente des plages décolorées, irrégulières et gaufrées.

Los síntomas típicos de esta enfermedad son enroscamiento, arrugamiento, hinchazón y decoloración de las hojas afectadas. El color varía de rojo a púrpura en las hojas más jóvenes, hasta amarillo en el follaje viejo. Los síntomas son notorios tan pronto como las hojas emergen. Después de volverse cloróticas, las hojas mueren y caen. Los árboles enfermos son conspicuos por su decoloración y temprana defoliación. Las ramillas terminales y algunas veces los capullos y frutos jóvenes se ven también afectados. Los frutos más viejos muestran áreas decoloradas, irregulares y arrugadas en la cáscara.

DISTRIBUTION: Af., A., Aust., Eur., N.A., C.A., S.A.

214

Prunus domestica + Taphrina pruni (Fuckel) Tul. (Exoascus pruni Fuckel) =

ENGLISH	Bladder Plum, Plum Pockets
FRENCH	Pochette, Maladie des Pochettes
SPANISH	Vesícula de ciruela, Bolsas de ciruela
PORTUGUESE	
ITALIAN	Bozzacchioni, Lebbia del susino
RUSSIAN	
SCANDINAVIAN	Blommepunge, Pungsjuka, Plommepung
GERMAN	Narrentaschenkrankheit, Taschenkrankheit der Pflaumen
DUTCH	Hongerpruimen
HUNGARIAN	
BULGARIAN	мехурки, рошкови, бебици
TURKISH	Ceb hastalığı
ARABIC	
PERSIAN	Ânbâne âloo انبانی آلو
BURMESE	
THAI	
NEPALI	Alubakhara kuhine rog आलु वखरा कुहिने रोग
HINDI	
VIETNAMESE	
CHINESE	
JAPANESE	Hukuromi-byô ふくろみ病

The first evidence of this disease appears on the fruit as small, whitish spots or blisters.
These blisters enlarge rapidly on the fruit as they develop and soon the entire fruit is involved.
The flesh becomes spongy and the young bud ceases development, turns brown and withers,
leaving a hollow cavity.

Le prime symptomas de iste maladia se manifesta in le fructo como minute maculas o pustulas
quasi blanc. Iste pustulas cresce rapidemente durante le developpamento del fructo usque a
involver tote le tempore in breve tempore. Le pulpa deveni spongiose e le button juvene non
se developpa plus, deveni brun, marcesce, e lassa un cavitate vacue.

Le premier indice de cette maladie paraît sous forme de petites taches ou d'ampoules blanchâtres. Ces ampoules s'agrandissent rapidement sur le fruit qu'elles ont bientôt fait d'envahir
tout entier. La chair devient spongieuse, et le jeune bourgeon cesse de se développer, brunit
et dépérit pour ne laisser qu'un trou.

La primer evidencia de esta enfermedad aparece en el fruto como pequeñas manchas o ampollas
blanquecinas. Estas ampollas crecen rápidamente en el fruto a medida que se desarrollan y
pronto todo el fruto se ve atacado. La carne se vuelve esponjosa y el capullo tierno cesa su
desarrollo, se vuelve de color café, y se marchita, dejando una cavidad hueca.

DISTRIBUTION: A., Eur.

Prunus spp. + Tranzchelia discolor (Fckl.) Tranz. & Litv. (Tranzchelia pruni-spinosae (Pers.) Diet.) =

ENGLISH	Rust
FRENCH	Rouille du Prunier
SPANISH	Roya, Polvillo,
PORTUGUESE	Ferrugem
ITALIAN	Ruggine del susino
RUSSIAN	ржавчина
SCANDINAVIAN	Blommerust, Plommonrost
GERMAN	Rost, Zwetschgenrost, Pflaumenrost, Zwetschenrost
DUTCH	Roest
HUNGARIAN	
BULGARIAN	ръжда
TURKISH	
ARABIC	
PERSIAN	Zánge áloo رنگ آلو
BURMESE	
THAI	
NEPALI	
HINDI	
VIETNAMESE	Rỉ lá
CHINESE	
JAPANESE	Kassabi-byð 褐さび病

Symptoms appear on leaves, fruit and twigs. On the leaves, the disease is first evident as pale yellowish spots on both surfaces. Later these spots become bright yellow. On the under-surface, rarely the uppersurface, the cinnamon-brown lesions appear. The fruit is rarely infected. When present, the affected areas appear first as water-soaked dark green spots. The spots then become sunken and are deeper green than the surrounding tissues. Later they turn to a deep yellow. On the twigs, the lesions break through the bark and the bark areas split open lengthwise.

Le symptomas se manifesta in le folios, fructos, e ramettos. In le folios, le maladia se presenta al initio como pallide maculas jalnette in tote le duo superfacies del folios. Plus tarde, iste maculas deveni jalne brillante. In le superfacie inferior, e rarmente in le superior, le lesiones brun como cannella se manifesta. Le fructo es rarmente afflicte. Si le infection es presente, le areas afflicte se manifesta al initio como aquose maculas verde obscur. Le maculas alora deveni depresse e es verde plus obscur que le texito circumjacente. Plus tarde, illos deveni jalne obscur. In le ramettos, le lesiones penetra le cortice le qual se finde in direction longitudinal.

Les symptômes apparaissent sur les feuilles, le fruit et les rameaux. Sur les feuilles, la maladie se manifeste d'abord sous forme de taches jaunâtre pâle sur les deux faces. Par la suite, les taches deviennent jaune vif. Des lésions brun cannelle se forment sur la face inférieure, rarement sur la face supérieure. Le fruit est rarement infecté. Lorsqu'il l'est, les parties atteintes se présentent d'abord sous forme de taches délavées vert foncé. Puis, les taches deviennent déprimées et d'un vert plus intense que celui des tissus voisins. Plus tard, ces taches deviennent jaune foncé. Sur les rameaux, les lésions percent l'écorce, et des portions de celle-ci se fendent sur la longueur.

Los síntomas aparecen en hojas, fruto y ramillas. En las hojas, la enfermedad se nota primero como manchas amarilloso claro en ambas superficies de las hojas. Luego estas manchas se vuelven de color amarillo brillante. En el envés, pero casi nunca en el haz, aparecen las lesiones de color café canela. La fruta casi nunca se ve afectada. Las áreas afectadas, cuando están presentes, aparecen primero como manchas grandes, de color verde obscuro y empapadas de agua. Luego las manchas se vuelven hundidas y son de un verde más intenso que el de los tejidos de alrededor. Más tarde se vuelven de un color amarillo intenso. En las ramillas las lesiones atraviesan la corteza y estas áreas de la corteza de rajan a lo largo.

DISTRIBUTION: Af., A., Aust., Eur., N.A., C.A., S.A.

216

Prunus persica (P. domesticus, Prunus spp.) + Xanthomonas pruni (Smith) Dowson =

ENGLISH   Bacterial Leaf Spot, Bacterial Canker
FRENCH   Tache Bactérienne, Bactériose, Taches Bactériennes Criblées
SPANISH   Cancrosis, Mancha foliar bacteriana, Mancha negra, Cancro bacterial
PORTUGUESE
ITALIAN   Batteriosi del susino
RUSSIAN   **пятнистость дырчатая бактериальная**
SCANDINAVIAN
GERMAN   Fleckenbakteriose
DUTCH
HUNGARIAN
BULGARIAN   **бактериална съчмянка**
TURKISH
ARABIC
PERSIAN
BURMESE
THAI
NEPALI
HINDI
VIETNAMESE
CHINESE
JAPANESE   Senkô-saikin-byô せん孔細菌病

The first foliage symptom occurs on the underside of the leaf as a small interveinal, circular or irregular, light-colored area. As the disease develops, the uppersurface is invaded and the spots enlarge, becoming angular and purple, brown or black. The necrotic spots fall out, giving the leaves a ragged hole appearance. The fruit symptoms are numerous, small, circular, brown spots followed by cracking. Early infection causes misshapen fruit. On the twigs inconspicuous cankers occur.

Le prime symptoma foliar occurre in le superfacie inferior como un minute area circular o irregular, de color clar, inter le venas. Quando le maladia se avantia, le superfacie superior es invase e le maculas cresce e deveni angular e de color purpuree, brun, o nigre. Le maculas necrotic se separa e cade, lassante perforationes in le folios. Le symptomas in le fructos es numerose minute maculas circular brun, e postea fissuaras. Infection precoce produce fructos deforme. In leramettos canceres inobservabile occurre.

Sur le feuillage, le premier symptôme paraît à la face inférieure sous forme d'une petite plage claire, circulaire ou irrégulière, entre les nervures. A mesure que la maladie progresse, la face supérieure se couvre de taches qui s'agrandissent en devenant anguleuses et pourprées, brunes ou noires. La tache nécrotique tombe, laissant un trou à bords déchirés. Les fruits se couvrent de nombreuses petites taches circulaires brunes et se fendillent. Une infection précoce produit des fruits difformes. Des chancres presque imperceptibles se forment sur les ramilles.

El primer síntoma del follaje aparece en el envés de las hojas, como un área entre las venas, circulares o irregulares, de color claro. A medida que avanza la enfermedad, el haz es invadido y las manchas se agrandan, volviéndose angulares y de color púrpura, café o negras. Las manchas necróticas se caen, dándole a las hojas una apariencia de agujero desgarrado. Los síntomas del fruto son numerosas manchas pequeñas, circulares, de color café, seguidas por rajaduras. La infección temprana causa malformación del fruto. En las ramillas aparecen cancros incospicuos.

DISTRIBUTION: Af., A., Aust., Eur., N.A., C.A., S.A.

Pyrus spp. + Gymnosporangium sabinae (Dicks.) Wint. =

ENGLISH	Rust
FRENCH	Rouille Grillagée du Poirier
SPANISH	Roya
PORTUGUESE	
ITALIAN	Ruggine del pero
RUSSIAN	ржавчина
SCANDINAVIAN	Gitterost, Päronrost, Gelérost
GERMAN	Gitterost, Birnengitterrost, Rost
DUTCH	Pereroest
HUNGARIAN	
BULGARIAN	ръжда
TURKISH	Armut memeli pası
ARABIC	
PERSIAN	
BURMESE	
THAI	
NEPALI	
HINDI	
VIETNAMESE	
CHINESE	
JAPANESE	

This disease occurs on leaves, twigs and occasionally on the fruit. On the leaves, small pale yellow spots appear on the upper surface. These gradually enlarge and turn to an orange color. The tissue of the leaf is not killed, but is somewhat swollen. On some varieties the tissue dies early and brown necrotic areas result. The final size of the lesions varies with the variety and number of spots on a leaf.

Iste maladia occurre in folios, ramettos, e, a vices, in le fructos. In le folios, minute maculas jalne e pallide se manifesta in le superfacie superior. Istos cresce gradualmente e deveni orange. Le texito del folio non mori, ma deveni assatis tumide. In alicun varietates, le texito mori prematurmente e le resultato es areas necrotic brun. Le dimension final del lesiones se varia secundo le varietate e le numero del maculas in un folio.

Cette maladie se manifeste sur les feuilles, les ramilles, et parfois le fruit. Sur les feuilles, de petites taches jaune pâle apparaissent à la face supérieure. Elles s'agrandissent graduellement et deviennent orangées. Les tissus de la feuille ne meurent pas, mais sont un peu renflés. Chez certaines variétés, les tissus meurent tôt, ce qui entraîne la formation de plages nécrotiques brunes. La grandeur finale des lésions varie selon la variété et le nombre de taches sur une feuille.

Esta enfermedad aparece en hojas, ramillas y ocasionalmente en el fruto. En las hojas aparecen pequeñas manchas amarillo pálido en el haz. Estas gradualmente se agrandan y se vuelven de un color anaranjado. El tejido de la hoja no muere, pero está un poco hinchado. En algunas variedades el tejido muere pronto y resultan áreas necróticas de color café. El tamaño final de las lesiones varía con la variedad y el número de manchas en una hoja.

DISTRIBUTION: Af., A., Eur., N.A.

218

Pyrus communis + Venturia pirini Aderh. =

ENGLISH	Scab
FRENCH	Tavelure du Poirier
SPANISH	Sarna, Roña del peral
PORTUGUESE	
ITALIAN	Ticchiolatura, Brusone de pero
RUSSIAN	**парша**
SCANDINAVIAN	Pæreskurv, Päronskorv
GERMAN	Schorf, Birnenschorf
DUTCH	Pereschurft
HUNGARIAN	Körtefa-varasodás
BULGARIAN	**струпясване**
TURKISH	Armut kara lekesi
ARABIC	
PERSIAN	
BURMESE	
THAI	
NEPALI	
HINDI	
VIETNAMESE	
CHINESE	
JAPANESE	Kurohosi-byô 黒星病

This disease is usually evident first on the underside of the leaf. The diseased area first shows a faint olive-green spot somewhat darker than the surrounding normal leaf green. The color deepens with age and becomes almost black on the uppersurface. The leaf blade may become curled, dwarfed and distorted as a result of heavy infection. On the fruit, early infection is often overlooked. Later, the lesions on the enlarging fruit have the same general appearance as on the leaves. The fungus also attacks the twigs and causes serious damage to them.

Iste maladia usualmente se manifesta primo in le superfacie inferior del folio. Le area afflicte al initio monstra un macula verde como olivas e pallide, assatis plus obscur que le verde normal del folio. Le color se obscura post qualque tempore e deveni quasi nigre in le superfacie superior. Le lamina foliar pote devenir crispate, nano, e distorte a causa del infection sever. In le fructo, infection precoce sovente non es remarcate. Plus tarde, durante le developpamento del fructo, le lesiones es similar a illos in le folios. Le fungo attacca anque le ramettos e es multo damnose a illos.

D'habitude, cette maladie apparaît d'abord à la face inférieure des feuilles. La partie malade présente au début une tache olive mal définie, un peu plus foncée que le vert normal qui l'entoure. Peu à peu, la couleur devient plus foncée, presque noire, à la face supérieure. Le limbe peut cesser de croître, se recroqueviller et se déformer suite à une infection grave. Sur le fruit, une infection précoce passe souvent inaperçue. Plus tard, les lésions sur le fruit en croissance ont le même aspect que les lésions sur les feuilles. Le champignon s'attaque également aux ramilles qu'il endommage gravement.

Esta enfermedad es generalmente evidente primero en el envés de la hoja. El área enferma muestra primero una mancha verde-olivo claro un poco más obscura que el verde normal alrededor. El color se acentúa con la edad y se vuelve casi negro en el haz. Toda la hoja puede enroscarse, achaparrarse y malformarse como resultado de una infección fuerte. En el fruto, el principio de la infección puede pasar sin notarse. Más tarde, las lesiones en el fruto que va creciendo tienen la misma apariencia general que en las hojas. El hongo también ataca las ramillas y les causa serio daño.

DISTRIBUTION: Af., A., Aust., Eur., N.A., S.A.

Ribes spp. + Pseudopeziza ribis Kleb. (Gloeosporium ribis Montagne et Desmaziers) =

ENGLISH	Anthracnose, Leaf, Stem and Fruit Spot
FRENCH	Anthracnose du Groseillier, Gléosporiose
SPANISH	Antracnosis, Mancha de la hoja, del tallo y del fruto
PORTUGUESE	
ITALIAN	Seccume delle foglie del ribes
RUSSIAN	антракноз, "мухосед" (на листьях и ягодах)
SCANDINAVIAN	Skivesvamp, Bladfallsjuka
GERMAN	Blattfallkrankheit, Anthraknose
DUTCH	Bladvalziekte der Aalbes
HUNGARIAN	Ribiszke pszeudopezizás levélhullása
BULGARIAN	антракноза
TURKISH	
ARABIC	
PERSIAN	
BURMESE	
THAI	
NEPALI	
HINDI	
VIETNAMESE	
CHINESE	
JAPANESE	Tanso-byô 炭そ病

All parts of the plant's current season's growth are susceptible to this disease. Spotting of the leaves is the most pronounced symptom. Usually numerous dark brown to black dots appear scattered at random on either or both surfaces of the leaf. They may appear at any time during the growing season. The spots enlarge, become more angular in outline and are bordered by a purplish area. Shortly after severe infection, the entire leaf turns yellow and defoliation occurs. On the stem of the young shoots, the superficial lesions are usually larger than on the leaves and light brown in color.

Tote le partes del accrescimento del planta in le saison presente es susceptibile a iste maladia. Maculas in le folios es le symptoma le plus remarcabile. Usualmente numerose maculas brun obscur o nigre se manifesta disperse al hasardo in un o ambes del superfacies del folios. Illos se manifesta per tote le periodos del saison. Le maculas cresce, deveni de contorno plus angular, e monstra un bordo purpuree. Post infection sever, le folio rapidemente deveni jalne e defoliation occurre. In le pedunculo de plantas juvene, le lesiones superficial es usualmente plus grande que illos in le folios e de color brun clar.

Toutes les parties de la pousse annuelle sont sujettes à cette maladie. Les taches sur les feuilles constituent le symptôme le plus marqué. D'habitude, de nombreux points bruns et noirs apparaissent éparpillés au hasard sur l'une ou l'autre ou sur les deux faces des feuilles. Ces points peuvent apparaître a tout moment de la saison de croissance. Les taches s'agrandissent, deviennent plus anguleuses et sont entourées d'une zone violacée. Peu après une infection grave, les feuilles jaunissent complètement, et la défoliation survient. Sur les tiges des jeunes pousses, les lésions superficielles sont d'habitude brun clair et plus étendues que sur les feuilles.

Todas las partes de la planta en la época de crecimiento son susceptibles a esta enfermedad. El síntoma más pronunciado son las manchas de las hojas. Generalmente aparecen numerosos puntos de color café obscuro a negro, diseminados al azar en una o en ambas superficies de las hojas. Pueden aparecer en cualquier momento durante el crecimiento. Las manchas se agrandan, es vuelven más angulares en su forma y se rodean de un área purpúrea. Poco después de una infección severa la hoja entera se vuelve de color amarillo y ocurre una defoliación. En el tallo de brotes jóvenes, las lesiones superficiales son generalmente más grandes que en las hojas y de color café claro.

DISTRIBUTION: A., Aust., Eur., N.A.

220

Ricinus communis + Alternaria ricini (Yoshii) Hansford =

ENGLISH	Seedling light, Spot and Rot of Leaves and Seed Pods
FRENCH	Alternariose des Feuilles, des Tiges et des Capsules
SPANISH	Tizón de las plántulas, Mancha y podredumbre de hojas y vainas
PORTUGUESE	Mancha de Alternaria
ITALIAN	
RUSSIAN	альтернариоз
SCANDINAVIAN	
GERMAN	Alternaria Blatt-, Stengel- und Samenflecken
DUTCH	
HUNGARIAN	
BULGARIAN	
TURKISH	
ARABIC	
PERSIAN	
BURMESE	
THAI	Bi-Mhai ใบไหม้
NEPALI	
HINDI	
VIETNAMESE	Đốm đen trên lá
CHINESE	
JAPANESE	Hagare-byô 葉枯病

This fungus produces symptoms on seedlings, leaves and pods. Leaf lesions are irregular in outline, variable in size, brown, zonate, with a yellow halo. Defoliation can be extensive. The floral parts can be attacked at any age, and they develop a sooty appearance. The seed coats wilt and become purple to dark brown, and the seed is poorly filled. Infected seedlings are stunted and may be killed.

Iste fungo produce symptomas in plantulas, folios, e siliquas. Lesiones foliares de contorno irregular, de dimensiones varie, brun, zonate, con un halo jalne. Defoliation es sovente extensive. Le partes floral es attaccate per tote le stadios de crescimento e illos es fuliginose. Le tegumentos marcesce e deveni purpuree o brun obcur e le granos es mal implete. Plantulas infectate es nanos e a vices mori.

Ce champignon s'attaque aux semis, aux feuilles et aux capsules des plantes adultes. Les lésions sur les feuilles sont de forme irrégulière, d'étendue variable, brunes, zonées, entourées d'une áreole jaune. La défoliation peut être abondante. Les parties florales peuvent être attaquées à n'importe quel moment et présentent un aspect fuligineux. Les parois des graines se dessèchent et deviennent pourprées à brun foncé, et les graines sont chétives. Les plantules infectées deviennent rabougries et peuvent mourir.

Este hongo produce síntomas en las plántulas, hojas y vainas. Las lesiones en la hoja son de forma irregular, varían en tamaño, de color café, zonadas con un halo amarillo. La defoliación puede ser extensive. Las partes florales pueden ser atacadas a cualquier edad y desarrollan una apariencia tiznada. La cubierta de la semilla se marchita y se vuelve de color púrpura a café obscuro y a semilla se llena pobremente. Las plántulas infectadas se achaparran y pueden morir.

DISTRIBUTION: Af., A., Aust., Eur., N.A.

Ricinus communis + Melampsorella ricini De Toni (Melampsora ricini Noronha) =

ENGLISH	Rust	
FRENCH	Rouille	
SPANISH	Roya	
PORTUGUESE		
ITALIAN		
RUSSIAN		
SCANDINAVIAN		
GERMAN		
DUTCH		
HUNGARIAN		
BULGARIAN		
TURKISH		
ARABIC	Sadaa	صدأ
PERSIAN		
BURMESE		
THAI		
NEPALI		
HINDI	Ratua	रतुवा
VIETNAMESE		
CHINESE		
JAPANESE	Sabi-byô	さび病

Severe attacks of this disease cause the leaves to dry up and wither prematurely. The symptoms are characterized by many yellowish-orange spots or lesions on the undersurface of the foliage. The disease usually attacks the leaves on the lower parts of the plant and if infection occurs early in the season, these leaves fall prematurely.

Attaccos sever de iste maladia face le folios desiccar se e marcescer prematurmente. Le symptoma characteristic es multe maculas o lesiones orange-jalnette in le superfacie inferior del foliage. Le maladia usualmente attacca le folios in le partes inferior del planta e, si infection occurre in le prime parte del saison, iste folios cade prematurmente.

Les infections graves de cette maladie produisent le desséchement et le dépérissement prématurés des feuilles. Les symptômes paraissent sous forme de nombreuses taches jaune orange ou de lésions sur la face inférieure des feuilles. La maladie s'attaque d'habitude aux feuilles de la partie inférieure de la plante, et ces feuilles tombent prématurément si l'infection se produit au début de la saison.

Ataques severos de esta enfermedad causan que las hojas se sequen y marchiten prematuramente. Los síntomas son caracterizados por numerosas manchas de color anaranjado-amarilloso o lesiones en el envés del follaje. La enfermedad generalmente ataca las hojas en las partes más bajas de la planta y si la infección aparece temprano en la estación, las hojas atacadas se caen prematuramente.

DISTRIBUTION: Af., A., Eur.

Ricinus communis + Xanthomonas ricini (Yoshi & Takimoto) Dowson =

ENGLISH	Leaf Spot	
FRENCH		
SPANISH	Mancha foliar	
PORTUGUESE	Bacteriose da fôlha	
ITALIAN		
RUSSIAN		
SCANDINAVIAN		
GERMAN		
DUTCH		
HUNGARIAN		
BULGARIAN		
TURKISH		
ARABIC		
PERSIAN		
BURMESE		
THAI	Bi-Jut	ใบจุด
NEPALI		
HINDI		
VIETNAMESE		
CHINESE		
JAPANESE	Hanten-saikin-byô	斑点細菌病

The symptoms of this leaf spot disease are small, dark green, water-soaked spots which expand and become brownish, angular and necrotic. These spots coalesce to form larger necrotic areas and badly infected leaves become chlorotic and may drop. Occasionally the stems and succulent branches have minute oval or linear spots. With heavy infection when leaves first open, the main veins become infected causing the leaves to become yellow and wither.

Le symptomas de iste maladia foliar es minute maculas aquose, verde obscur, le quales cresce e deveni brunette, angular, e necrotic. Iste maculas coalesce usque a formar plus grande areas necrotic. Folios severmente infectate deveni chlorotic e pote cader. A vices, le pedunculos e ramos succulente monstra minute maculas oval o linear. Si il ha infection sever quando le folios se aperi, le venas principal es infectate, faciente le folios devenir jalne e marcescer.

Les symptômes de cette maladie des feuilles sont de petites taches vert foncé, détrempées que s'étendent et deviennent brunâtres, anguleuses et nécrotiques. Ces taches s'unissent pour former des plages nécrotiques étendues, et les feuilles gravement infectées deviennent chlorotiques et peuvent tomber. Parfois, les tiges et les branches tendres présentent de minuscules taches ovales ou linéaires. Si une infection grave se produit au moment où les feuilles paraissent, les nervures principales deviennent infectées et produisent le jaunissement et le dessèchement des feuilles.

Los síntomas de esta enfermedad causante de manchas foliares son pequeñas manchas de color verde obscuro, empapadas de agua, las cuáles crecen y se vuelven cafesosas, angulares y necróticas. Las manchas se unen para formar áreas necróticas mayores y las hojas severamente infectadas se vuelven cloróticas, pudiendo caerse. Ocasionalmente los tallos y ramas suculentas tienen manchas diminutas, ovaladas o lineales. Con infección severa, cuando las hojas acaban de abrirse, las venas principales se infectan cuasando que las hojas se vuelvan amarillas y se marchiten.

DISTRIBUTION: A., Af., Eur.

Rubus idaeus (Rubus ursinus var. loganobaccus) + Didymella applanata (Niessl.) Sacc. =

ENGLISH   Spur Blight
FRENCH   Brûlure des Dards, Didymella, Desséchement des Rameaux, Taches Violacées
SPANISH   Tizón de espuela
PORTUGUESE
ITALIAN
RUSSIAN   пятнистость пурпуровая, метельчатость
SCANDINAVIAN   Hindbærstængelsyge, Hallonskottsjuka, Bringebærskuddsyke, Stængelsyge
GERMAN   Rutenkrankheit, Himbeerrutensterben
DUTCH
HUNGARIAN   Didimellás vesszőfoltosság
BULGARIAN   сиво-виолетови петна около пъпките
TURKISH
ARABIC
PERSIAN
BURMESE
THAI
NEPALI
HINDI
VIETNAMESE
CHINESE
JAPANESE

The disease first appears on the new canes early in the season. Discolored areas, brown or purplish, appear on the shoot immediately below the leaf attachment. The spots enlarge and extend up and down the cane. The leaf arising from the node is often chlorotic in appearance. Leaflets fall off. Dark, enlongated lesions sometimes are found on the midrib and larger veins. The infected area on the cane remains dark brown to purplish until late in the season when it takes on a grayish color.

Le morbo se manifesta al initio in le novelle cannas in le prime parte del saison. Areas discolorate, brun o purpuree, se manifesta in le plantones juvene exactemente sub le attachamento foliar. Le folios cresce e se extende in alto e in basso per la canna. Sovente le folio que emerge ab le nodo sembla esser chlorotic. Foliettos se separa e cade. Lesiones obscur e elongate a vices se manifesta in le nervature central e in le venas plus grande. Le area infectate del canna resta brun obscur o purpuree usque al ultime parte del saison. Alora illo deveni quasi gris.

La maladie paraît d'abord sur les tiges nouvelles au début de la saison. Des plages décolorées, brunes ou pourprées, apparaissent sur les pousses, juste au dessous de l'attache de la feuille. Les taches grandissent et se répandent vers le haut et le bas de la tige. La feuille provenant de noeud présente souvent un aspect chlorotique. Les folioles tombent. Des lésions foncées, allongées se forment souvent sur la nervure médiane et sur les grosses nervures. La région infectée de la tige reste brun foncé à pourpré jusque tard dans la saison. A ce moment, là partie infectée devient grisâtre.

La enfermedad aparece primero en los tallos nuevos al principio de la estación. En los brotes, inmediatamente abajo del pegue de la hoja aparecen áreas decoloradas, de color café o purpúreo. Las manchas crecen se extienden hacia arriba y hacia abajo en los tallos. La hoja que sale del nudo es frecuentemente clorótica en su apariencia. Las hojillas se caen. Algunas veces se encuentran lesiones obscuras, alargadas, en la vena central y otras venas grandes. El área infectada en el tallo permanece de color café obscuro a purpúreo hasta tarde en la estación, y entonces adquiere un color grisáceo.

DISTRIBUTION: A., Aust., Eur., N.A.

224

Rubus idaeus (Rubus spp.) + Elsinoe veneta (Berk.) Jenk. =

ENGLISH	Cane Spot, Gray Bark, Anthracnose
FRENCH	Anthracnose
SPANISH	Antracnosis, Mancha del tallo, Corteza gris
PORTUGUESE	
ITALIAN	
RUSSIAN	антракноз
SCANDINAVIAN	Flekkskurv
GERMAN	Brennfleckenkrankheit, Anthraknose
DUTCH	Stengelziekte, Twijgsterfte bij Framboos
HUNGARIAN	Málna elzinoés vesszőfoltossága
BULGARIAN	
TURKISH	
ARABIC	
PERSIAN	
BURMESE	
THAI	
NEPALI	
HINDI	
VIETNAMESE	
CHINESE	
JAPANESE	

The disease first appears early in the season on the young shoots as small, purplish, slightly raised spots. As the cane grows, the spots enlarge and become lens-shaped or oval with a slightly raised purple edge. The lesions on the leaves are similar in appearance, with spotting on the blades first appearing as small indefinite purplish areas which later enlarge and turn brown. Infection on the fruit is rare, but when present, fruits are shrunken, brown and dry.

Le maladia se manifesta in le prime parte del saison in le plantones juvene como minute maculas purpuree e legiermente elevate. Quando le canna cresce, le maculas anque cresce e deveni lentiforme o oval con un bordo purpuree legiermente elevate. Le lesiones in le folios es similar con maculas in le laminas, le quales al initio se manifesta como minute areas purpuree indefinite que plus tarde cresce e deveni brun. Infection in le fructos es rar, ma si infection occurre, le fructos es contracte, brun, e desiccate.

La maladie apparaît au début de la saison sur les jeunes pousses, sous forme de petites taches violacées légèrement dressées. A mesure que la tige pousse, les taches grandissent et deviennent lenticulées ou ovales, avec des contours pourprés légèrement relevés. Les lésions sur les feuilles présentent un aspect similaire: les limbes se couvrent d'abord de petites taches mal définies, violacées, qui plus tard s'agrandissent et brunissent. L'infection du fruit est rare, mais, si elle se produit, les fruits sont ratatinés, bruns et secs.

La enfermedad aparece primero a principios de la estación en los tallos jóvenes como pequeños manchas de color purpúreo, un poco levantadas. A medida que los tallos crecen, las manchas se agrandan y adquieren forma de lente u ovaladas, con un borde purpúreo, un poco levantado. Las lesiones en las hojas son de apariencia similar, apareciendo las manchas primero como pequeñas áreas purpúreas indefinidas, que más tarde crecen y se vuelven de color café. La infección del fruto es rara, pero cuando aparece, los frutos se encojen, se vuelven de color café y se secan.

DISTRIBUTION: Eur., N.A., C.A.

Rubus spp. + <u>Gymnoconia</u> <u>peckiana</u> (Howe) Trotter (<u>Gymnoconia</u> <u>interstitialis</u> (Schl.) Lag.) =

ENGLISH	Orange Rust
FRENCH	Rouille Orangée
SPANISH	Roya anaranjada
PORTUGUESE	
ITALIAN	
RUSSIAN	ржавчина оранжевая
SCANDINAVIAN	
GERMAN	Oranger Rost
DUTCH	
HUNGARIAN	
BULGARIAN	
TURKISH	
ARABIC	
PERSIAN	
BURMESE	
THAI	
NEPALI	
HINDI	
VIETNAMESE	
CHINESE	
JAPANESE	

This disease can be identified shortly after new growth appears. The slender infected shoots develop a lighter green foliage than uninfected plants. The young leaves, before they unfold, are dotted with small, yellowish bodies which later turn black. Two or three weeks later, the lower surface of the leaves is covered with a bright orange-colored mass of spores, at first waxy, then powdery in texture. Usually most of the leaves on a shoot show the disease. The plants may become dwarfed, but remain alive for several years.

Iste maladia es identificate post le emergentia del plantones novelle. Le gracile plantones infectate monstra un foliage verde plus clar que normal. Le folios juvene, ante que illos se aperi, es coperte de minute structuras sporal jalne, le quales plus tarde deveni nigre. Intra duo o tres septimanas, le superfacie inferior del folios es coperte de un massa orange brillante de sporas, al initio similar a cera, plus tarde de textura pulverose. Usualmente le major parte del folios in un planton monstra le morbo. Sovente le plantas deveni nanos, ma resta vivente per alicun annos.

On peut constater cette maladie peu après le début de la nouvelle croissance. Les fines pousses infectées présentent un feuillage d'un vert plus clair que les plantes saines. Avant qu'elles ne s'étalent, les jeunes feuilles sont couvertes de petits corps jaunâtres qui plus tard deviennent noirs. Deux ou trois semaines après, la face inférieure des feuilles se couvre d'une masse de spores orange vif, tout d'abord de texture cireuse, puis poudreuse. D'habitude, la plupart des feuilles de la pousse présentent des indices de maladie. Les plantes peuvent devenir rabougries, mais survivent pendant plusieurs années.

Esta enfermedad puede ser identificada poco después de aparecer el crecimiento nuevo. Los brotes delgados, infectados, desarrollan un follaje de color verde más claro que las plantas sanas. Las hojas tiernas, antes de abrirse, se manchan con cuerpos amarillosos que luego se vuelven negros. Dos o tres semanas más tarde, el envés de las hojas está cubierto con una masa de esporas de color anaranjado brillante, al principio cerosas y luego polvorientas en su textura. Generalmente la mayoría de las hojas en un brote muestran la enfermedad. Las plantas pueden achaparrarse, pero permanecen vivas durante varios años.

DISTRIBUTION: A., N.A.

226

Saccharum _officinarum_ + _Cephalosporium_ _sacchari_ Butler =

ENGLISH   Wilt
FRENCH   Flétrissement dû à Cephalosporium
SPANISH   Marchitez
PORTUGUESE
ITALIAN   Avvizzimento della canna da zuchero
RUSSIAN   **увядание цефалосползлозное**
SCANDINAVIAN
GERMAN   Cephalosporium-Welke
DUTCH
HUNGARIAN
BULGARIAN
TURKISH
ARABIC
PERSIAN
BURMESE
THAI
NEPALI
HINDI   Mlani   मलानी
VIETNAMESE
CHINESE
JAPANESE

Symptoms are inconspicuous until plant is half grown. Either single canes or whole clumps gradually wither and continue to do so until harvest. Affected leaves turn yellow and dry up, appearing as if suffering from drought. Stems become light and hollow, the pith having a diffuse purple or dirty-red discoloration which is mostly longitudinally streaky.

Le symptomas non es remarcabile usque a medie-saison. O singule cannas o racemos gradual-mente marcesce continualmente usque al rendimento. Folios afflicte deveni jalne e se desicca como si a causa de siccitate. Cannas deveni legier e cavernose e le medulla monstra un dif-fuse discoloration purpuree o rubiette in strias longitudinal.

Tant que la plante n'a pas atteint la moitié de sa grosseur, les symptômes sont imperceptibles. Par la suite, des tiges isolées ou des touffes entières commencent à péricliter et continuent ainsi jusqu'à la récolte. Les feuilles atteintes jaunissent et se dessèchent, comme sous l'effet de la sécheresse. Les tiges deviennent légères et creuses, la moëlle exhibant une couleur de pourpre diffus ou de rouge sale, le plus souvent en bandes longitudinales.

Los síntomas son inconspicuos hasta que la planta ha alcanzado la mitad de su crecimiento. Ya sea un tallo solo o macollas enteras se marchitan gradualmente y continúan así hasta la cosecha. Las hojas afectadas se vuelven de color amarillo y se secan, aparentando como que están sufriendo de sequía. Los tallos se vuelven livianos y huecos, teniendo el centro una de-coloración de color púrpura difuso o rojo sucio, la cuál generalmente tiene las rayas longi-tudinalmente.

DISTRIBUTION: Af., A., Eur., N.A., C.A., S.A.

Saccharum officinarum + Cercospora koepkei Kreuger (Cercospora longipes Butl.) =

ENGLISH   Brown Spot
FRENCH   Taches Foliaires Jaunes, Foliaires Brunes
SPANISH   Mancha café de la vaina, Cercosporiosis
PORTUGUESE   Mancha Parda
ITALIAN   Macchie brune
RUSSIAN   пятнистость церкоспорозная бурая
SCANDINAVIAN
GERMAN   Gelbe Blattfleckenkrankheit, Braune Blattfleckenkrankheit
DUTCH
HUNGARIAN
BULGARIAN
TURKISH
ARABIC
PERSIAN
BURMESE
THAI   Bi-Jut-Leung   ใบจุดเหลือง
NEPALI
HINDI   Bhura parn daag   भूरा पर्ण दाग
VIETNAMESE   Đốm đỏ
CHINESE   Gan jùoh yèh pèan chìr ban bìng   甘蔗葉片赤斑病
JAPANESE   Yōhensekihan-byō   葉片赤斑病

Oval- to linear-shaped reddish-brown spots appear on the cane leaves. They are characteristic-ally surrounded by a narrow yellow halo. The centers of older spots become dry and straw-colored, surrounded by a red zone and the outer yellow halo. Spots may enlarge and coalesce to form large reddish-brown patches of irregular shape, appearing first on older leaves and progressing upward as the plant grows. Severely infected leaves die prematurely.

Oval o linear maculas brun-rubiette se manifesta in le folios del cannas. Illos es imbraciate per un stricte halo jalne characteristic. Le centros del maculas plus vetule deveni desiccate e de color paleate, imbraciate per un zona rubie e le halo jalne exterior. Maculas a vices cresce e coalesce usque a formar grande areas brun-rubiette de contorno irregular, le quales se manifesta al initio in folios plus vetule e postea se extende in alto quando le planta cresce. Folios severmente infectate mori prematurmente.

Des taches brun rougeâtre de forme ovale à linéaire apparaissent sur les feuilles. Ces taches sont entourées d'un halo jaune et mince, caractéristique. Le centre des vieilles taches se dessèche et devient paille, bordé par une zone rouge qu'entoure le halo jaune. Les taches peu-vent grossir et se fusionner, de façon à former de grandes plaques brun rougeâtre de forme irrégulière, qui apparaissent d'abord sur les vieilles feuilles et se propagent vers le haut à mesure que la plante croît. Les feuilles gravement atteintes meurent prématurément.

En las hojas de la caña aparecen manchas ovaladas o lineales, de color café-rojizo. Están característicamente rodeadas por un angosto halo amarillo. Los centros de las manchas más viejas se secan y adquieren un color de paja, rodeadas por una zona roja y el halo exterior amarillo. Las manchas pueden crecer y juntarse para formar grandes manchones café-rojizos de forma irregular, que aparecen primero en las hojas viejas y van progresando hacia arriba a medida que la planta crece. La hojas infectadas severamente mueren prematuramente.

DISTRIBUTION: Af., A., Aust., N.A., C.A., S.A.

228

Saccharum officinarum + Cercospora vaginae Krueger =

ENGLISH	Red Leaf-Sheath Spot
FRENCH	Taches Rouges sur les Gaines
SPANISH	Mancha roja del peciolo, Mancha roja de la cubierta de la hoja
PORTUGUESE	
ITALIAN	
RUSSIAN	**пятнистость церкоспорозная красная (на влагалищах), церкоспороз**
SCANDINAVIAN	
GERMAN	Rote Scheidenfleckigkeit
DUTCH	
HUNGARIAN	
BULGARIAN	
TURKISH	
ARABIC	
PERSIAN	
BURMESE	
THAI	Bi-Jut-Si-Dang ใบจุดสีแดง
NEPALI	
HINDI	
VIETNAMESE	
CHINESE	Gan jùoh yèh shau chìr ban bìng 甘蔗葉鞘赤斑病
JAPANESE	Yôsyôsekihan-byô 葉しょう赤斑病

The first symptom is the appearance of small, round, bright red spots on the upper leaf sheaths, sharply delimited from the normal green of surrounding tissue. Later they enlarge and coalesce to form irregular, bright red patches. These lesions extend through to the inner sheaths. In later stages the sooty fruiting structures may be produced on the affected areas, more abundantly on the inside of the sheaths.

Le prime symptoma es minute maculas rotunde rubie brillante in le vaginas foliar superior, acutemente delimitate ab le verde normal del texito circumjacente. Plus tarde illos cresce e coalesce usque a formar areas irregular rubie brillante. Iste lesiones penetra in le vaginas interior. In stadios subsequente, le structuras sporal fuliginose es formate in le areas afflicte, plus numerose in le interior del vaginas.

Le premier symptôme de cette maladie est l'apparition, sur les gaines du haut, de petites taches rondes rouge vif, qui forment un contraste frappant avec le vert normal des tissus voisins. Plus tard, ces taches s'agrandissent et peuvent se fusionner pour former des plaques irrégulières rouge vif. Ces lésions se propagent jusqu'aux gaines intérieures. A des phases subséquentes, des fructifications fuligineuses peuvent se former sur les parties atteintes, plus abondamment à l'intérieur des gaines.

El primer síntoma es la aparición de pequeñas manchas redondas, de color rojo brillante en las cubiertas superiores de las hojas, marcadamente delimitadas del verde normal del tejido de alrededor. Más tarde se agrandan y se juntan para formar manchones irregulares, de color rojo brillante. Etas lesiones se extienden a través hasta llegar a las oubiertas interiores de las hojas. En estados avanzados las estructuras fructíferas hollinientas pueden ser producidas en las áreas afectadas, más abundatemente en la parte interior de las cubiertas.

DISTRIBUTION: Af., A., Aust., N.A., C.A., S.A.

Saccharum officinarum (Gramineae) + Colletotrichum falcatum Went (Physalospora tucumanensis Speg., Glomerella tucumanensis (Speg.) Arx & Muller) =

ENGLISH Red Rot of Sugarcane, Anthracnose of Gramineae
FRENCH Maladie de la Morve Rouge
SPANISH Podredumbre colorado, Muermo roja, Pudrición roja de la caña, Podredumbre roja
PORTUGUESE Podridão Vermelha
ITALIAN Marciume rosso della canna da zucchero
RUSSIAN **гниль красная**
SCANDINAVIAN
GERMAN Rotfäule
DUTCH
HUNGARIAN
BULGARIAN
TURKISH
ARABIC
PERSIAN
BURMESE Oo-nee roga ကော်လက်တိုတြိုခမ် ဖဂါလ်ကေတမ်
THAI Nao-Si-Dang เน่าสีแดง
NEPALI Ukhu Ko rate rog उखुको राते रोग
HINDI Lal viglan लाल विगलन
VIETNAMESE Đỗ thân - Khô lá
CHINESE Gan jùoh chìr fǔ bìng 甘蔗赤腐病
JAPANESE Akagusare-byô 赤腐病

This disease can be seen on any part of the plant, but principally on standing stalks, plant cuttings and leaf midribs. The first external symptoms are the drooping, withering and yellowing of the upper leaves. This is followed by the death of the entire top of the plant. If the fungus gains entrance to the stalk, rot will spread with tissues becoming dull red, interrupted by whitish patches elongated at right angles to the long stalk axis. Often the entire planted sett is rotted, taking on various shades of red, brown or gray.

Iste maladia es visibile in tote le partes del planta, ma principalmente in cannas erecte, in nervaturas foliar, e in plantones. Le prime symptomas externe es marcescentia, languor, e jalnessa del folios superior. Le morte de tote le planta super le radices seque. Si le fungo penetra le canna, putrefaction se extende e le texito deveni rubie mat interrupte per areas quasi blanc elongate a angulos recte al axe longe del cannas. Sovente tote leplantation putresce e deveni rubie, brun, o gris.

Cette maladie peut s'observer sur n'importe quelle partie de la plante, mais principalement sur les tiges dressées, les boutures et les nervures principales des feuilles. Les premiers symptômes externes sont l'affaissement, le dépérissement et le jaunissement des feuilles du haut, suivis de la mort de toute la partie supérieure de la plante (cime). Si le champignon vient à pénétrer dans la tige, la pourriture envahit les tissue, qui deviennent rouge terne, avec, çà et là, des plaques allongées formant angle droit avec le grand axe de la tige. Souvent, la bouture entière est pourrie et prend des teintes variées de rouge, de brun ou de gris.

La enfermedad puede verse en cualquier parte de la planta, pero principalmente en los tallos erectos, esquejes de la planta y en las venas centrales de las hojas. El primer síntoma externo es el decaimiento, marchitez y amarillamiento de las hojas superiores. Esto es seguido por la muerte de toda la parte superior de la planta, (corona). Si el hongo logra entrar dentro del tallo, la podredumbre se disemina y los tejidos se vuelven de color rojo opaco, interrumpidos por machones blanquecinos alargados en ángulos a la derecha de los axis de los tallos largos. Frecuentemente la planta entera se pudre, adquiriendo varios tonos de rojo, café o gris.
DISTRIBUTION: Af., A., Aust., N.A., C.A., S.A.

230

Saccharum officinarum + Cytospora sacchari Butler =

ENGLISH   Sheath Rot
FRENCH
SPANISH   Podredumbre de la cubierta
PORTUGUESE   Podridão da Bainha
ITALIAN
RUSSIAN
SCANDINAVIAN
GERMAN
DUTCH
HUNGARAN
BULGARIAN
TURKISH
ARABIC
PERSIAN
BURMESE
THAI
NEPALI
HINDI   Stamb canker   स्तम्भ कॅंकर
VIETNAMESE
CHINESE   Gan jùoh shau ku bìng   甘蔗鞘枯病
JAPANESE   Sayagare-byồ   さや枯病

The disease is primarily one of the sheaths, but the fungus also infects seed cuttings, stubble pieces and young shoots and stalks of standing canes. Infection first appears as irregular brick-red to brownish-red patches on the leaf sheath near the ground. These areas usually enlarge until most of the sheath is covered. Fruiting bodies cover the diseased areas as black masses which, in moist weather, exude amber droplets.

Iste maladia occurre principalmente in le vaginas, ma le fungo attacca anque plantones, barbas, e pedunculos erecte. Infection se manifesta al initio como areas rubie como bricca o rubie-brunette in le vagina foliar presso le terra. Iste areas cresce usualmente usque a coperir le major parte del vagina. Structuras sporal in massas nigre coperi areas morbide e, in tempore humide, illos exsuda minute guttas de color de ambra.

C'est avant tout une maladie des gaines, mais le champignon infecte aussi les boutures, le chaume, les jeunes pousses et les cannes dressées. L'infection se manifeste au début sous forme de plaques irrégulières rouge brique a rouge brunâtre sur la gaine de la feuille près du sol. Ces plaques s'agrandissent en général jusqu'à recouvrir la plus grande partie de la gaine. Des masses noires de fructifications recouvrent les parties malades; à l'humidité, ces masses exsudent des gouttelettes ambrées.

La enfermedad ataca principalmente las cubiertas, pero el hongo también afecta los esquejes, rastrojos, brotes tiernos y tallos de las cañas erectas. La infección aparece primero como manchones irregulares de color rojo-ladrillo a rojo-cafesoso en la cubierta de la hoja cerca del suelo. Estas áreas generalmente se agrandan hasta que casi toda la cubierta se ve atacada. Los cuerpos fructíferos cubren las áreas enfermas como masas negras, las cuáles exudan gotitas de color ámbar en tiempo húmedo.

DISTRIBUTION: Af., A., N.A., C.A., S.A.

Saccharum officinarum + Helminthosporium stenospilum Drechs. (Drechslera stenospila (Drechsler) Subram. & Jain, Cochliobolus stenoospilus (Carp.) Mats. & Yama) =

ENGLISH	Brown Stripe
FRENCH	Rayures Brunes, Maladie des Bandes Brunes
SPANISH	Raya café
PORTUGUESE	Estrias Pardas
ITALIAN	Striscie brune
RUSSIAN	полосатость гельминтоспориозная бурая
SCANDINAVIAN	
GERMAN	Helminthosporium-Braunstreifigkeit
DUTCH	
HUNGARIAN	
BULGARIAN	
TURKISH	
ARABIC	
PERSIAN	
BURMESE	
THAI	Bi-Kheit-Si-Numtan ใบขีดสีน้ำตาล
NEPALI	
HINDI	
VIETNAMESE	Đốm sọc nâu trên lá
CHINESE	Gan jùoh heh tíao bìng 甘蔗褐條病
JAPANESE	Katuzyô-byô 褐条病

Minute, watery spots appear on the young leaves. As the infection matures it turns reddish, assumes an elongated shape, then brownish-red and indefinite stripes. Surrounding the brown linear stripe is a definite yellowish halo. When infection is severe the lesions coalesce, giving the older leaves a prematurely dried appearance.

Minute maculas aquose se manifesta in le folios juvene. Quando le infection se matura, illo deveni rubiette de contorno elongate, postea rubie-brunette in strias definite. Circum le brun stria linear es un halo jalne definite. Si le infection es sever, le lesiones coalesce, facente le folios plus vetule parer prematurmente desiccate.

De minuscules taches aqueuses apparaissent sur les jeunes feuilles. En mûrissant, l'infection devient rougeâtre, prend une forme allongée, puis vire au rouge brunâtre en raies définies. Un halo jaunâtre défini entoure la raie linéaire brune. Quand l'infection est grave, les lésions confluent, ce qui donne aux vieilles feuilles l'aspect de feuilles séchées prématurément.

En las hojas jóvenes aparecen manchas diminutas, empapadas de agua. A medida que la infec-ción madura, se vuelve rojiza, adquiere una forma alargada y luego se vuelve de color rojo caf-esoso y en rayas definidas. Rodeando la raya café lineal hay un halo amarilloso definido. Cuando la infeccion es severa, las lesiones se juntan, dándole a las hojas más viejas una apariencia se-ca prematura.

DISTRIBUTION: Af., A., Aust., N.A., C.A., S.A.

232

Saccharum officinarum + Leptosphaeria michotii (West.) Sacc. =

ENGLISH   Leaf Spot
FRENCH
SPANISH   Mancha foliar
PORTUGUESE
ITALIAN
RUSSIAN
SCANDINAVIAN
GERMAN
DUTCH
HUNGARIAN
BULGARIAN
TURKISH
ARABIC
PERSIAN
BURMESE
THAI
NEPALI
HINDI
VIETNAMESE
CHINESE
JAPANESE

This disease produces characteristic ring-spot lesions on otherwise healthy leaves. The spots, always near the midrib, usually are surrounded by yellow margins. Severe infection results in the entire plant being infected and loss of leaves.

Iste maladia produce characteristic lesiones anellate in folios que es in altere respectos salubre. Le maculas, semper presso le nervatura central, usualmente es imbraciate per margines jalne. Infection sever face tote le planta esser morbide e causa defoliation.

La maladie produit des lésions caractéristiques en forme d'anneaux sur des feuilles saines par ailleurs. Les taches, toujours situées près de la nervure mediane, sont généralement bordées de jaune. Quand l'infecton est grave, toute la plante est envahie et perd ses feuilles.

Esta enfermedad produce lesiones características de manchas en forma de anillo sobre hojas que, de otra manera, estarían sanas. Las manchas, siempre cerca de la vena central, generalmente están rodeadas por márgenes amarillos. Una infección severa da como resultado la infección de toda la planta y pérdida de las hojas.

DISTRIBUTION: Af., A., Eur., S.A., W.I.

Saccharum officinarum + Leptosphaeria sacchari v. Breda de Haan (Helminthosporium sacchari Butl.) =

ENGLISH	Ring Spot, Eyespot, Florida Ring Spot
FRENCH	Taches Foliaires, Maladie des Taches Rondes
SPANISH	Mancha de anillo, Mancha circular, Mancha de ojo, Mancha linear, Mancha anular
PORTUGUESE	Mancha anular, Mancha ocular
ITALIAN	Macchie anulari, Macchie ocellate
RUSSIAN	**пятнистость гельминтоспориозная, гельминтоспориоз**
SCANDINAVIAN	
GERMAN	Ringfleckenkrankheit
DUTCH	
HUNGARIAN	
BULGARIAN	
TURKISH	
ARABIC	
PERSIAN	
BURMESE	
THAI	
NEPALI	Ukhu Ko rato thople rog उखुको रातो पोप्ले रोग
HINDI	Drik bindu द्रिक बिंदु
VIETNAMESE	Đốm thuần
CHINESE	Gan jùoh yěan dǐan bìng 甘蔗眼點病 , Gan jùoh luén ban bìng 甘蔗輪斑病
JAPANESE	Rinhan-byô 輪斑病

This disease occurs principally on the leaf blades but may be on the sheaths and stems. It usually affects the older leaves. On the blades, the spots may be abundant, at first dark green to brown with narrow, yellowish borders and somewhat elongate-oval in shape. As they enlarge they become more irregular in outline, with some coalescing to form large reddish-brown patches. Sometimes the centers of the older spots turn straw-colored with well-defined reddish margins and at times the entire lesion remains reddish-brown. Black fruiting bodies are usually present on older spots. Stalk lesions vary from oval to circular to irregular in shape.

Iste maladia occurre principalmente in le laminas foliar, ma anque in le vaginas e a vices in le pedunculos. Usualmente illo affice le folios plus vetule. In le laminas, le maculas es a vices numerose, al initio verde obscur o brun con stricte bordos jalnette e de contorno assatis elongate-oval. A mesura que illos cresce, illos deveni plus irregular de contorno e alicunes coalesce usque a formar grande areas brun-rubiette. A vices le centros del maculas plus vetule deveni paleate con margines ben definite brun-rubiette. Structuras sporal nigre usualmente se manifesta in maculas plus vetule. Lesiones in le cannas se varia ab oval usque a circular o irregular de contorno.

Cette maladie se manifeste principalement sur le limbe des feuilles, mais on peut aussi l'observer sur les gaines et parfois sur les tiges. Elle s'attaque généralement aux vieilles feuilles. Sur les limbes, les taches peuvent être abondantes, d'abord vert foncé à brunes, avec des bords jaunâtres étroits, et de forme ovale quelque peu allongée. A mesure que les taches grossissent, leurs contours deviennent plus irréguliers et certaines se fusionnent pour former de grandes plaques brun rougeâtre. Quelquefois, le centre des vieilles taches prend une couleur paille avec des marges rougeâtres bien définies, et parfois toute la lésion demeure brun rougeâtre. Sur les vieilles taches se trouvent généralement des fructifications noires. Les lésions de la tige sont de forme ovale à ronde ou irrégulière.

La enfermedad aparece principalmente sobre la hoja, pero puede aparecer en las cubiertas y algunas veces en los tallos. Generalmente ataca las hojas más viejas. En las hojas las manchas pueden ser abundantes, al principio de color verde obscuro a café, con bordes amarillosos y algunas veces de forma alargada-ovalada. A medida que se agrandan se vuelven de forma más irregular, juntándose algunas para formar grandes parches café-rojizos. Algunas veces los centros de las manchas más viejas se vuelven de color paja con márgenes rojizos bien definidos y a veces toda la lesión es de color café-rojizo. Cuerpos fructíferos negros están generalmente presentes en las manchas más viejas. Las lesiones del tallo varían en su forma de ovalada a circular o irregular.

DISTRIBUTION: Af., A., Aust., N.A., C.A., S.A.

234

Saccharum spp. + Phaeocytostroma sacchari (Ell. & Ev.) Sutton (Pleocyta sacchari (Massee) Petrak & Sydow) =

ENGLISH   Rind Disease, Sour Rot
FRENCH
SPANISH   Enfermedad de la corteza, Podredumbre ácida
PORTUGUESE
ITALIAN
RUSSIAN
SCANDINAVIAN
GERMAN
DUTCH
HUNGARIAN
BULGARIAN
TURKISH
ARABIC
PERSIAN
BURMESE
THAI
NEPALI
HINDI
VIETNAMESE
CHINESE   Gan jùoh wùei pi bìng   甘蔗外皮病
JAPANESE

This disease is recognized by the numerous black, coiled masses of hair-like fruiting structures of the fungus which issue from pustules, breaking through the stalks. Internal tissue becomes dark in color and has a sour odor. Leaf sheaths and blades near the joints may be affected, resulting in premature yellowing and dying. The black, coiled fruiting structures also appear on them.

Iste maladia es identificate per le numerose nigre massas inrolate de structuras sporal capillar del fungo, le quales emana ab le pustulas penetrante le pedunculos. Texito interne deveni de color obscur e ha un odor acerbe. Vaginas e laminas foliar presso le nodos es a vices afflicte, causante jalnessa prematur e morte. Le structuras sporal inrolate nigre se manifesta anque in illos.

Cette maladie est reconnaissable aux nombreuses masses noires torsadées des fructifications chevelues du champignon, qui sortent de pustules et traversent la tige. Les tissus internes se mélanisent et exhalent une odeur de suri. Les gaines et les limbes des feuilles au voisinage des noeuds peuvent être atteints, ce qui entraîne un jaunissement et un dessèchement précoces. Il s'y forme également des fructifications noires en torsades.

Esta enfermedad se reconoce por las numerosas masas negras, enroscadas de estructuras fructíferas del hongo, que parecen de pelo y las cuáles brotan de pústulas que atraviesan los tallos. El tejido interior se vuelve de color obscuro y tiene un olor ácido. Las cubiertas y las hojas cerca del os nudos pueden ser afectadas, dando como resultado un amarillamiento prematuro y muerte. Las estructuras fructíferas negras, enroscadas también aparecen sobre éstos.

DISTRIBUTION: Af., A., Aust., Eur., N.A., C.A., S.A.

Saccharum spp. + Puccinia kuehnii Butler =

ENGLISH	Leaf Rust	
FRENCH	Rouille	
SPANISH	Roya de la hoja	
PORTUGUESE		
ITALIAN	Ruggine	
RUSSIAN	**ржавчина**	
SCANDINAVIAN		
GERMAN	Rost	
DUTCH		
HUNGARIAN		
BULGARIAN		
TURKISH		
ARABIC		
PERSIAN		
BURMESE		
THAI	Ra-Snim-Lek	ราสนิมเหล็ก
NEPALI		
HINDI	Ratua	रतुवा
VIETNAMESE		
CHINESE	Gan jùoh shǐou bìng	甘蔗銹病
JAPANESE	Sabi-byō	さび病

The earliest symptoms on the leaves are minute, elongated, yellowish spots which are visable on both leaf surfaces. They increase in size, mainly in length, and turn brown to orange-brown in color, meanwhile gaining a slight but definite pale yellow-green halo. The lesions rapidly take on a pustular appearance mainly on the lower leaf surface. When the disease is severe, considerable numbers of lesions occur on a leaf, coalescing to form large, irregular, necrotic areas and resulting in premature death of even young leaves.

Le prime symptomas in le folios es minute maculas jalne elongate, le quales es visibile in tote le duo superfacies foliar. Illos cresce, principalmente in longitude, e deveni brun o brun-orange, monstrante in plus un halo definite verde-jalne pallide. Le lesiones rapidemente deveni pustular, specialmente in le superfacie inferior foliar. Quando le infection es sever, numerose lesiones occurre in cata folio, coalescente usque a formar grande areas irregular necrotic le quales causa morte prematur mesmo de folios juvene.

Les premiers symptômes sur les feuilles sont de minuscules taches jaunâtres allongées, visibles sur les deux faces. Ces taches s'accroissent, surtout en longueur, et prennent une couleur brune à brun orangé, cependant qu'elles s'entourent d'un halo léger, mais défini, de couleur gris jaunâtre pâle. Les lésions ne tardent pas à prendre l'aspect de pustules, surtout à la face inférieure des feuilles. Quand la maladie est grave, il se forme sur la feuille un nombre considérable de lésions qui se fondent pour produire de grandes plages nécrotiques irrégulières, ce qui entraîne la mort prématurée des feuilles même jeunes.

Los primeros síntomas en las hojas son diminutas manchas alargadas, amarillosas que son visibles en ambas superficies de la hoja. Estas aumentan en tamaño, especialmente en el largo y se vuelven de color café a café-anaranjado, mientras van adquiriendo un leve, pero definitivo, halo amarillo verdoso. Las lesiones rápidamente adquieren una apariencia de pústula principalmente en el envés de la hoja. Cuando la enfermedad es severa, un considerable número de lesiones aparecen en la hoja, se juntan para formar grandes áreas necróticas, irregulares y resultan en la muerte prematura de aún las hojas jóvenes.

DISTRIBUTION: Af., A., Aust., N.A., W.I., S.A.

236

Saccharum officinarum + Ustilago scitaminea Sydow =

ENGLISH   Smut
FRENCH   Charbon
SPANISH   Carbón, Tizón
PORTUGUESE   Carvão
ITALIAN   Carbone
RUSSIAN   головня стеблей
SCANDINAVIAN
GERMAN   Stengelbrand
DUTCH
HUNGARIAN
BULGARIAN
TURKISH
ARABIC
PERSIAN
BURMESE   Kyat-kho-hmo-ah-hnan roga   ၊   အာ ၀ကၬ ဂု ၀ က ၆ တ ၄ ၌ ယ ၁း
THAI   Ra-Khamouw-Dum   ราเขมาคำ
NEPALI   Ukhu Ko kalo tuppo   उखुको कालो टुप्पो
HINDI   Kand   कंद
VIETNAMESE   Thối đen lá ruột
CHINESE   Gan jùoh hei swày bỉng   甘 蔗 黑 穗 病
JAPANESE   Kuroho-byõ   黒穂病

The characteristic symptom of this disease is the production of a whip-like structure from the
apex of the infected stalk. It varies in size from a few inches to several feet, with the shorter
ones straight or slightly curved and the long ones usually doubled back. The membranous
covering eventually ruptures, exposing a layer of black soot-like spores which are disseminated
by the wind, leaving only the core.

Le symptomas characteristic de iste maladia es le formation de un structura como un flagello
ab le apice del canna infectate, le qual se varia in dimension ab alicun centimetros usque a
multe decimetros--le structuras plus curte essente recte o legiermente curvate e le longes
usualmente retro-curvate. Le copetura membranacee al fin se rumpe, lassante un strato de
sporas nigre como fuligine, le quales rapidemente es disseminate per le ventos, lassante
solmente le parte interior.

Le symptôme caractéristique de cette maladie est l'apparition, à l'apex de la canne infectée,
d'une sorte de fouet. La longueur des fouets varie de quelques pouces à plusieurs pieds, les
plus petits se tenant droits ou légèrement courbés, tandis que les longs sont généralement
repliés. La membrane qui les recouvre se déchire par la suite, mettant ainsi à nu une couche
de spores noires fuligineuses que le vent dissémine, en ne laissant que le trognon.

El síntoma característico de esta enfermedad es la producción de una estructura en forma de
látigo desde el ápice del tallo infectado. Varía en tamaño desde unas pocas pulgadas hasta
varios piés, con los más cortos erectos o un poco encorvados, y los largos generalmente dobla-
dos hacia atrás. La cubierta membranosa eventualmente se rompe, dejando expuesta una capa
de esporas negras que parecen hollín, las cuáles son diseminadas por el viento dejando sólo
el corazón.

DISTRIBUTION: Af., A., Aust., Eur., C.A., S.A.

Saccharum officinarum + Xanthomonas albilineans (Ashby) Dowson =

ENGLISH	Leaf Scald
FRENCH	Brûlure de la Feuille de la Canne a sucre
SPANISH	Estriado de la hoja, Escaldadura de la hoja
PORTUGUESE	Escaldadura
ITALIAN	Batteriosi
RUSSIAN	ожог листьев бактериальный
SCANDINAVIAN	
GERMAN	Bakterieller Blattbrand
DUTCH	
HUNGARIAN	
BULGARIAN	
TURKISH	
ARABIC	
PERSIAN	
BURMESE	
THAI	Bi-Luak โรคใบลวก
NEPALI	
HINDI	
VIETNAMESE	
CHINESE	Gan jùoh bór tíao bìng 甘蔗白條病
JAPANESE	Sirosuzi-byô 白条病

This disease has two distinct phases (1) the chronic form and (2) the acute form. (1) The chief symptoms are narrow whitish stripes on the leaves and leaf sheaths, stunted stalks with stiff upward and inward curling leaves scalded at the tips, whitish color of leaves and profuse development of side shoots. (2) Affected plants suddenly wilt and die as if killed by drought and are symptomless for the most part.

Iste maladia ha duo phases distinctive (1) le forma chronic e (2) le forma acute. (1) le symptomas principal es stricte strias quasi blanc in le folios e in le vaginas foliar; pedunculos de crescimento impedite e con folios rigide, inrolante se in alto e ad intro, con escaldatura in le extremitates; color quasi blanc in le folios; e developpamento extensive de crescentia lateral. (2) plantas afflicte subito marcesce e mori como a causa de siccitate. Symptomas usualmente non es remarcabile.

Cette maladie a deux phases distinctes (1) la forme chronique et (2) la forme aiguë. (1) Les principaux symptôme sont d'étroites raies blanchâtres sur les feuilles et les gaines, des tiges rabougries avec des feuilles raides à bout échaudé, dont la face supérieure s'enroule vers l'intérieur ou l'extérieur, des feuilles blanchâtres et des pousses latérales à profusion. (2) Les plantes atteintes se flétrissent soudainement et meurent comme si elles avaient succombé à la sécheresse, sans le plus souvent présenter d'autres symptomês.

Esta enfermedad tiene dos fases distintas (1) la forma crónica y la forma aguda, (2). Los síntomas principales de la primera son rayas angostas blanquecinas en las hojas y cubiertas de las hojas, tallos achaparrados con hojas erectas y enroscadas hacia adentro, escaldadas en las puntas, color blanquecino de las hojas y desarrollo profuso de brotes laterales. En la segunda, las plantas se marchitan repentinamente y mueren como si hubieran muerto a causa de la sequía y la mayoría no tiene síntomas.

DISTRIBUTION: Af., A., Aust., N.A., S.A., W.I.

238

Saccharum officinarum + Xanthomonas rubrilineans (Lee et al.) Starr & Burkh. (Pseudomonas rubrilineans (Lee et al.) Starr & Burkh.) =

ENGLISH	Red Stripe
FRENCH	Maladie Bactérienne des Stries Rouges, Rayure Rouge de la Canne a Sucre
SPANISH	Raya roja, Estriado de la hoja, Rayado bacterial
PORTUGUESE	Estrias Vermelhas, Podridão do Tôpo
ITALIAN	Batteriosi
RUSSIAN	полосатость бактериальная красная
SCANDINAVIAN	
GERMAN	Rotstreifigkeit
DUTCH	
HUNGARIAN	
BULGARIAN	
TURKISH	
ARABIC	
PERSIAN	
BURMESE	
THAI	
NEPALI	
HINDI	Lal Dhari लाल धारी
VIETNAMESE	Sọc đỏ - Thối đọt
CHINESE	Gan jùoh chỉr tỉao bỉng 甘蔗赤條病
JAPANESE	Akasuzi-byô 赤条病

Leaf stripes are at first water-soaked, narrow lesions. They elongate rapidly and turn red to maroon in color. Often the stripes are bordered by water-soaked or yellow zones. They are usually continuous with a uniformly necrotic and colored area. Later the stripes coalesce, forming alternate red and chlorotic stripes. Exudate is formed on the necrotic areas and red staining occurs in the vascular bundles. The upper portion of the stalk tissue rots and the upper leaves are killed.

Le strias foliar es al initio aquose lesiones stricte. Illos se elonga rapidemente e deveni de color rubie o castanie. Souvente le strias es bordate per zonas jalne o aquose. Usualmente illos es continue con un area uniformemente necrotic e colorate. Plus tarde le strias coalesce in strias alternativemente rubie e chlorotic. Exsudato se forma in le areas necrotic e rubification occurre in le fasces vascular. Le portion superior del texito del pedunculo putresce, e le folios superior mori.

Les rayures des feuilles sont d'abord des lésions étroites, délavées. Elles s'allongent rapidement et prennent une couleur rouge à marron. Souvent, les rayures sont bordées par des zones jaunes ou d'aspect détrempé. Elles sont généralement continues et présentent une plage uniformément nécrotique et colorée. Plus tard, les rayures se fusionnent, créant un alternance de raies rouges et de raies chlorotiques. Un exsudat se forme sur les plages nécrotiques et les faisceaux vasculaires se colorent en rouge. La partie supérieure des tissus de la tige pourrit et les feuilles du haut sont détruites.

Las rayas de las hojas aparecen primero como lesiones angostas, empapadas de agua. Estas se alargan rápidamente y se vuelven de color marrón. Frecuentemente las rayas están bordeadas por zonas empapadas de agua o de color amarillo. Estas son generalmente contínuas, con un área necrótica uniforme y coloreada. Más tarde las rayas se juntan, formando rayas alternas cloróticas y rojas. En las áreas necróticas se forma una exudación y los cilindros vasculares se tiñen de rojo. La parte superior del tejido del tallo se pudre y las hojas superiores mueren.

DISTRIBUTION: Af., A., Aust., N.A., C.A., S.A.

Saccharum officinarum + Xanthomonas rubrisubalbicans (Christopher & Edgerton) Bergey =

ENGLISH	Mottled Stripe
FRENCH	Maladie des Rayures Marbrées, Moucheture Bactérienne
SPANISH	Estriado de la hoja, Raya moteada
PORTUGUESE	
ITALIAN	Striscia variegata
RUSSIAN	**крапчатость бактериальная, полосатость крапчатая**
SCANDINAVIAN	
GERMAN	Bakterielle Scheckung
DUTCH	
HUNGARIAN	
BULGARIAN	
TURKISH	
ARABIC	
PERSIAN	
BURMESE	
THAI	
NEPALI	
HINDI	
VIETNAMESE	
CHINESE	
JAPANESE	

Leaf stripes are linear with somewhat irregular margins and centers. The color is predominantly red although frequently white areas or white margins occur. When the streaks coalesce, mottled red and white bands are formed across the leaf blade. Top rot is not associated with this disease.

Le strias foliar es linear con assatis irregular margines e centros. Le color es principalmente rubie ma areas o margines blanc occurre frequentemente. Quando le strias coalesce, bandas maculate rubie e blanc se forma per la lamina foliar. Putrefaction apical non se associa con iste maladia.

Les raies de la feuille sont linéaires avec des marges et un centre quelque peu irréguliers. Elles sont le plus souvent rouges, bien qu'il se présente souvent des plages ou des marges blanches. Quand les stries se fusionnent, il se forme à travers le limbe des bandes marbrées de rouge et de blanc. La pourriture de la tête n'a rien à voir avec cette maladie.

Las rayas en las hojas son lineales con márgenes y centros un poco irregulares. El color es predominantemente rojo, aunque frecuentemente tienen áreas o márgenes blancos. Cuando las rayas se juntan, se forman bandas moteadas de rojo y blanco a lo ancho de la hoja. La podredumbre del extremo no está asociada con esta enfermedad.

DISTRIBUTION: Af., Aust., Eur., N.A., C.A., S.A.

Saccharum officinarum + Xanthomonas vasculorum (Cobb) Dowson =

ENGLISH	Gummosis
FRENCH	Gommose Bactérienne de la Canne a Sucre
SPANISH	Estriado de la hoja, Gomosis
PORTUGUESE	
ITALIAN	Gommosi della canna da zucchero
RUSSIAN	увядание бактериальное, болезнь кобба, гоммоз
SCANDINAVIAN	
GERMAN	Schleimkrankheit, Cobb'sche Krankheit, Gummikrankheit
DUTCH	
HUNGARIAN	
BULGARIAN	
TURKISH	
ARABIC	
PERSIAN	
BURMESE	
THAI	
NEPALI	
HINDI	Gondarti गोनदरती
VIETNAMESE	
CHINESE	
JAPANESE	Gomu-byô ゴム病

In its early stages this disease is primarily a leaf disease showing pale green to yellow stripes flecked with reddish dots, regular in outline when young and diffused in outline as it becomes older. The longitudinal streaks enlarge and turn red to brown as the leaves mature. In advanced stages of the disease, the upper leaves of the older canes develop linear stripes while the older leaves show red blotches and brown streaks. Dwarfing of the plant, necrotic pockets in the stalk tissue and honey-yellow exudate in the conductive tissue of the stalk and veins are characteristic.

In le stadios initial iste maladia es principalmente un morbo foliar con strias verde pallide o jalne maculate per punctos rubiette, de contorno regular al initio e de contorno plus diffuse quando le morbo se avantia. Le strias longitudinal cresce e deveni rubie o brun quando le folios se matura. In stadios avantiate del morbo le folios superior del cannas plus vetule forma strias linear, e le folios plus vetule monstra pustulas rubie e strias brun. Nanismo del planta, caviatates necrotic in le texito del pedunculo, e exsudato jalne como melle in le texitos conductive del pedunculo e del venas es characteristic.

Dans ses premières phases, cette maladie est avant tout une maladie des feuilles montrant des bandes vert pâle à jaunes, mouchettées de points rougeâtres, à contours réguliers au début et plus imprécis avec le temps. Les stries longitudinales s'agrandissent et deviennent rouges à brunes à mesure que les feuilles vieillissent. Aux phases avancées de la maladie, les feuilles supérieures des vieilles cannes forment des stries linéaires, alors que les vieilles feuilles montrent des éclaboussures rouges et des stries brunes. Le rabougrissement de la plante, des poches nécrotiques dans les tissus de la tige et un exsudat jaune miel dans les tissus conducteurs de la tige et des nervures sont des points caractéristiques.

En sus primeras etapas esta enfermedad es principalmente una enfermedad foliar que muestra rayas de color verde pálido a amarillo moteado con puntos rojizos, de forma angular cuando están jóvenes y de forma difusa a medida que se vuelven más viejas. En estados avanzados de la enfermedad, las hojas superiores de los tallos más viejos desarrollan rayas lineales, mientras que las hojas mas viejas muestran manchones rojos y rayas de color café. El acha-parramiento de la planta, cavidades necróticas en el tejido del tallo y exudaciones de color amarillo-miel en los tejidos conductivos del tallo y venas son característicos.

DISTRIBUTION: Af., A., Aust., N.A., C.A., S.A., W.I.

Secale cereale (Gramineae) + Claviceps purpurea (Fr.) Tul. =

ENGLISH	Ergot
FRENCH	Ergot du seigle
SPANISH	Cornezuelo del centeno
PORTUGUESE	
ITALIAN	Chiodo segalino, Segala cornuta, Sclerozi cornuto, Sprone, Mal dell sclerogio
RUSIAN	спорынья
SCANDINAVIAN	Meldrøjer, Mjöldryga, Meldrøjersvampen
GERMAN	Mutterkorn Roggen, Sclerotium-Krankheit
DUTCH	Moederkoren
HUNGARIAN	
BULGARIAN	мораво рогче
TURKISH	Çavdar mahmuzu
ARABIC	
PERSIAN	
BURMESE	
THAI	
NEPALI	
HINDI	
VIETNAMESE	
CHINESE	
JAPANESE	Bakkaku-byð 麦角病

The first symptom of this disease is the sticky exudate which appears in the spikes. A blue-black, compact, hard mass of the fungus develops next, in place of the kernel. These hard bodies resemble the kernel but are longer and darker and are very conspicuous. On a single spike there may be a few or several of these hard, black fungus bodies.

Le symptoma prime de iste maladia es le exsudato viscose le qual se manifesta in le spicas. Un dense massa dur nigre-blau del fungo se forma alora in loco del grano. Iste corpores dur resimila le grano ma es plus elongate, plus obscur, e multo remarcabile. In un sol spica se forma a vices plure structuras sporal nigre e dur.

Le premier symptôme de cette maladie est l'exsudat collant qui apparaît sur les épis. Il se forme ensuite, à la place du grain, une masse bleu foncé, compacte et dure, du champignon. Ces corpuscules durs ressemblent aux grains, mais ils sont plus longs, plus foncés et très voyants. Il peut y avoir sur le même épi un nombre plus ou moins grand de ces masses fongueuses noires et dures.

El primer síntoma de esta enfermedad es la exudación pegajosa que aparece en las espigas. Luego se desarrolla una masa dura del hongo, compacta, de color azul negro, que sustituye al grano. Estos cuerpos duros semejan granos, pero son más largos y duros y muy conspicuos. En una sola espiga puede haber uno o varios de estos cuerpos duros y negros del hongo.

DISTRIBUTION: Af., A., Aust., Eur., N.A., S.A.

Secale cereale (Gramineae) + Puccinia rubigo-vera (DC.) Wint. =

ENGLISH	Brown Leaf Rust, Leaf Rust
FRENCH	Rouille des Feuilles, Rouille Brune
SPANISH	Roja, Foliar café, Roya Foliar
PORTUGUESE	
ITALIAN	Ruggine bruna
RUSSIAN	ржавчина бурая
SCANDINAVIAN	Rugens brunrust, Rågbrunrost, Brunrost på råg
GERMAN	Braunrost, Roggenbraunrost
DUTCH	Bruine Roest
HUNGARIAN	A rozs levélrozsdája
BULGARIAN	
TURKISH	
ARABIC	
PERSIAN	
BURMESE	
THAI	
NEPALI	
HINDI	
VIETNAMESE	
CHINESE	
JAPANESE	Akasabi-byō 赤さび病

The symptoms of this disease are small, oval, orange-brown pustules on both surfaces of the leaves. The pustules break as the disease develops and powdery masses of reddish-brown spores are exposed. Elongated pustules may develop on the stem also. Toward maturity of the plant, small, elongated, dark gray pustules appear which do not immediately break through the epidermis.

Le symptomas de iste maladia es minute, oval pustulas brun-orange in tote le duo superfacies del folios. Le pustulas se finde quando le morbo se avantia e massas pulverose de sporas brun-rubiette es discoperte. Pustulas elongate pote formar se anque in le pedunculos. Verso maturation del planta minute pustulas elongate gris obscur se manifesta, le quales non immediatemente penetra le epidermis.

Les symptômes de cette maladie sont de petites pustules ovales, brun orangé, sur les deux faces des feuilles. Pendant que la maladie évolue, les pustules se brisent et exposent des masses poudreuses de spores brun rougeâtre. Des pustules allongées peuvent se former aussi sur la tige. Lorsque la plante arrive à maturité, on voit apparaître de petites pustules allongées gris foncé, qui ne traversent pas immédiatement l'épiderme.

Los síntomas de esta enfermedad son pequeñas pústulas ovaladas, de color café-anaranjado en ambas superficies de las hojas. La pústula se revienta a medida que la enfermedad se desarrolla, dejando expuestas masas polvorientas de esporas café-rojizas. En el tallo pueden desarrollarse también pústulas alargadas. Hacia el maduramiento de la planta aparecen pequeñas pústulas alargadas, de color gris obscuro, las cúales atraviesan la epidermis.

DISTRIBUTION: Af., A., Aust., Eur., N.A., C.A., S.A.

Secale cereale + Urocystis occulta (Wallr.) Rabh. =

ENGLISH	Stalk Smut
FRENCH	Charbon de la Tige
SPANISH	Tizón de bandera
PORTUGUESE	
ITALIAN	Carbone del culmo, Tarlo del gambo della segala
RUSSIAN	головня стеблей
SCANDINAVIAN	Rugens stængelbrand, Stråsot på råg
GERMAN	Stengelbrand des Roggens, Streifenstengelbrand
DUTCH	Stengelbrand van de Rogge
HUNGARIAN	
BULGARIAN	стъблена главня
TURKISH	Sap sürmesi
ARABIC	
PERSIAN	
BURMESE	
THAI	
NEPALI	
HINDI	
VIETNAMESE	
CHINESE	
JAPANESE	Karakuroho-byô から黒穂病

Diseased plants have a darker green color than healthy ones, and traces of lighter green streaks may be observed on the upper leaves. Soon these plants develop the characteristic symptoms of the disease. These are long, lead-colored stripes on leaves, sheathes and stems; the stripes soon become black and break through the tissue, exposing dusty masses of spores. Affected parts may be twisted and distorted and the leaves split along the stripes. Affected plants are usually smaller than normal ones.

Plantas morbide monstra un color verde plus obscur que normal. Vestigios de strais verde plus clar a vices se manifesta in le folios superior. In breve tempore iste plantas manifesta le symptomas characteristic del maladia. Istos es strias elongate de color de graphite in folios, vaginas, e pedunculos. Le strias rapidemente deveni nigre e penetra le texito, discoperiente massas pulverose de sporas. Partes afflicte es a vices distorte e le folios se finde al longitude del strias. Plantas afflicte es usualmente minor que plantas normal.

Les plantes malades sont d'un vert plus foncé que les plantes saines, et l'on peut observer sur les feuilles du haut des traces de stries d'un vert plus léger. Bientôt, ces plantes présentent les symptômes caractéristiques de la maladie. Ce sont de longues raies couleur de plomb sur les feuilles, les gaines et les tiges; ces raies noircissent rapidement et brisent les tissus, de façon à exposer des masses poussiéreuses de spores. Les parties atteintes peuvent être tordues et difformes, et les feuilles se fendent le long des raies. Les plantes atteintes sont ordinairement plus petites que les plantes saines.

Las plantas enfermas tienen un color verde más obscuro que las plantas sanas y rasgos de rayas de un verde más claro que peuden observarse en las hojas superiores. Pronto estas plantas desarrollan los síntomas característicos de la enfermedad. Estas son rayas largas, de color plomo que aparecen sobre las hojas, cubiertas y tallos; las rayas pronto se vuelven de color negro y penetran a través del tejido, dejando expuestas masas de esporas polvorientas. Las partes afectadas pueden estar torcidas y mal formadas y las hojas se parten a lo largo de las rayas. Las plantas atacadas son generalmente más pequeñas que las normales.

DISTRIBUTION: Af., A., Aust., Eur., N.A., C.A., S.A.

<u>Sesamum</u> <u>indicum</u> + <u>Alternaria</u> <u>sesami</u> (Kawamura) Mohanty & Behere =

ENGLISH	Damping-off, Leaf Spot, Stem Spot, Pod Spot
FRENCH	
SPANISH	Mancha parda de la hoja, Mal del talluelo, Mancha foliar
PORTUGUESE	
ITALIAN	
RUSSIAN	
SCANDINAVIAN	
GERMAN	
DUTCH	
HUNGARIAN	
BULGARIAN	
TURKISH	
ARABIC	
PERSIAN	
BURMESE	
THAI	
NEPALI	
HINDI	
VIETNAMESE	
CHINESE	
JAPANESE	Kokuhan-byô　黒斑病

This disease causes a variety of symptoms since it causes pre-and post-emergence damping-off and stem, leaf and pod spot. On the leaves, the symptoms are brown, irregular spots with concentric zonations on the upper surface which later coalesce. With severe infection, the leaves fall prematurely. Stem lesions are less conspicuous even though they frequently involve the whole length of the stem. Infected seed lead to reduced emergence.

Iste maladia produce varie symptomas perque illo causa putrefaction o ante o post le emergentia e maculation del pedunculos, folios, e siliquas. In le folios, le symptomas es maculas irregular brun con zonationes concentric al superfacie superior, le quales plus tarde coalesce. In infection sever, le folios cade prematurmente. Lesiones in le pedunculos non es tanto remarcabile, ben que frequentemente illos se extende per tote le longitude. Granos infectate causa emergentia reducte (in le anno sequente).

Cette maladie étale un variété de symptômes, puisqu'elle cause une fonte des semis, de pré-émergence et de postémergence, ainsi qu'une tache des tiges, des feuilles et des capsules. Sur les feuilles, les symptômes sont des taches brunes irrégulières avec, à la face supérieure, des ceintures concentriques qui se fusionnent ensuite. Dans les cas graves, les feuilles tombent prématurément. Bien qu'elles se présentent souvent tout le long de la tige, les lésions y sont moins apparentes. L'infection des semences réduit l'émergence.

Esta enfermedad causa una variedad de síntomas ya que causa mal del talluelo en la pre y post emergencia y manchas en el tallo, hojas y vainas. En las hojas los síntomas son manchas irregulares, de color café, con zonas concéntricas en el haz, que luego se juntan. Con infección severa, las hojas se caen prematuramente. Las lesiones del tallo son menos conspicuas, aunque frecuentemente todo el largo del mismo. La semilla infectada causa una emergencia reducida.

DISTRIBUTION: Af., A., N.A., S.A.

Sesamum indicum + Xanthomonas sesami Sabet & Dowson (Pseudomonas sesami Malkoff) =

ENGLISH	Bacterial Leaf Spot
FRENCH	Taches Foliaires Bactériennes
SPANISH	Mancha angular de la hoja, Mancha foliar bacterial
PORTUGUESE	
ITALIAN	
RUSSIAN	пятнистость бактериалсная, бактериоз
SCANDINAVIAN	
GERMAN	Bakterielle Blattfleckenkrankheit
DUTCH	
HUNGARIAN	
BULGARIAN	черно гниене бактериоза
TURKISH	Susam bakteriyel solgunluğu
ARABIC	
PERSIAN	
BURMESE	
THAI	
NEPALI	
HINDI	
VIETNAMESE	
CHINESE	Hú má shìh jìunn ban dǐan bǐng 胡 蔴 細 菌 斑 點 病
JAPANESE	Hanten-saikin-byô 斑点細菌病

This disease causes a leaf spot which affects all aerial parts of the plant. Blackish-brown spots coalesce and may extend along the entire length of the stem. The angular spots on the leaves are delimited by the veins. Infected seed pods turn black.

Iste maladia produce un macula foliar que affice tote le partes aeree del planta. Maculas brunnigre coalesce e a vices se extende per tote le longitude del pedunculo. Le maculas angular foliar es delimitate per le venas. Siliquas infectate deveni nigre.

Cette maladie cause une tache des feuilles qui affecte toutes les parties aériennes de la plante. Les taches brun noirâtre se fusionnent et peuvent couvrir la tige sur toute sa longueur. Sur les feuilles, les taches angulaires sont délimitées par les nervures. Les capsules infectées noircissent.

Esta enfermedad causa una mancha foliar que afecta todas las partes aéreas de la planta. Las manchas café-negruzcas se juntan y pueden extenderse a todo lo largo del tallo. Las manchas angulares en las hojas están delimitadas por las venas. Las vainas infectadas se vuelven negras.

DISTRIBUTION: Af., A., Eur., N.A., S.A.

Solanum melongena + Phomopsis vexans (Sacc. & Syd.) Harter =

ENGLISH	Fruit Rot, Phomopsis Blight, Tip-over
FRENCH	Brûlure Phomopsienne, Pourriture des Fruits, Taches Foliaires
SPANISH	Podredumbre del fruto, Tizón de phomopsis
PORTUGUESE	Podridão sêca de Phomopsis, Câncro de Phomopsis
ITALIAN	
RUSSIAN	гниль плодов сухая и пятнистость (листьев) фомопсисная
SCANDINAVIAN	
GERMAN	Phomopsis-Fruchtfäule, Blattfleckenkrankheit
DUTCH	
HUNGARIAN	
BULGARIAN	
TURKISH	
ARABIC	
PERSIAN	
BURMESE	
THAI	
NEPALI	
HINDI	Phomopsis angmaari फोमोपसिस अंगमारी
VIETNAMESE	
CHINESE	
JAPANESE	Katumon-byô 褐斑病

This pathogen causes damping-off of seedlings and a stem canker, leaf blight and fruit rot of older plants. The foliage is attacked at any time, the lesions usually showing first on the leaves near the ground. The spots are clearly defined, circular, gray to brown, and have a light-colored center. Cankers often form at the base of the stem, either as a constriction or as a gray dry-rot. The spots on the fruit are pale sunken areas that finally may include the whole fruit.

Iste pathogeno causa putrefaction del plantulas e, in plantas plus vetule, cancere del pedunculos, necrosis foliar, e putrefaction del fructos. Le foliage es attaccate per tote le stadios de crescimento, le lesiones usualmente manifestante se in le folios presso le terra al initio. Le maculas es ben definite, circular, gris, o brun, e monstra un centro de color clar. Canceres sovente se forma al pede del pedunculo, o como un constriction o como un putrefaction sic. Le maculas in le fructos es areas pallide depresse le quales a vices se extende usque a involver tote le fructo.

Ce pathogène cause la fonte des semis et un chancre de la tige, ainsi que la brûlure des feuilles et la pourriture du fruit chez les plantes adultes. Les feuilles sont attaquées à n'inporte quel moment, les lésions apparaissant d'ordinaire en premier lieu sur les feuilles situées près du sol. Les taches sont nettement définies, circulaires, grises à brunes, avec un centre plus pâle. Des chancres se forment souvent à la base de la tige, soit sous forme d'étranglement ou de pourriture sèche grise. Les taches sur le fruit consistent en des plages déprimées et pâles qui peuvent finalement envahir tout le fruit.

Este patógeno causa mal del talluelo en las plántulas y cancro en el tallo, tizón en las hojas y podredumbre del fruto en plantas viejas. El follaje es atacado en cualquier época, las lesiones generalmente apareciendo primero en las hojas cercanas al suelo. Las manchas están claramente definidas, son circulares, de color gris a café y tienen centros de color más claro. Los cancros frecuentemente se forman en la base del tallo, ya sea como una constricción o como una podredumbre seca. Las manchas en los fructos son áreas pálidas y hundidas que finalmente pueden incluir todo el fruto.

DISTRIBUTION: Af., A., Aust., N.A., C.A., S.A.

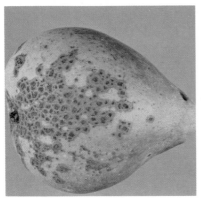

Fig. 218. Pear-Scab; ARS

Fig. 219. Currant-Anthracnose; Cornell Univ.

Fig. 220. Castorbean-Seedling Blight; R. G. Orellana, ARS

Fig. 221. Castorbean-Rust; ARS

Fig. 222. Castorbean-Leaf Spot; Charles A. Thomas, ARS

Fig. 223. Raspberry-Spur Blight; A. J. Braun, N.Y. St. Agric. Exp. Station

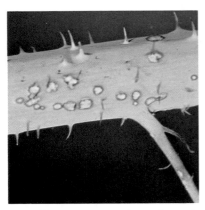

Fig. 224. Raspberry-Cane Spot; A. J. Braun, N.Y. St. Agric. Exp. Station

Fig. 225. *Rubus* spp.-Orange Rust; ARS

Fig. 226. Sugarcane-Wilt; H. Koike, ARS, Houma, La.

PLATE 25

Fig. 227. Sugarcane-Brown Spot; H. Koike, ARS, Houma, La.

Fig. 228. Sugarcane-Sheath Spot; Charles Y. Yang, Taiwan, Republic of China

Fig. 229. Sugarcane-Red Rot; H. Koike, ARS, Houma, La.

Fig. 230. Sugarcane-Sheath Rot; H. Koike, ARS, Houma, La.

Fig. 231. Sugarcane-Brown Stripe; H. Koike, ARS, Houma, La.

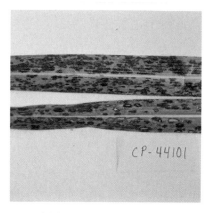

Fig. 232. Sugarcane-Leaf Spot; Benjamin Waite, El Salvador

Fig. 233. Sugarcane-Eyespot; H. Koike, ARS, Houma, La.

Fig. 234. Sugarcane-Rind Disease; H. Koike, ARS, Houma, La.

Fig. 235. Sugarcane-Leaf Rust; H. Koike, ARS, Houma, La.

PLATE 26

Fig. 236. Sugarcane-Smut; H. Koike, ARS, Houma, La.

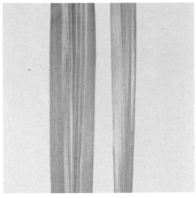

Fig. 237. Sugarcane-Leaf Scald; H. Koike, ARS, Houma, La.

Fig. 238. Sugarcane-Red Stripe; H. Koike, ARS, Houma, La.

Fig. 239. Sugarcane-Mottled Stripe; Benjamin Waite, El Salvador

Fig. 240. Sugarcane-Gummosis; H. Koike, ARS, Houma, La.

Fig. 241. Rye-Ergot; Dept. of Pl. Path., Purdue Univ.

Fig. 242. Rye-Brown Leaf Rust; J. A. Browning

Fig. 243. Rye-Flag Smut

Fig. 244. Sesame-Leaf Spot; Charles A. Thomas, ARS

PLATE 27

Fig. 245. Sesame-Bacterial Leaf Spot; Charles A. Thomas, ARS

Fig. 246. Eggplant-Fruit Rot; ARS

Fig. 247. Eggplant-Leaf Spot

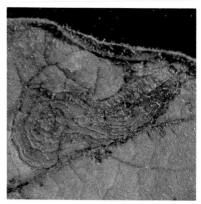

Fig. 248. Potato-Early Blight; Arden F. Sherf, Cornell Univ.

Fig. 249. Potato-Leaf Blotch; Dept. of Pl. Path., West Va.

Fig. 250. Potato-Black Dot; F. Manzer, U. of Me.

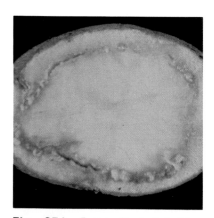

Fig. 251. Potato-Bacterial Ring Rot; Arden F. Sherf, Cornell Univ.

Fig. 252. Potato-Black Leg; ARS

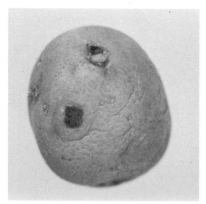

Fig. 253. Potato-Skin Spot; Robert Goth, ARS

PLATE 28

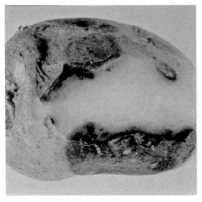

Fig. 254. Potato-Canker; Clark Livingston, Colo. St. Univ.

Fig. 255. Potato-Pink Rot; Clark Livingston, Colo. St. Univ.

Fig. 256. Potato-Late Blight; Robert Goth, ARS

Fig. 257. Potato-Bacterial Wilt; ARS

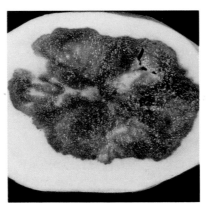

Fig. 258. Potato-Leak; Arden F. Sherf, Cornell Univ.

Fig. 259. Potato-Violet Root Rot

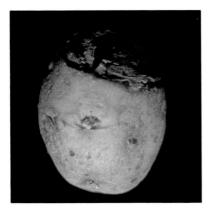

Fig. 260. Potato-Silver Scurf; Clark Livingston, Colo. St. Univ.

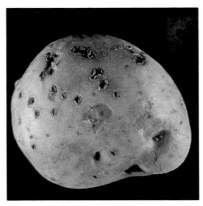

Fig. 261. Potato-Powdery Scab; ARS

Fig. 262. Potato-Scab; F. Manzer, U. of Me.

PLATE 29

Fig. 263. Potato-Wart; Dept. of Pl. Path., West Va.

Fig. 264. Potato-Wilt; Clark Livingston, Colo. St. Univ.

Fig. 265. Sorghum-Leaf Spot; Richard A. Fredrickson, Texas A&M Univ.

Fig. 266. Sorghum-Anthracnose; Richard A. Fredrickson, Texas A&M Univ.

Fig. 267. Sorghum-Zonate Leaf Spot; Richard A. Fredrickson, Texas A&M Univ.

Fig. 268. Sorghum-Charcoal Rot; Richard A. Fredrickson, Texas A&M Univ.

Fig. 269. Sorghum-Leaf Rust; Richard A. Fredrickson, Texas A&M Univ.

Fig. 270. Sorghum-Downy Mildew; Richard A. Fredrickson, Texas A&M Univ.

Fig. 271. Sorghum-Long Smut; Richard A. Fredrickson, Texas A&M Univ.

PLATE 30

Fig. 272. Sorghum-Loose Kernel Smut; Richard A. Fredrickson, Texas A&M Univ.

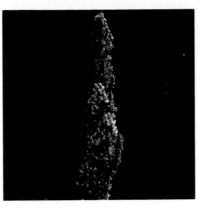

Fig. 273. Sorghum-Covered Kernel Smut; Richard A. Fredrickson, Texas A&M Univ.

Fig. 274. Sorghum-Bacterial Streak; Richard A. Fredrickson, Texas A&M Univ.

Fig. 275. Tea-Brown Blight

Fig. 276. Tea-Gray Blight; Venkata Ram, India

Fig. 277. Tea-Red Blight

Fig. 278. Tea-Root Disease; Venkata Ram, India

Fig. 279. Cacao-Canker; Ernest P. Imle, ARS

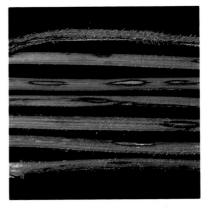

Fig. 280. Clover-Northern Anthracnose; E. W. Hanson, ARS

PLATE 31

Fig. 281. Clover-Downy Mildew

Fig. 282. Clover-Sclerotinia Wilt; E. W. Hanson, ARS

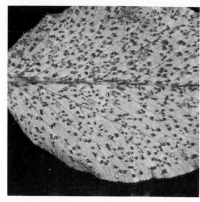

Fig. 283. Clover-Rust; E. W. Hanson, ARS

Fig. 284. Wheat-Eyespot Foot Rot; Dept. of Pl. Path., Wash. St. Univ.

Fig. 285. Wheat-Yellow Slime

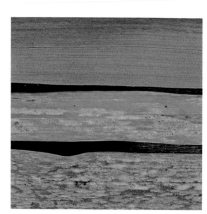

Fig. 286. Wheat-Powdery Mildew; R. A, Kilpatrick, ARS

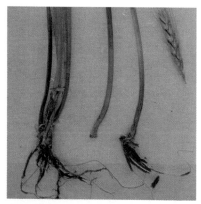

Fig. 287. Wheat-Root and Culm Rot; Dept. of Pl. Path., Wash. St. Univ.

Fig. 288. Wheat-Snow Mold; Don M. Huber, Purdue Univ.

Fig. 289. Wheat-Take-All; Dept. of Pl. Path., Univ. of Wisc.

PLATE 32

Fig. 290. Wheat-Stripe; Earl D. Hansing, Dept. Pl. Path., Kansas St. Univ.

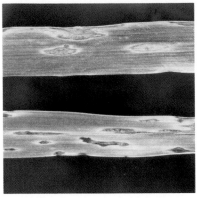

Fig. 291. Wheat-Glume Blotch; A. S. Williams, Univ. of Kentucky

Fig. 292. Wheat-Basal Glume Rot

Fig. 293. Wheat-Yellow Rust; R. A. Kilpatrick, ARS

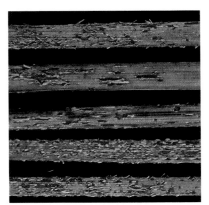

Fig. 294. Wheat-Stem Rust; J. A. Browning

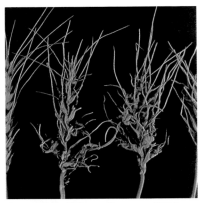

Fig. 295. Wheat-Downy Mildew; R. A. Kilpatrick, ARS

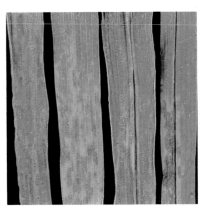

Fig. 296. Wheat-Speckled Leaf Blotch; R. A. Kilpatrick, ARS

Fig. 297. Wheat-Rough-spored Bunt; Earl D. Hansing, Dept. of Pl. Path., Kansas St. U.

Fig. 298. Wheat-Smooth-spored Bunt; Dept. of Pl. Path., U. of Wisc.

PLATE 33

Fig. 299. Wheat-Flag Smut; Dept. of Pl. Path., Wash. St. Univ.

Fig. 300. Wheat-Black Chaff; A. S. Williams, Univ. of Kentucky

Fig. 301. Elm-Dutch Elm Disease; Dept. of Botany & Pl. Path., U. of Md.

Fig. 302. Boradbean-Gray Mold Shoot Blight

Fig. 303 Broadbean-Rust; Charles Y. Yang, Taiwan, Republic of China

Fig. 304. Grape-White Rot

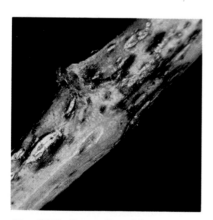

Fig. 305. Grape-Dead Arm; Austin C. Goheen, ARS

Fig. 306. Grape-Anthracnose; ARS

Fig. 307. Grape-Bacterial Blight; Austin C. Goheen, ARS

PLATE 34

Fig. 308. Grape-Black Rot; ARS

Fig. 309. Grape-Leaf Rust

Fig. 310. Grape-Downy Mildew; ARS

Fig. 311. Grape-Root Rot

Fig. 312. Grape-Powdery Mildew; ARS

Fig. 313. Corn-Bacterial Wilt; Arden F. Sherf, Cornell Univ.

Fig. 314. Corn-Glume Mold; Dept. of Pl. Path., Univ. of Illinois

Fig. 315. Corn-Dry Rot of Ears; Dept. of Pl. Path., Univ. of Illinois

Fig. 316. Corn-Gibberella Blight; Mauricio Manzano, El Salvador

PLATE 35

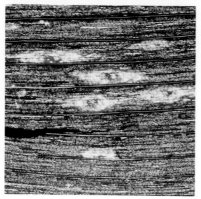

Fig. 317. Corn-Leaf Spot; Dept. of Pl. Path., Univ. of Illinois

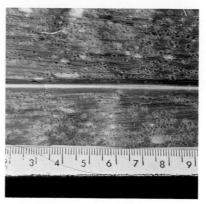

Fig. 318. Corn-Southern Leaf Blight; Dept. of Pl. Path., Univ. of Illinois

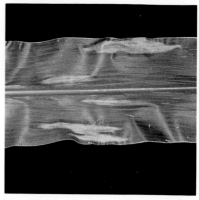

Fig. 319. Corn-Leaf Blight; Dept. of Pl. Path., Univ. of Illinois

Fig. 320. Corn-Brown Spot; Dept. of Pl. Path., Univ. of Illinois

Fig. 321. Corn-Rust; Mauricio Manzano, El Salvador

Fig. 322. Corn-Rust; Mauricio Manzano, El Salvador

Fig. 323. Corn-Leaf Rust; J. Stanley Melching, Pl. Dis. Res. Lab., ARS

Fig. 324. Corn-Head Smut; Dept. of Pl. Path., Univ. of Illinois

Fig. 325. Corn-Smut; Arden F. Sherf, Cornell Univ.

PLATE 36

Solanum melongena (Solanum spp., Lycopersicon esculentum, Capsicum spp.) + Stemphylium
solani Weber =

ENGLISH   Leaf Spot of Eggplant, Gray Leaf Spot of Tomato
FRENCH
SPANISH   Mancha foliar
PORTUGUESE
ITALIAN
RUSSIAN
SCANDINAVIAN
GERMAN
DUTCH
HUNGARIAN
BULGARIAN
TURKISH
ARABIC
PERSIAN
BURMESE
THAI
NEPALI
HINDI
VIETNAMESE
CHINESE
JAPANESE

The disease is limited to the leaf blades, rarely on the stems, and is absent on the fruit. The
plant can become infected from the earliest seedling stage to the mature plant. Only a few
spots may be present on the oldest leaves as there may be infection in varying degrees of
severity until the entire plant is defoliated. The gray spots appear first as small, circular,
sunken brownish to black lesions. Some of the spots are surrounded by a yellow halo.

Le maladia es limitate al laminas foliar, rarmente se trova in le petiolos, e nunquam in le
fructos. Le planta pote suffrer infection ab le stadios initial de plantulas usque al planta maturate. A vices solmente pauc maculas se manifesta in le folios plus vetule perque, ante que
tote le planta suffre defoliation, le severitate de infection pote variar se. Le maculas gris
se manifesta al initio como minute lesiones circular depresse brunette o nigre. Alicunes del
maculas es imbraciate per un halo jalne.

Cette maladie se limite au limbe des feuilles; elle s'attaque rarement aux tiges et n'atteint pas
le fruit. La plante peut s'infecter depuis les premiers stades de plantule jusqu'à maturité. Il
peut n'y avoir que de rares taches sur les vieilles feuilles, comme il peut y avoir des infections de divers degrés de gravité jusqu'à ce que toute la plante soit défoliée. Les taches
grises se présentent d'abord sous forme de petites lésions circulaires, déprimées, brunâtres
à noires. Certaines taches sont entourées d'un halo jaune.

La enfermedad está limitada a las hojas, raramente ataca los tallos y nunca los frutos. La
planta puede infectarse desde la temprana edad, hasta que es planta madura. Solamente unas
pocas manchas pueden estar presentes en las hojas más viejas ya que puede haber infección
en diferentes grados de severidad, hasta que toda la planta está defoliada. Las manchas grises
aparecen primero como pequeñas lesiones circulares, hundidas, de color cafesoso a negro.
Algunas de las manchas están rodeadas por un halo amarillo.

DISTRIBUTION: Af., A., Aust., Eur., N.A., C.A.

248

Solanum tuberosum (Lycopersicon esculentum and others) + Alternaria solani Sorauer =

ENGLISH   Early Blight
FRENCH   Brûlure Alternarienne, Alternariose, Maladie des Taches Brunes et Taches Noires
SPANISH   Mancha negra de la hoja, Alternariosis, Lancha temprana, Tizón temprano
PORTUGUESE   Pinta preta, Queima das fôlhas
ITALIAN   Alternaria, Nebbia, Seccume primaverile dell patate
RUSSIAN   альтернариоз, пятнистость бурая, гниль ранняя, пятнистость краевая
SCANDINAVIAN   Kartoffelbladpletsyge, Torrfläcksjuka
GERMAN   Dörrfleckenkrankheit, Alternaria-Fäule, Hartfäule, Knollenbefall
DUTCH   Alternaria-ziekte, Droogrot, Alternariarot der Aardappelen
HUNGARIAN
BULGARIAN   кафяви листни петна
TURKISH   Erken yaprak yanıklığı
ARABIC   Lafha Moubakera   لفحـمبكرة
PERSIAN
BURMESE   Set-waing-yit-pyauk roga   အာ ၁လ၈ာ ၁ ၆၄ ၛ၊ ၁းဈၛ၎င၆
THAI
NEPALI   Alu Ko rato-dadhuwa rog   आलुको रातो दद्दुवा रोग
HINDI   Agete angmaari   अगेटे अंगमारी
VIETNAMESE   Bệnh úa sớm
CHINESE   Mǎ líng shǔ shìa yìh bìng   馬鈴薯夏疫病
JAPANESE   Natueki-byô   夏疫病

This disease develops in the field after the plant has passed its stage of greatest vigor, and in storage. Lesions of various sizes and shapes are usually sunken with regular raised borders. Internally, the lesions are shallow and sharply set off from the healthy tissues by a layer of purplish-brown, metallic-hued cork. A yellowish discoloration of the flesh is sometimes produced just beyond the edge of the lesion. On the foliage, lesions are brown to black and usually show concentric rings and are generally oval in shape. At times, they are vein-limited. The spots enlarge as the leaf weakens and dies.

Iste maladia se forma in le campo post que le planta ha passate su stadio de vigor plus grande, e in immagasinage. Lesiones de contornos e de dimensiones varie usualmente es depresse con bordos regular e elevate. Internemente, le lesiones es nonprofunde e acutemente distincte ab le texito salubre per un strato de corco, brun-purpuree con un lustro metallic. Un discoloration jalne del pulpa a vices se forma un pauc ultra le margine del lesion. In le foliage, lesiones es brun o nigre e usualmente monstra anellos concentric e es generalmente de contorno oval. A vices illos es delimitate per le venas. Le maculas cresce quando le folio es infirmate e mori.

Cette maladie survient dans le champ, après que la plante a passé son stade le plus vigoureux, ainsi qu'en entrepôt. Les lésions, de dimensions et de formes diverses, sont généralement déprimées, avec des rebords régulièrement élevés. A l'intérieur, les lésions sont peu profondes et nettement séparées des tissus sains par une couche de liège brun pourpré, à reflets métalliques. Un jaunissement de la chair se produit quelquefois juste au delà de bord de la lésion. Sur le feuillage, les lésions sont brunes à noires, ovales, et présentent généralement des anneaux concentriques. Parfois, elles sont limitées par les nervures. Les taches s'agrandissent pendant que les feuilles s'affaiblissent et meurent.

Esta enfermedad se desarrolla en el campo después que la planta ha pasado su época de mayor vigor y también durante el almacenaje. Las lesiones de varios tamaños y formas son generalmente hundidas con bordes regulares levantados. Internamente, las lesiones son poco profundas y bien fijadas en el tejido sano por una capa de corcho de color café purpúreo, matizado metálico. Algunas veces se produce una decoloración amarillenta en la carne, un poco abajo del borde de la lesión. En el follaje, las lesiones son de color café a negro y generalmente muestran anillos concéntricos y son generalmente ovaladas en su forma. A veces, están limitadas por las venas. Las manchas crecen a medida que la hoja se debilita y muere.
DISTRIBUTION: Af., A., Aust., Eur., N.A., C.A., S.A.

Solanum tuberosum + Cercospora concors (Casp.) Sacc. =

ENGLISH	Leaf Blotch
FRENCH	Taches Foliaires, Cercosporose
SPANISH	Manchón foliar
PORTUGUESE	
ITALIAN	
RUSSIAN	церкоспороз, пятнистость желтая
SCANDINAVIAN	
GERMAN	Cercospora-Blattfleckenkrankheit
DUTCH	Viltvlekken
HUNGARIAN	
BULGARIAN	
TURKISH	
ARABIC	
PERSIAN	
BURMESE	
THAI	
NEPALI	
HINDI	
VIETNAMESE	
CHINESE	
JAPANESE	

The symptoms of this disease appear when the plants begin to blossom. The lower leaves show obscurely defined, pale spots, 3-5 mm. in diameter. These spots sometimes coalesce to form larger areas. Often the entire leaf slowly turns yellow and dies.

Le symptomas de iste maladia se manifesta al comenciamento de floration. Le folios inferior monstra maculas pallide e mal definite, 3-5 mm de diametro. Iste maculas a vices coalesce usque a formar areas plus grande. Sovente tote le folio deveni jalne e mori.

Les symptômes de cette maladie apparaissent quand les plantes commencent à fleurir. Les feuilles du bas montrent des taches pâles, mal définies, de 3 à 5 mm de diamètre. Quelquefois, ces taches se joignent pour couvrir une surface plus grande. Souvent, la feuille entière jaunit lentement et meurt.

Los síntomas de esta enfermedad aparecen cuando la planta comienza a florecer. Las hojas inferiores muestran manchas obscuramente definidas, pálidas, de 3-5 mm. de diámetro. Estas manchas a veces se juntan para formar grandes áreas. Frecuentemente toda la hoja lentamente se vuelve amarilla y muere.

DISTRIBUTION: Af., A., Eur., N.A.

250

Solanum tuberosum (Lycopersicon esculentum) + Colletotrichum atramentarium (Berk. & Br.)
Taubenh. =

ENGLISH	Black Dot Disease, Root Rot
FRENCH	Dartrose, Anthracnose
SPANISH	Antracnosis, Enfermedad de punto negro
PORTUGUESE	
ITALIAN	Dartrosi della patata, Puntulatura nera
RUSSIAN	антракноз
SCANDINAVIAN	
GERMAN	Fusskrankheit, Anthraknose, Blattdürre
DUTCH	Zwarte Spikkel
HUNGARIAN	
BULGARIAN	
TURKISH	
ARABIC	
PERSIAN	
BURMESE	Ah-net-set-roga ကၡလက်တို့ကြိုင်ခခ်အက်ကြၠၦငၧတရၧယၥ
THAI	
NEPALI	
HINDI	
VIETNAMESE	Héo lá
CHINESE	
JAPANESE	Tanso-byô 炭そ病

The first sign in the field is yellowing of leaflets, beginning at the top. Upward rolling of the margins may occur and the leaves gradually droop and die. In the stem just above the soil-line down to the first tuber the cortex is broken down, sometimes with an amethyst coloration of the tissue. Similar lesions are more abundant on the stolons. Small tubers shrivel while larger tubers may just be reduced in size. Other mature tubers bear fungus fruiting bodies, scattered or in patches.

In le campo le prime symptoma es le jalnessa de foliettos, comenciante in le extremitates. Inrolamento in alto del margines pote occurrer e le folios gradualmente marcesce e mori. In le pedunculos, le cortices ab le terra usque al prime tubere es corrupte, a vices con un coloration como amethysto del texito. Lesiones similares plus numerose in le stolones. Minute tuberes se crispa ma le tuberes plus grande es a vices solmente impedite in lor crescimento. Altere tuberes maturate porta structuras sporal del fungo o disperse o in gruppamentos.

Le premier indice dans le champ est un jaunissement du bout des folioles. Les marges peuvent se retrousser petit à petit, et les feuilles pendent et meurent. Depuis le ras du sol jusqu'au premier tubercule, le cortex se désintègre parfois en conférant aux tissus la couleur de l'améthyste. Des lésions semblables sont plus abondantes sur le stolon. Les petits tubercules se recroquevillent, alors que les gros tubercules perdent simplement du volume. D'autres tubercules mûrs portent les fructifications du champignon, soit dispersées, soit en plaques.

El primer síntoma en el campo es el amarillamiento de las hojillas, comenzando desde arriba. Las hojas pueden enroscarse hacia arriba, decayéndose prematuramente y muriendo. En el tallo un poco arriba de la línea del suelo y abajo hasta el primer tubérculo, la corteza se desagrega algunas veces mostrando una coloración de color amatista en el tejido. Lesiones similares son más abundantes en los estolones. Los tubérculos pequeños se arrugan, mientras que los grandes pueden sólo reducirse en su tamaño. Otros tubérculos maduros tienen cuerpos fructíferos del hongo, diseminados o en manchones.

DISTRIBUTION: Af., A., Aust., Eur., N.A., C.A., S.A..

Solanum tuberosum + Corynebacterium sepedonicum (Spieck. & Kotth.) Skapt & Burkh. =

ENGLISH	Bacterial Ring Rot
FRENCH	Flétrissure Bactérienne, Bactériose Annulaire
SPANISH	Pudrición anular, Pudrición bacterial anular
PORTUGUESE	
ITALIAN	Batteriosi
RUSSIAN	**гниль клубней кольцевая**
SCANDINAVIAN	
GERMAN	Bakterienringfäule, Bakterielle Ringfäule, Ringbakteriose
DUTCH	
HUNGARIAN	
BULGARIAN	
TURKISH	Patates halka çürcüklüğü
ARABIC	
PERSIAN	
BURMESE	
THAI	
NEPALI	Alu Ko rato- chakke rog आलुको रातो चक्के रोग
HINDI	
VIETNAMESE	Thối vòng - Chết tươi
CHINESE	Mǎ líng shǔ lúen fǔ bìng 馬鈴薯輪腐病
JAPANESE	Wagusare-byô 輪腐病

The first evidence of this disease is usually a slight wilting of the leaves fairly late in the growing season. Affected leaves become pale green, mottled and chlorotic, then they develop a marginal necrosis and finally die. Symptoms on the tubers vary considerably. Mildly infected tubers develop a yellowish, cheesy rot of the vascular tissue. More advanced stages show a breakdown of the vascular ring and ragged cracks on the surface. This stage is followed by invasion of soft bacterial rot.

Le prime symptoma de iste maladia es usualmente un legier marcescentia del folios assatis tarde in le saison. Folios afflicte deveni verde clar, maculate, e chlorotic. Alora, illos monstra un necrosis marginal e mori. Symptomas in le tuberes se varia grandemente. Tuberes legiermente infectate monstra un putrefaction jalne como caseo in le texito vascular. In le stadios avantiate, destruction del anello vascular occurre e fissuras corrugate in le superfacie se manifesta. Postea seque blande putrefaction bacterial.

Le premier indice de cette maladie est habituellement un léger flétrissement des feuilles passablement tard dans la saison. Les feuilles atteintes deviennent vert pâle, marbrées et chlorotiques, puis forment une nécrose marginale et finissent par mourir. Les symptômes sur les tubercules varient considérablement. Les tubercules légèrement infectés produisent une pourriture des tissus vasculaires jaune et caséeuse. Les phases plus avancées présentent une détérioration de l'anneau vasculaire et des crevasses déchirées à la surface. Cette phase est suivie d'une invasion bactérienne du type pourriture molle.

La primera evidencia de esta enfermedad es generalmente una leve marchitez de las hojas bastante tarde en la época de crecimiento. Las hojas afectadas se vuelven de color verde pálido, moteadas y cloróticas; luego desarrollan una necrosis marginal y finalmente mueren. Los síntomas en los tubérculos varían considerablemente. Los tubérculos infectados poco desarrollan una pudrición amarillosa, con textura como de queso en el tejido vascular. Los estados más avanzados muestran una desintegración del anillo circular y rajaduras desiguales en la superficie. Este estado es seguido por la invasión de la pudrición bacterial suave.

DISTRIBUTION: A., Eur., N.A., C.A., S.A.

Solanum tuberosum + Erwinia phytophthora (Appel) Bergey et al. (Erwinia atroseptica (Van Hall) Jennison) =

ENGLISH	Black Leg
FRENCH	Jambe Noire
SPANISH	Pierna negra, Pudrición suave del fruto
PORTUGUESE	
ITALIAN	
RUSSIAN	
SCANDINAVIAN	Stengelbakteriose, Sortbensyge, Stjälkbakterios
GERMAN	Schwarzbeinigkeit
DUTCH	Zwartbenigheid
HUNGARIAN	
BULGARIAN	
TURKISH	Patates siyah çürüklüğü, Patates dip yanıklığı
ARABIC	
PERSIAN	
BURMESE	
THAI	
NEPALI	
HINDI	
VIETNAMESE	Thối den gốc
CHINESE	
JAPANESE	

This fungus affects seed pieces, stems and tubers. Affected plants turn yellow, wilt and eventually die. The stems turn an inky-black color. A conical dark-colored rot frequently develops at the stem end of the infected tubers.

Iste fungo attacca patatas de semination, pedunculos, e tuberes. Plantas afflicte deveni jalne, marcesce, e al fin mori. Le pedunculos deveni nigre obscur. Putrefaction sovente se manifesta in le tuberes infectate presso le pedunculo.

Ce pathogène attaque la semence, les tiges et les tubercules. Les plantes attaquées jaunissent, se flétrissent et finissent par mourir. Les tiges prennent une couleur noire d'encre. Souvent, une pourriture foncée se produit au point d'attache du tubercule malade.

Este hongo ataca los pedazos de semilla, tallos y tubérculos. Las plantas afectadas se vuelven de color amarillo, se marchitan y eventualmente mueren. Los tallos se vuelven de color de tinta negra. Frecuentemente se desarrolla una podredumbre de color obscuro en el centro, al final del tallo de los tubérculos infectados.

DISTRIBUTION: Af., A., Aust., Eur., N.A., C.A., S.A.

Solanum tuberosum + Oospora pustulans Owen & Wakef. =

ENGLISH	Skin Spot
FRENCH	Tache de la Pelure, Oosporose, "Skin Spot"
SPANISH	Mancha de la cáscara
PORTUGUESE	
ITALIAN	
RUSSIAN	парша бугорчатая, ооспороз клубней
SCANDINAVIAN	Blæreskurv
GERMAN	Oospora-Flekkenkrankheit, Tüpfelfleckigkeit
DUTCH	
HUNGARIAN	
BULGARIAN	
TURKISH	
ARABIC	
PERSIAN	
BURMESE	
THAI	
NEPALI	
HINDI	
VIETNAMESE	
CHINESE	
JAPANESE	

The disease first appears as minute, reddish or brownish spots on the surface of the tuber. This discoloration often is found on the roughened points scattered over the tuber's surface. They may extend quite rapidly to adjacent tissue, becoming deeper in color and being associated with the development of an abnormal corky growth. This area may constitute a more or less irregular scab-like crust over the surface, or more frequently may become deeply cracked and furrowed.

Le maladia se manifesta al initio como minute maculas rubiette o brunette al superfacie del tubere. Iste discoloration sovente se trova in le punctos asperate disperse super le superfacie del tubere. Illos pote extender se rapidemente al texito cirumjacente, deveniente de color plus obscur e associante se con le crescimento de un corco anormal. Iste area pote formar un plus o minus irregular crusta scaliose super le superfacie, o plus frequentement, pote finder se profundemente e devenir cannellate.

Cette maladie s'annonce par de très petites taches rougeâtres ou brunâtres à la surface du tubercule, surtout aux endroits rugueux dispersés ça et là. Les taches peuvent s'étendre très rapidement aux tissus adjacents, devenir plus foncées et se trouver impliquées dans la formation anormale d'une couche de liège. Cette couche peut former à la surface une croûte scabieuse plus ou moins régulière. Le plus souvent, elle se marque de fissures et de sillons profonds.

La enfermedad aparece primero como manchas diminutas, rojizas o cafesosas, en la superficie del tubérculo. Esta decoloración frecuentemente se encuentra en los puntos ásperos esparcidos sobre la superficie del tubérculo. Estas pueden extenderse rápidamente al tejido adyacente, volviéndose de color más intenso y estando asociadas con el desarrollo de un crecimiento corchoso anormal. Esta área puede constituir una costra más o menos irregular como de roña sobre la superficie, o más frecuentemente puede volverse muy rajada y arrugada.

DISTRIBUTION: Aust., Eur., N.A.

254

Solanum tuberosum + Phoma solanicola Prill. & Delacr. =

ENGLISH	Canker
FRENCH	Pourriture Phoméene, Gangrène
SPANISH	Pudrición de la raíz, Cancro
PORTUGUESE	
ITALIAN	
RUSSIAN	фомоз, гниль клубней фомозная
SCANDINAVIAN	
GERMAN	Pustelfäule
DUTCH	
HUNGARIAN	
BULGARIAN	
TURKISH	
ARABIC	
PERSIAN	
BURMESE	
THAI	
NEPALI	
HINDI	
VIETNAMESE	
CHINESE	
JAPANESE	

This disease is more severe on some varieties than on others and is often found in association with powdery scab. When not associated with the latter, the rot is slate-colored, dry and powdery. On the surface of the tuber, the lesion is brown to gray at first, becoming dark with age. When found on the lower stems, brown lesions enlarge until the older decayed tissue turns whitish, the stalks breaking over at the point of advanced lesions.

Iste maladia causa infection plus sever in alicun varietates que in alteres e illo occurre sovente in assocation con crustas pulverose. Si le crustas non es presente, le putrefaction es de color ardesiose, desiccate, e pulverose. Al superfacie del tubere le lesion es, al initio, brun o gris e deveni obscur post qualque tempore. Si le infection occurre in le pedunculos inferior, lesiones brun cresce usque a facer le texito putrefacte plus vetule devenir quasi blanc e le pedunculos finder se al punctos del lesiones avantiate.

Cette maladie atteint certaines variétés plus gravement que d'autres et s'associe souvent à la gale poudreuse. Quand elle se présente suele, elle produit une pourriture couleur ardoise, sèche et poudreuse. Sur le tubercule, les lésions sont d'abord brunes à grises et deviennent foncées avec le temps. Les lésions brunes situées à la base des tiges s'agrandissent jusqu'à ce que les tissus pourris deviennent blanchâtres. Les tiges se brisent au niveau des lésions.

Esta enfermedad es más severa en algunas variedades que en otras y frecuentemente se encuentra asociada con la roña polvorienta. Cuando no está asociada con está, la podredumbre es de color pizarra, seca y polvorienta. Sobre la superficie del tubérculo la lesión es de color café a gris primero, volviéndos obscura con la edad. Cuando se encuentra en los tallos inferiores, las lesiones color café se agrandan hasta que el tejido más viejo podrido se vuelve blanquecino, y los tallos se quiebran en el punto donde están las lesiones avanzadas.

DISTRIBUTION: Aust., Eur., N.A., S.A.

Solanum tuberosum + Phytophthora erythroseptica Pethybridge =

ENGLISH	Pink Rot, Watery Rot, Wilt
FRENCH	Pourriture Rose, Pourriture Humide
SPANISH	Podredumbre rosada, Pudrición de la raíz, Marchitez, Pudrición acuosa
PORTUGUESE	
ITALIAN	Marciume delle patate
RUSSIAN	**гниль клубней розовая**
SCANDINAVIAN	
GERMAN	Rotfäule, Rosafäule
DUTCH	Roodrot
HUNGARIAN	
BULGARIAN	
TURKISH	
ARABIC	
PERSIAN	
BURMESE	
THAI	
NEPALI	
HINDI	
VIETNAMESE	
CHINESE	
JAPANESE	Huhai-byô 腐敗病

Although wilt and stem decay may be seen, rot of tubers is the most important expression of this disease. Decay usually begins at the stem end and may progress rapidly throughout the length of the tuber. The affected tissues are firm and resilient but exude water under pressure. The almost normal-colored tissue turns pink when exposed to the air. Later the color darkens and becomes purplish-brown to black. Externally, the tubers have a dull brownish or black dead appearance, and the lenticels stand out as black spots.

Ben que on vide marcescentia e putrefaction de pedunculos, le symptoma le plus importante de iste marcescentia es putrefaction de tuberes. Putrescentia usualmente comencia in le pedunculo e sovente se extende rapidemente per tote le longitude del tubere. Le texitos afflicte es dur e elastic, ma sub pression illos exsuda aqua. Si on los expone al aere, le texitos de color quasi normal deveni rosee. Plus tarde le color es obscurate a brun-purpuree o nigre. Externemente, le tuberes ha un mat e morte apparentia brun o nigre, e le lenticellas se manifesta como distincte maculas nigre.

Il peut y avoir pourriture et flétrissement de la tige, mais la pourriture du tubercule est le symptôme le plus remarquable de cette maladie. La pourriture débute généralement au point d'attache et progresse rapidement tout le long du tubercule. Les tissus atteints sont fermes et élastiques, mais laissent échapper de l'eau lorsqu'on les presse. Lorsqu'ils sont exposés à l'air, les tissus de couleur presque normale virent au rose. Plus tard, la couleur se fonce et devient de brun pourpré à noire. Extérieurement, les tubercules sont de couleur brunâtre terne ou noire et semblent morts, et les lenticelles émergent comme des points noirs.

Aunque puede verse una marchitez y podredumbre del tallo, la expresión más importante de esta enfermedad es la pudrición de los tubérculos. La pudrición generalmente comienza en la base del tallo y puede progresar rápidamente a través del largo del tubérculo. Los tejidos afectados son firmes y elásticos pero exudan agua al apretarlos. El tejido de color casi normal se vuelve rosado cuando es expuesto al aire. Luego el color se obscurece y se vuelve café-purpúreo a negro. Externamente, los tubérculos tienen un color cafesoso opaco a negro, una apariencia muerta y los lenticelas se muestran como manchas negras.

DISTRIBUTION: A., Aust., Eur., N.A.

Solanum tuberosum (Lycopersicon esculentum) + Phytophthora infestans (Mont.) De Bary =

ENGLISH	Late Blight, Downy Mildew, Potato Blight, Tomato Blight
FRENCH	Mildiou
SPANISH	Hielo, Pudrición de la raíz, Tizón tardio, Fitoftora, Lancha tardía
PORTUGUESE	Requeima, Mildio, Fitoftora, Peste preta, Crestamento
ITALIAN	Peronospora, Marciume
RUSSIAN	фитофтороз
SCANDINAVIAN	Kartoffelskimmel, Brunröta, Potettørråte, Bladmögel
GERMAN	Kraut- und Knollenfäule, Braunfäule, Kartoffelkrautfäule
DUTCH	Aardappelziekte, Aardappelplaag
HUNGARIAN	Burgonyavész
BULGARIAN	мана
TURKISH	Patates mildiyösü
ARABIC	Lafha Motaakhera لفحة متأخرة
PERSIAN	Sēfēdākē dōrōōghēyē sēbzâmēne سفیدک دروغی سیب زمینی
BURMESE	Laung-mee, Ywet-pok, Pin-nar-kya roga
THAI	Bi-Mhai ใบไหม้
NEPALI	Alu Ko dadhuwa rog आलुको दद्‌ुवा रोग
HINDI	Vilambit angmaari विलम्बित अंगमारी
VIETNAMESE	Bệnh úa muộn
CHINESE	Mǎ líng shǔ yìh bìng 馬鈴薯疫病
JAPANESE	eki-byð 疫病

Infected leaves have, at first, a water-soaked appearance. In wet weather a white fungus growth develops on the water-soaked areas, particularly on the undersides of the leaves. In dry weather, the water-soaked areas turn brown and dry up, killing the infected leaf areas. Infected petioles and stems turn brown and are sometimes covered with the white fungus growth. Tubers develop a shallow, reddish-brown dry rot. In wet weather other organisms invade the tubers and cause their complete breakdown in the form of soft rot.

Folios afflicte ha al initio un apparentia multo aquose. Sub conditiones humide, un strato blanc fungal se forma al areas aquose; specialmente in le superfacie inferior del folios. Sub conditiones sic, le areas aquose deveni brun e se desicca, causante le morte del areas foliar afflicte. Petiolos e pedunculos infectate deveni brun e a vices es coperte del strato blanc fungal. Tuberes developpa un putrefaction sic brun-rubiette nonprofunde. In tempore humide, altere organismos invade le tuberes e causa collapso total per putrefaction blande.

Les feuilles infectées apparaissent au début comme imbibées d'eau. Par temps humide, un mycélium blanc recouvre les plages délavées, particulièrement sous les feuilles. Par temps sec, les parties détrempées brunissent et sèchent, ce qui entraîne la mort des plages atteintes. Les pétioles et les tiges infectés brunissent et son parfois recouverts d'un mycélium blanc. Les tubercules produisent une pourriture sèche, brun rougeâtre, peu profonde. Par temps humide, d'autres organismes envahissent le tubercule et provoquent une désagrégation complète sous forme de pourriture molle.

Las hojas infectadas tienen al principio una apariencia de estar empapadas en agua. En tiempo húmedo se desarrolla un crecimiento fungoso blanco en las áreas empapadas de agua, especialmente en el envés de las hojas. En el tiempo seco, las áreas empapadas de agua se vuelven de color café y se secan, matando las áreas de la hoja que están infectadas. Los pecíolos y tallos infectados se vuelven de color café y algunas veces están cubiertos con el crecimiento blanco del hongo. Los tubercúlos desarrollan una pudrición poco profunda, café-rojiza y seca. En tiempo húmedo otros organismos invaden los túberculos y causan su completa desintegración en la forma de una pudrición suave.

DISTRIBUTION: Af., A., Aust., Eur., N.A., C.A., S.A.

Solanum tuberosum (Nicotiana tabacum) + Pseudomonas solanacearum E. F. Smith (Bacterium
solanacearum E. F. Smith) =

ENGLISH	Bacterial Wilt of Potato, Granville Wilt of Tobacco, Slime Disease
FRENCH	Bactériose Vasculaire, Pourriture Brune
SPANISH	Podredumbre anular, Dormidera, Marchitez bacterial, Enfermedad de limo
PORTUGUESE	Murcha bacteriana, Murchadeira, Bacteriose
ITALIAN	Avvizzimento batterico
RUSSIAN	гниль бурая, увядание бактериальное, бактериоз слизистый
SCANDINAVIAN	
GERMAN	Braunfäule, Bakterielle Welke, Schleimkrankheit
DUTCH	Bacterieel-ringrot de Aardappelen
HUNGARIAN	
BULGARIAN	
TURKISH	
ARABIC	Affan Bony عـفـن بنى
PERSIAN	
BURMESE	
THAI	Hiaw-Chaow เหี่ยวเฉา
NEPALI	
HINDI	
VIETNAMESE	Thối năm - Chết tươi
CHINESE	Mǎ líng shǔ ching ku bìng 馬鈴薯青枯病
JAPANESE	Aogore-byō 青枯病

The first symptom is the wilting of leaves during the heat of the day with recovery at night,
but eventually plants fail to recover and they die. Vascular bundles are stained brown, giving
a streaked appearance and the foliage has a bronze tint. A white, slimy mass of bacteria ooze
from vascular bundles which are broken or cut.

Le prime symptoma es marcescentia de folios durante le parte calide del die, con recuperation
nocturne. Al fin, le plantas non se recupera e illos mori. Fasces vascular es tintate brun,
causante striation, e le foliage es bronzate. Un massa blanc viscose de bacterios se exsuda
ab le fasces vascular si on los rumpe o talia.

Le premier symptôme est le flétrissement des feuilles durant la chaleur du jour. Les feuilles
reprennent leur apparence normale durant la nuit pendant un certain temps, mais la plante
finit par mourir. Les faisceaux vasculaires sont teintés de brun, ce qui confère au feuillage
un aspect strié et bronzé. Une masse de bactéries blanchâtre et visqueuse exsude des faisceaux vasculaires quand ces derniers sont brisé ou coupés.

El primer síntoma es la marchitez de las hojas durante las horas calurosas del día, las cuáles
se recuperan en la noche, pero eventualmente ya no se recuperan y mueren. Los tubos vasculares están manchados de café, dándole una apariencia rayada y el follaje tiene un tinte
bronceado. De los tubos vasculares quebrados o cortados emana una masa de bacterias blanca
y limosa.

DISTRIBUTION: Af., A., Eur., N.A., C.A., S.A.

258

<u>Solanum</u> <u>tuberosum</u> (Gramineae) + <u>Pythium</u> <u>vexans</u> De Bary =

ENGLISH   Water Rot and Leak of Potatoes, Seedling Disease of Gramineae
FRENCH
SPANISH   Pucrición de agua, goteo
PORTUGUESE
ITALIAN
RUSSIAN
SCANDINAVIAN
GERMAN
DUTCH
HUNGARIAN
BULGARIAN   сечене
TURKISH
ARABIC
PERSIAN
BURMESE
THAI
NEPALI
HINDI
VIETNAMESE
CHINESE
JAPANESE

The most characteristic symptom is the extremely watery nature of the affected tissues. When pressure is applied, a yellowish to brown liquid is readily given off. Another characteristic is the granular nature of diseased tissues. Externally, the affected tissues appear turgid and show a discoloration ranging from metallic gray in red varieties to brown shades in white and dark-skinned varieties. Internally, the creamy color tissue, upon cutting, turns tan or slightly reddish, brown and finally black.

Le symptoma le plus characteristic es le essentia grandemente aquose del texito afflicte. Sub presion, un liquor jalnette o brun se exsuda facilemente. Anque symptomatic es le character granular del texitos morbide. Externemente, le texitos afflicte es turgide e monstra discoloration que se varia ab gris metallic in varietates rubie usque a tintas brun in varietates que ha un pelle blanc o obscur. Internemente, le texito de color de crema deveni, post section, bronzate, brun, e, al fin, nigre.

Le symptôme le plus caractéristique est l'aspect très délavé des tissus atteints. Quand on applique une pression, un liquide jaune à brun exsude facilement. Une autre caractéristique est la texture granuleuse, des tissus malades. A l'extérieur, les tissus affectés paraissent turgescents, et leur couleur va du gris métallique chez les variétés rouges, aux teintes de brun chez les variétés blanches et celles dont la peau est foncée. Sectionnés, les tubercules laissent voir les tissus intérieurs crémeux, qui bientôt deviennent couleur chamois ou légèrement rougeâtres, puis bruns et finalement noirs.

El síntoma más característico es la extrema naturaleza acuosa de los tejidos afectados. Cuando se hace presión, sale fácilmente un líquido amarilloso o café. Otra característica es la naturaleza granular de los tejidos enfermos. Externamente, los tejidos afectados parecen túrgidos y muestran una decoloración que varía de gris metálico en las variedades rojas a tonos café en las variedades blancas y de cáscara obscura. Internamente, el tejido de color cremoso, al cortarlo se vuelve café bien claro o algo rojizo, café y finalmente negro.

DISTRIBUTION: Af., A., Aust., Eur., N.A.

Solanum tuberosum (and other hosts) + Rhizoctonia crocorum (Pers.) DC. ex Fr. (Helico-
basidium purpureum Pat.) =

ENGLISH   Violet Root Rot
FRENCH   Rhizoctone Violet
SPANISH   Pudrición radicular violeta
PORTUGUESE
ITALIAN   Mal vinato
RUSSIAN   гниль красная
SCANDINAVIAN
GERMAN   Violetter Wurzeltöter
DUTCH   Violet Wortelrot, Rhizoctonia Rot
HUNGARIAN
BULGARIAN
TURKISH   Çökerten
ARABIC
PERSIAN
BURMESE
THAI
NEPALI
HINDI
VIETNAMESE
CHINESE
JAPANESE

This disease affects the roots and underground parts of the plant. Lesions tend to be most abundant at the stem end of tubers. They are at first small and violet in color; later they may enlarge and increase in number until the entire tuber is involved. Shortly after a lesion develops, the epidermis breaks and exposes ashen-gray tissue which becomes dry and powdery. At times, enough roots are affected to produce yellowing and wilting in above-ground parts.

Iste maladia afflige le radices e le partes hypogee del planta. Lesiones es usualmente le plus numerose in le tuberes presso le pedunculo. Illos es al initio minute e de color violette; plus tarde illos cresce e deveni plus numerose usque a involver tote le tubere. Post que un lesion se forma, le epidermis rapidemente se finde e discoperi texito gris cinerose le qual deveni desiccate e pulverose. A vices le sytema radical es tanto afflicte que le partes epigee suffre jalnessa e marcescentia.

Cette maladie attaque les racines et les parties souterraines de la plante. Les lésions sont généralement beaucoup plus nombreuses au point d'attache du tubercule. Au début, elles sont petites et violettes; plus tard, elles peuvent augmenter en nombre et en étendue au point d'envahir tout le tubercule. Peu de temps après la formation d'une lésion, la pelure se brise et laisse voir des tissus gris cendré qui deviennent secs et poudreux. Quelquefois, il y a suffisamment de racines atteintes pour provoquer le jaunissement et le flétrissement des parties aériennes de la plante.

Esta enfermedad afecta las raíces y partes subterráneas de la planta. Las lesiones tienden a ser más abundantes en la base del tallo de los tubérculos. Primero son pequeñas y de color violeta; más tarde pueden agrandarse y aumentar en su número hasta que todo el tubérculo está dañado. Poco después de que la lesión se desarrolla, le epidermis se raja, dejando expuestos tejidos de color gris-ceniza, los cuáles se vuelven secos y polvorientos. A veces suficientes raíces son afectadas para producir amarillamiento y marchitez en las partes áereas.

DISTRIBUTION: Af., A., Aust., Eur., N.A., C.A., S.A.

<u>Solanum</u> <u>tuberosum</u> + <u>Spondylocladium</u> <u>atrovirens</u> Harz =

ENGLISH	Silver Scurf
FRENCH	Gale Argentée de la Pomme de Terre
SPANISH	Costra plateada, Mancha plateada, Caspa plateada
PORTUGUESE	Sarna prateada
ITALIAN	Tigna argentata
RUSSIAN	парша серебристая
SCANDINAVIAN	Sølvskurv
GERMAN	Silberflecken
DUTCH	Zilverschurft
HUNGARIAN	
BULGARIAN	
TURKISH	
ARABIC	
PERSIAN	
BURMESE	
THAI	
NEPALI	
HINDI	
VIETNAMESE	
CHINESE	Mǎ líng shǔ yín já bìng　　馬 鈴 薯 銀 痂 病
JAPANESE	Ginka-byð 銀か病

This disease is characterized by a gray, smooth, leathery appearance of the skin, usually near the stem end of the affected tubers and is more noticeable, because of a silvery sheen, when the tubers are wet. No rot occurs and damage involves reduction in size, malformation and severe shriveling in storage. Many tubers tend to be wedge-shaped or lopsided.

Iste maladia se characterisa per un apparition gris, rase, e coriacee del pelle del tuberes afflicte, usualmente presso le pedunculo. Iste symptoma deveni plus remarcabile a causa de un politura argentee quando le tuberes es molliate. Putrefaction non occurre, ma le tuberes suffre damno a causa de diminution de grandor, malformation, e crispation sever durante immagasinage. Multe tuberes es cuneiforme o asymmetric.

Cette maladie se caractérise par l'aspect de cuir lisse et gris que prend la pelure, généralement près du point d'attache des tubercules atteints, et qu'on remarque davantage à cause des reflets d'argent qu'émettent ces tubercules lorsqu'ils sont humides. Il n'y a pas de pourriture, et les dommages se résument à une réduction de volume, des malformations, et beaucoup de ratatinement en entrepôt. Plusieurs tubercules deviennent cunéiformes ou asymétriques.

Esta enfermedad se caracteriza por una apariencia cueruda, gris y suave de la cáscara, generalmente cerca de la base del tallo de los tubérculos afectados y es más notoria debido a un brillo plateado cuando los tubérculos están mojados. No ocurre pudrición y el daño consiste en reducción del tamaño, malformación y arrugamiento severo en el almacenaje. Muchos tubérculos tienden a ser cuneiformes o más pesados de un lado que de otro.

DISTRIBUTION: Af., A., Aust., Eur., N.A., C.A., S.A.

Solanum tuberosum + Spongospora subterranea (Wallr.) Lagerh. =

ENGLISH	Powdery Scab, Corky Scab
FRENCH	Gale Poudreuse, Gale Spongieuse
SPANISH	Roña, Sarna polvorienta, Roña corchosa, Sarna polvosa, Roña polvorienta
PORTUGUESE	
ITALIAN	Crosta, Rogna, Scabbia polverulenta
RUSSIAN	парша порошистая
SCANDINAVIAN	Pulverskurv, Pulverskorv, Vorteskurv
GERMAN	Räude, Pulverschorf der Kartoffel, Geschwülste, Schwammschorf
DUTCH	Poederschurft
HUNGARIAN	
BULGARIAN	брашнеста, прашеста, краста
TURKISH	
ARABIC	
PERSIAN	
BURMESE	
THAI	
NEPALI	
HINDI	Choorni scab चूरनी स्काव
VIETNAMESE	
CHINESE	
JAPANESE	Hunzyôsôka-byô 粉状そうか病

There are no foliar symptoms to this disease. Infected tubers are more or less covered with small circular depressions which are filled with brown powdery spores. Loose papery remnants of skin adhere to the margins of the lesions.

Iste maladia ha nulle symptomas foliar. Tuberes infectate es coperte, plus o minus, de minute depressiones circular le quales· contine sporas pulverose brun. Fragmentos disligate de pelle adhere al margines del lesiones.

Cette maladie ne présente aucun symptôme sur les feuilles. Les tubercules infectés sont plus ou moins couverts de petites dépressions circulaires remplies de spores poudreuses brunes. Des débris de pelure lâches et minces comme du papier adhèrent à la marge des lésions.

No hay síntomas foliares de esta enfermedad. Los tubérculos infectados están más o menos cubiertos con pequeñas depresiones circulares, las cuáles están llenas de esporas polvorientas, de color café. Restos sueltos de cáscara que semeja papel se adhieren a los márgenes de las lesiones.

DISTRIBUTION: Af., A., Aust., Eur., N.A., S.A.

Solanum tuberosum + Streptomyces scabies (Thaxt.) Wakes & Henrici (Actinomyces scabies (Thaxt.) Gussow) =

ENGLISH	Scab
FRENCH	Gale Commun, Gale Ordinaire, Gale Profonde
SPANISH	Roña, Sarna común, Sarna morena, Actinomicosis, Sarna de América
PORTUGUESE	Sarna comun
ITALIAN	Scabbia commune, Rogna dei tuberi, Rogna
RUSSIAN	парша обыкновенная, парша актиномикозная
SCANDINAVIAN	Kartoffelskurv, Vanlig potatisskorv, Flatskurv, Almindelig skurv
GERMAN	Gewöhnlicker Kartoffelschorf, Schorf, Gewöhnlicker Schorf
DUTCH	Gewone Schurft, Aardappelschurft
HUNGARIAN	
BULGARIAN	корковидна краста
TURKISH	
ARABIC	Garabaaddy جــرب عــا دى
PERSIAN	
BURMESE	
THAI	Scab สะแกป
NEPALI	Alu Ko dade rog आलुको दडे रोग
HINDI	Scab स्काव
VIETNAMESE	Bệnh ghẻ trên củ
CHINESE	Mǎ líng shǔ chong já bìng 馬鈴薯瘡痂病
JAPANESE	

This disease is characterized by raised corky areas, varying in size and shape. The color is grayish-white to dark tan. The scabby areas can appear pitted or raised depending on the variety. In severe infections the size and shape of the tubers may be changed.

Iste maladia se characterisa per areas elevate similar a corcole quales se varia e in dimension e in contorno. Le color es blanc-gris o bronzate obscur. Le areas crustose pote esser o depresse o elevate secundo le varietate. In infection sever, le dimension e le contorno del tuberes a vices suffre alteration.

Cette maladie est caractérisée par des aires liégeuses soulevées de dimensions et de formes variables. La couleur varie du blanc grisâtre au beige foncé. Les aires scabieuses peuvent paraître trouées ou surélevées selon la variété. Dans les cas graves, la grosseur et la forme du tubercule peuvent être modifiées.

Esta enfermedad se caracteriza por áreas levantadas corchosas, que varían de tamaño y forma. El color es blanco-grisáceo a crema obscuro. Las áreas roñosas peuden aparecer picados o levantados dependiendo de la variedad. En infecciones severas el tamaño y forma de los tubérculos peude cambiar.

DISTRIBUTION: Af., A., Aust., Eur., N.A., S.A.

Solanum tuberosum + Synchytrium endobioticum (Schilb) Perc. =

ENGLISH	Wart Disease, Black Scab
FRENCH	Tumeur Verruqueuse, Gale Verruqueuse, Gale Noire, Maladie Verruqueuse
SPANISH	Verreuga, Enfermedad de verruga, Roña negra
PORTUGUESE	
ITALIAN	Rogna nera, Rogna verrucolosa dei tuberi
RUSSIAN	рак
SCANDINAVIAN	Kartoffelbrok, Potatiskräfta, Potetkreft
GERMAN	Krebs, Kartoffelkrebs
DUTCH	Wratziekte, Aardappelwratziekte
HUNGARIAN	
BULGARIAN	
TURKISH	
ARABIC	
PERSIAN	
BURMESE	
THAI	
NEPALI	Alu Ko ainjeru आलुको अंजेरी
HINDI	Massa मसा
VIETNAMESE	
CHINESE	
JAPANESE	Gansyu-byô がんしゅ病

Although the fungus may attack all parts of the plant, the most severe effects are produced on the tubers. It enters the eyes and causes abnormal growth that is warty and cauliflower-like. If infection takes place early, the entire tuber may become a spongy mass. Similar warts may appear on stolons and other underground parts. Young warts usually have the same color as the affected tissues, later turning light brown and finally dark brown or black. The black color is associated with decay.

Ben que le fungo pote attaccar tote le partes del planta, le effectos le plus vitiose se manifesta in le tuberes. Le infection penetra per le oculos e produce crescentia anormal, verrucose e similar a cuale flor. Si infection occurre in le prime parte del saison, tote le tubere a vices deveni un massa spongiose. Verrucas similar occurre anque in stolones e in altere partes hypogee. Verrucas juvene usualmente es del mesme color que le texitos afflicte o, plus tarde, deveni brun clar e, al fin, brun obscur o nigre. Le color nigre se associa con putrescentia.

Bien que le champignon puisse attaquer toutes les parties de la plante, les méfaits les plus graves portent sur le tubercule. Le champignon pénètre par les germes et provoque une végétation anormale verruqueuse en forme de chou-fleur. Si l'infection débute tôt, le tubercule entier peut se transformer en une masse spongieuse. Des verrues semblables peuvent apparaître sur les stolons ou sur d'autres parties souterraines. Les jeunes verrues sont généralement de la même couleur que les tissus affectés, puis virent au brun clair et deviennent finalement de brun foncé à noires. La couleur noire va de pair avec la pourriture.

Aunque el hongo peude atacar todas las partes de la planta, los efectos más severos se producen en los tubérculos. Entra a través de los ojos y causa crecimiento anormal que es verrugoso y parece coliflor. Si la infección aparece temprano, todo el tubérculo puede convertirse en una masa esponjosa. Verrugas similares peuden aparecer en los estolones y otras partes subterráneas. Las verrugas jóvenes generalmente tienen el mismo color que los tejidos afectados, volviéndose más tarde de color café claro y finalmente café obscuro o negro. El color negro está asociado con la pudrición.

DISTRIBUTION: Af., A., Eur., N.A., S.A.

264

Solanum tuberosum (and others) + Verticillium alboatrum Reinke & Berth. =

ENGLISH	Potato Wilt, Blight of Tomato, Verticillium Hadromycosis
FRENCH	Flétrissure Verticillienne, Maladie Jaune, Verticilliose
SPANISH	Marchitez de la papa, Verticilosis, Verticillium hadromycosis, Tizón
PORTUGUESE	Verticillium
ITALIAN	Tracheo-verticilliosi
RUSSIAN	увядание вертициллезное, вертициллез
SCANDINAVIAN	Krankimmel, Vissnesjuka, Kransskimmel
GERMAN	Welkekrankheit
DUTCH	Ringvuur, Verticillium-ziekte
HUNGARIAN	
BULGARIAN	вертицилийно увяхване
TURKISH	Verticillium solgunluğu
ARABIC	
PERSIAN	Bōōtёhmḗrёyё sёbzảmḗne بوته میری سیب زمینی
BURMESE	
THAI	
NEPALI	
HINDI	
VIETNAMESE	
CHINESE	
JAPANESE	

Vines wilt, turn yellow and die prematurely. The internal woody portion of the stem turns a cinnamon-brown color and can be observed by peeling the stem or cutting it at an angle at about groundlevel. Infected tubers show a brown ring that is visible if the stem end is removed. Sometimes a pink discoloration may develop around the eyes or as blotches on the surface. In storage these pink blotches turn brown and dry up.

Vites marcesce, deveni jalne, e mori prematurmente. Le portion interne lignose del pedunculo deveni de color brun de cannella -- lo que es observabile si on remove le cortice del pedunculo o lo seca a un angulo presso le solo. Tuberes infectate monstra un anello brun que on vide quando on remove le portion presso le pedunculo. A vices discoloration rosee se forma circum le oculos o como pustulas al superfacie. In immagasinage, iste pustulas rosee deveni brun e se desicca.

Les fanes se flétrissent, jaunissent et meurent prématurément. La partie ligneuse de la tige devient brun canelle, coloration que l'on peut observer en pelant la tige ou en la coupant en biais près du sol. Les tubercules infectés laissent voir un anneau brun lorsqu'on les coupe au point d'attache. Quelquefois, une teinte rosée apparaît autour des germes ou sous forme d'éclaboussures à la surface. En entrêpot, ces éclaboussures roses brunissent et sèchent.

Las guías se marchitan, se vuelven amarillas y mueren prematuramente. La porción maderosa interna del tallo se vuelve de color café-canela y peude observarse al pelar el tallo o al cortarlo en un ángulo a más o menos el nivel del suelo. Los tubérculos infectados muestran un anillo café que es visible si la base del tallo se quita. Algunas veces puede desarollarse una decoloración rosada alrededor de los ojos o como manchones en la superficie. En el almacenaje estos manchones rosados se vuelven de color café y se secan.

DISTRIBUTION: Af., A., Aust., Eur., N.A., C.A., S.A.

Sorghum vulgare (Zea mays) + Cercospora sorghi Ell. & Ev. =

ENGLISH	Leaf Spot, Gray Leaf Spot
FRENCH	Cercosporose, Taches Foliaires, Trachéomycose
SPANISH	Mancha gris, Mancha de la hoja, Mancha foliar
PORTUGUESE	
ITALIAN	
RUSSIAN	церкоспороз
SCANDINAVIAN	
GERMAN	Cercospora-Blattfleckenkrankheit
DUTCH	
HUNGARIAN	
BULGARIAN	
TURKISH	
ARABIC	
PERSIAN	
BURMESE	
THAI	Bai-Jut ใบจุด
NEPALI	
HINDI	
VIETNAMESE	
CHINESE	Gao líang lsўy luén bìng 高粱紫輪病
JAPANESE	Sirin-byð 紫輪病

This disease is characterized by lesions which are usually reddish-purple, but in some crop varieties they are tan. When these spots are small, they resemble those of other leaf spot diseases on this host, but as they enlarge they become long and narrow, being limited some-what by the leaf veins. The long, narrow spots may come together and then kill large areas of the leaves. As the spots enlarge, they usually become covered with a grayish fuzz made up of the fruiting structures of the fungus.

Iste maladia se characterisa per lesiones usualmente purpuree-rubiette, ma in alicun varie-tates bronzate. Quando iste maculas es minute, illos es similar a altere maladias de maculas foliar que afflige iste plantahospite. Ma quando illos cresce, illos deveni elongate e stricte, aliquanto limitate per le venas foliar. Le maculas elongate e stricte a vices coalesce e causa morte de grande areas del folios. Quando le maculas cresce, illos es usualmente coperte de un villositate gris del structuras sporal del fungo.

Cette maladie est caractérisée par des lésions généralement rouge pourpré, mais qui, chez certaines variétés cultivées, sont couleur chamois. Quand ces taches sont petites, elles res-semblent à d'autres taches du feuillage sur cet hôte, mais à mesure qu'elles s'agrandissent, elles deviennent longues et étroites et sont quelque peu limitées par les nervures de la feuille. Les taches longues et étroites peuvent se fusionner et tuer de grandes aires de la feuille. En s'agrandissant, les taches se couvrent généralement d'un duvet grisâtre: ce sont les fructi-fications du champignon.

Esta enfermedad se caracteriza por lesiones que generalmente son rojizo-púrpura, pero en al-gunas variedades de cultivos son de color crema. Cuando estas manchas son pequeñas se aseme-jan a las de otras enfermedades de manchas foliares en esta hospedera, pero a medida que se agrandan se vuelven largas y angostas, estando limitadas un poco por las venas de las hojas. Las manchas largas y angostas peuden juntarse y luego matar grandes áreas en las hojas. A medida que las manchas crecen, generalmente se cubren con una peluza gris hecha con las es-tructuras fructíferas del hongo.
DISTRIBUTION: Af., A., Aust., Eur., N.A., C.A., S.A.

Sorghum vulgare (Zea mays, Triticum spp., Secale spp., Medicago spp., Trifolium pratense, Melilotus indica, Glycine max) + Colletotrichum graminicola (Ces.) Wilson (Colletotrichum cereale Manns) =

ENGLISH Red Stalk Rot, Anthracnose, Red Leaf Spot, Seedling Blight
FRENCH Anthracnose
SPANISH Podredumbre roja del tallo, Antracnosis, Mancha foliar roja, Tizón
PORTUGUESE Antracnose
ITALIAN
RUSSIAN
SCANDINAVIAN
GERMAN
DUTCH
HUNGARIAN
BULGARIAN
TURKISH
ARABIC
PERSIAN
BURMESE Ywet-nee-pyauk roga ကော်လက်တိုတြိုင်ခမ် ဂရမိနိကိုလ
THAI Bi-Mhai ใบไหม้
NEPALI Junelo Ko rate rog जुनेलोको राते रोग
HINDI Lal Parn Daag लाल परण दाग
VIETNAMESE
CHINESE Gao líang tán dzǔ bìng 高粱炭疽病
JAPANESE Tanso-byô 炭そ病

This fungus causes spots to develop on the leaves when plants are still in the seedling stage and later spreads to other leaves as they appear. At first the spots are small, circular to elliptical and are reddish-purple on most varieties. Later the spots enlarge and may unite to form areas of dead tissue. As the disease develops the centers of the spots fade to a grayish-straw-color surrounded by a reddish-purple border. The leaf midribs are often strikingly discolored. In many of the spots, a blackish fungus growth appears.

Iste fungo face maculas formar se al folios quando plantas es ancora multo juvene e plus tarde extender se al nove folios quando illos emerge. Al initio, le maculas es minute, circular o elliptic, e purpuree-rubiette in le major parte de varietates. Plus tarde le maculas cresce e a vices se uni usque a formar areas de texito morte. Quando le maladia se avantia, le centros del maculas se discolora a un color paleate-gris imbraciate per un margine purpuree-rubiette. Sovente le nervaturas central del folios es extraordinarimente discolorate. In multe maculas un fungo nigre se manifesta.

Ce champignon provoque l'apparition de taches sur les feuilles des plantules. Par la suite, ces taches se propagent à d'autres feuilles, dès que celles-ci apparaissent. Au début, les taches sont petites, circulaires à elliptiques, et rouge pourpré chez la plupart des variétés. Ensuite, les taches s'agrandissent et peuvent s'unir pour former des plages nécrotiques. Pendant que la maladie évolue, le centre des taches se ternit, devient paille grisâtre et s'entoure d'une marge rouge pourpré. Les nervures principales sont souvent décolorées de façon frappante. Dans plusieurs taches, on peut voir croître le champignon noirâtre.

Este hongo causa manchas que se desarrollan en las hojas cuando las plantitas están aún muy pequeñas y luego ataca otras hojas a medida que éstas van apareciendo. Al principio las manchas son pequeñas, circulares o elípticas y son rojizas-púrpura en la mayoría de variedades. Más tarde las manchas se agrandan y pueden unirse para formar áreas de tejido muerto. A medida que la enfermedad se desarrolla los centros de las manchas disminuyen su color y se tornan color paja grisácea, rodeadas por una orilla rojiza-púrpura. Las venas centrales de la hoja frecuentemente están marcadamente decoloradas. En muchas de las manchas aparece un crecimiento negruzco del hongo.
DISTRIBUTION: Af., A., Aust., Eur., N.A., C.A., S.A.

Sorghum vulgare (Gramineae) + Gloeocercospora sorghi Bain & Edgerton =

ENGLISH	Zonate Leaf Spot
FRENCH	Gloeocercosporose, Taches Foliaires
SPANISH	Mancha zonal, Mancha foliar zonada
PORTUGUESE	
ITALIAN	
RUSSIAN	пятнистость глеоцеркоспорозная, пятнистость медная, глеоцеркоспороз
SCANDINAVIAN	
GERMAN	Gloeocercospora-Blattfleckenkrankheit
DUTCH	
HUNGARIAN	
BULGARIAN	
TURKISH	
ARABIC	
PERSIAN	
BURMESE	
THAI	Bi-Jut-Zonate ใบจุดโซเนท
NEPALI	
HINDI	
VIETNAMESE	
CHINESE	Gao lĭang lúen ban bĭng 高粱輪斑病
JAPANESE	Hyômon-byô ひょう紋病

This disease is very conspicuous on the leaves as reddish-purple bands of tissue alternating with tan or straw-colored areas, forming a zonate pattern. The spots often occur along the margins of the leaves, forming semi-circular patterns, or they may occur on other parts of the leaf where they are more nearly circular and show more strikingly their irregular border. These irregular spots or blotches vary greatly in size. Sometimes they unite to cover most of the leaf surface.

Iste maladia es multo remarcabile in le folios como bandas purpuree-rubiette de texito in alternation con areas bronzate o paleate in un designo zonate. Le maculas sovente se trova al margines del folios in designos semicircular o in altere portiones del folios, in qual caso illos es plus circular e monstra plus evidentemente lor bordos irregular. Iste maculas o pustulas irregular se varia grandemente in dimension. A vices illos coalesce usque a coperir le major parte del superfacie foliar.

Cette maladie est très voyante sur les feuilles. Des bandes rouge pourpré alternent avec des aires beiges ou paille pour former des bariolages bien dessinés. Les taches apparaissent souvent le long de la marge des feuilles de façon à former des demi-cercles, ou bien elles apparaissent à d'autres endroits de la feuille où elles deviennent presque rondes et montrent de façon plus frappante leurs marges irrégulières. La grosseur de ces taches irrégulières ou éclaboussures varie beaucoup. Quelquefois, elles s'unissent pour recouvrir presque toute la surface de la feuille.

Esta enfermedad es muy conspicua en las hojas como bandas rojizas purpúreas de tejido alternando con áreas de color crema o color paja, formando un diseño zonado. Las manchas frecuentemente aparecen a lo largo de los márgenes de las hojas formando diseños semicirculares, o pueden aparecer en otras partes de la hoja donde son casi circulares y muestran más marcadamente sus bordes irregulares. Estas manchas irregulares o manchones varían mucho en tamaño. Algunas veces se unen para cubrir casi toda le superficie de la hoja.

DISTRIBUTION: Af., A., N.A., C.A., S.A.

Sorghum vulgare (Zea mays, Gossypium spp., Corchorus spp., Glycine max) + Macrophomina phaseoli (Maubl.) Ashby (Sclerotium bataticola Taubenhaus) =

ENGLISH	Charcoal Rot
FRENCH	Pourriture Charbonneuse des Tiges
SPANISH	Podredumbre de charcol
PORTUGUESE	
ITALIAN	
RUSSIAN	гниль стеблей угольная (сухая)
SCANDINAVIAN	
GERMAN	Schwarze Stengelfäule
DUTCH	
HUNGARIAN	
BULGARIAN	
TURKISH	
ARABIC	
PERSIAN	
BURMESE	
THAI	
NEPALI	
HINDI	
VIETNAMESE	
CHINESE	Gao líang tàn fǔ bìng 高粱炭腐病
JAPANESE	

Symptoms due to this disease usually do not become evident until the plants approach maturity. Close examination at this time reveals many poorly filled spikes with lightweight kernels, a premature ripening and drying of entire stalks, and the presence of many stalks on the ground. These stalks are often soft and discolored at the base, with the pith disintegrated. The entire stalk appears shredded. An abundant mold-like growth of a pink or white fungus is usually present.

Symptomas de iste maladia usualmente non se manifesta ante le maturation del plantas. Examination precise revelara multe spicas implete con granos legier, maturation ante le tempore ordinari, desiccation de pedunculos integre, e cadita de plure pedunculos al solo. Iste pedunculos es sovente molle e discolorate al pede, con le medulla disintegrate. Tote le pedunculo pare esser fragmentate. Un copiose massa como mucor de un fungo rosee o blanc usualmente es presente.

Les symptômes de cette maladie ne sont généralement pas visibles jusqu'à ce que les plantes approchent de la maturité. A ce moment, un examen minutieux permet de déceler beaucoup d'épis chétifs à grains légers, une maturation précipitée, le dessèchement de tiges entières, ainsi que plusieurs tiges étendues sur le sol. Ces tiges sont souvent molles et décolorées à la base, et leur moëlle est désintégrée. La tige entière apparaît dechiquetée. On y trouve generalement un abondant mycelium de champignon blanc ou rose qui ressemble a une moisissure.

Los síntomas debido a esta enfermedad generalmente no se vuelven evidentes hasta que las plantas se acercan a la madurez. Un examen de cerca en esta época revela muchas espículas pobremente llenadas, que pesan poco, una maduración prematura y los tallos enteros se secan, habiendo muchos de ellos en el suelo. Estos tallos frecuentemente son suaves y decoloridos en la base, con la médula desintegrada. El tallo entero aparece hecho trizas. Un abundante crecimiento parecido al moho de hongos rosados o blancos está generalmente presente.

DISTRIBUTION: A., Aust., Eur., N.A.

Sorghum spp. + Puccinia purpurea Cooke =

ENGLISH	Leaf Rust
FRENCH	Rouille, Rouge des Feuilles
SPANISH	Roya de la hoja
PORTUGUESE	Ferrugem
ITALIAN	
RUSSIAN	ржавчина
SCANDINAVIAN	
GERMAN	Rost
DUTCH	
HUNGARIAN	
BULGARIAN	
TURKISH	
ARABIC	
PERSIAN	
BURMESE	Than-chi roga ပတ်စိ: နိ: ယ၁:ပျြုဇ ၇ံ:ယ၁:
THAI	Ra-Snim-Leck ราสนิมเหล็ก
NEPALI	
HINDI	Ratua रतुवा
VIETNAMESE	
CHINESE	Gao liang shiou bing 高梁銹病
JAPANESE	Sabi-byō さび病

This disease appears on the leaves as raised pustules or blisters covered with a brownish coating that eventually breaks open and allows the dark chestnut-brown spores to escape. The pustules occur on both the upper and lower surfaces of the leaf. Before the pustules appear, small purple, red or tan spots may be seen at the points where infection is developing. As the pustules develop, the colored regions around them become larger and large areas of the leaves may be destroyed.

Iste maladia se manifesta al folios como pustulas elevate coperte de un revestimento brunette le qual al fin se finde e vomita le sporas castanie obscur. Le pustulas se trova a tote le duo superfacies del folios. Ante que le pustulas se manifesta, minute maculas purpuree, rubie, o bronzate es visibile in le punctos ubi le infection se developpa. A mesura que le pustulas se forma, le regiones colorate circum illos cresce e grande areas del folios es sovente destructe.

Cette maladie apparaît sur les feuilles sous forme de pustules ou de vésicules soulevées recouvertes d'une enveloppe brunâtre qui, éventuellement, se brise et laisse échapper des spores châtain foncé. Les pustules s'observent sur les deux faces de la feuille. Avant l'apparition des pustules, de petites taches pourprées, rouges ou beiges s'observent au niveau de l'infection. Pendant l'évolution des pustules, les aires colorées qui les entourent s'agrandissent, et de grandes plages de la feuille peuvent être détruites.

Esta enfermedad aparece en las hojas como pústulas o ampollas levantadas cubiertas con una capa cafesosa que eventualmente se abre y permite que las esporas de color café castaño obscuro se escapen. Las pústulas aparecen tanto en el haz como en el envés de la hoja. Antes de que las pustulas aparezcan, pueden verse pequeñas manchas de color púrpura, rojas o crema en los puntos en donde se está desarrollando le infección. A medida que las pústulas se desarrollan, las regiones coloreadas de alrededor se vuelven más grandes y grandes áreas de las hojas pueden ser destruídas.

DISTRIBUTION: Af., A., Aust., Eur., N.A., C.A., S.A.

270

Sorghum vulgare (Zea mays) + Sclerospora sorghi (Kulk.) Weston & Uppal =

ENGLISH	Downy Mildew
FRENCH	Sclérosporose
SPANISH	Mildiu, Mildiú lanoso
PORTUGUESE	
ITALIAN	
RUSSIAN	склеорспороз
SCANDINAVIAN	
GERMAN	Sklerosporose
DUTCH	
HUNGARIAN	
BULGARIAN	
TURKISH	
ARABIC	
PERSIAN	
BURMESE	
THAI	
NEPALI	
HINDI	Mriduromil Phaphundi मृदुरोमिल फाफुन्दी
VIETNAMESE	
CHINESE	Gao líang lwù jìunn bìng 高梁露菌病
JAPANESE	Siraga-byô 白髪病

The symptoms of this disease vary greatly since the disease development is influenced by its environment. Generally, white to yellow streaking of the leaves and development of the white fungus growth on the streaks, followed by necrosis and browning of the streaks are characteristic leaf symptoms. Dwarfing, retarded development of the plants and poorly formed ears are typical where infection occurs early.

Le symptomas de iste maladia se varia grandemente perque le developpamento del maladia es affecte per le ambiente. Usualmente le symptomas characteristic in le folios es striation blanc o jalne e le apparition de un fungo blanc in le strias, e, al fin, necrosis e imbrunimento del strias. Nanismo, obstruction de crescimento, e spicas mal formate es typic se infection occurre in le prime parte del saison.

Les symptômes de cette maladie varient beaucoup parce que son évolution est influencée par l'environnement. Habituellement, des bandes blanches ou jaunâtres s'observent sur les feuilles, et le mycélium blanc du champignon croît sur les bandes, ce qui entraîne la nécrose et le brunissement à cet endroit. Ce sont là les symptômes caractéristiques sur la feuille. Le nanisme, le retard de croissance et la malformation des épis indiquent une infection précoce.

Los síntomas de esta enfermedad varían mucho, ya que el desarrollo de la enfermedad está influenciado por el medio ambiente. Generalmente, los síntomas foliares característicos son las rayas blancas a amarillas en las hojas y desarrollo de un crecimiento blanco en las rayas, seguido por necrosis y color café de las rayas. El achaparramiento, desarrollo retardado de las plantas y panojas pobremente formadas son típicos cuando la infección aparece temprano.

DISTRIBUTION: Af., A., Eur.

Sorghum vulgare + Sorosporium filiferum Kühn (Tolyposporium ehrenbergii (Kühn) Pat. =

ENGLISH	Long Smut
FRENCH	Charbon des Sorgho
SPANISH	Carbón largo
PORTUGUESE	
ITALIAN	Carbone del sorgo
RUSSIAN	головня соцветий крупнопузырчатая
SCANDINAVIAN	
GERMAN	Tolyposporium-Brand
DUTCH	
HUNGARIAN	
BULGARIAN	
TURKISH	Baş rastığı
ARABIC	
PERSIAN	
BURMESE	
THAI	
NEPALI	
HINDI	Deergh Kand दीर्घ कंद
VIETNAMESE	
CHINESE	Gao lǐang cháng hei swài bìng 高粱長黑穗病
JAPANESE	

This fungus attacks the floral structures and frequently only a small portion of the spike. The fungus fruiting structures are long, cylindrical, slightly curved and rupture at the base to release the brownish-green spore balls.

Iste fungo attacca solmente le structuras floral e, frequentemente, un minute portion del spica. Le structuras sporal del fungo es elongate, cylindric, legiermente curvate; illos se rumpe a basso e libera le sporas verde-brunette.

Le champignon attaque les parties florales et souvent seulement une petite portion de l'épi. Les fructifications du champignon sont longues, cylindriques, légèrement enroulées et se brisent à la base pour laisser échapper les boules de spores vert brunâtre.

Este hongo ataca las estructuras florales y frecuentemente sólo a una pequeña parte de la panoja. Las estructuras fructíferas del hongo son largas, cilíndricas, un poco encurvadas y se rompen en la base para dejar salir las bolas verde-cafesosas de esporas.

DISTRIBUTION: Af., A., Eur.

272

Sorghum vulgare + Sphacelotheca cruenta (Kühn) Potter =

ENGLISH	Loose Kernel Smut	
FRENCH	Charbon Nu de Sorgho	
SPANISH	Carbón suelto del grano	
PORTUGUESE		
ITALIAN	Carbone polveroso del sorgo, Carbone della rachide	
RUSSIAN	головня завязей мелкопузырчатая	
SCANDINAVIAN		
GERMAN	Kornbrand	
DUTCH		
HUNGARIAN		
BULGARIAN	праховита главня	
TURKISH	Darı açik rastığı	
ARABIC		
PERSIAN		
BURMESE		
THAI		
NEPALI		
HINDI	Shlath Kand	रलाथ कंद
VIETNAMESE		
CHINESE		
JAPANESE	Hadakakuroko-byô	裸黒穂病

This disease is identified by the early appearance of the diseased spikes and dwarfing of the plants of most varieties. Generally all the flowers in a spike are infected. The infected floral parts tend to elongate and proliferate. The membrane ruptures early, releasing the powdery black spore mass. The central portion of the floral structure persists after the spores are discharged.

Iste maladia se identifica per le apparition prematur del spicas morbide e per nanismo del plantas in le major parte de varietates. Usualmente, tote le flores de un spica es infectate. Le partes floral infectate propende a elongar se e proliferar. Le membrana se finde rapidemente, delivrante le massa pulverose nigre de sporas. Le portion central del structura floral persiste post que le sporas es discargate.

Cette maladie se distingue par l'apparition précoce d'épis malades et par le nanisme des plantes chez la plupart des variétés. Généralement, toutes les fleurs de l'épi sont infectées. Les parties florales infectées tendent à s'allonger et à proliférer. La membrane se brise tôt et laisse échapper une masse poudreuse de spores noires. La partie centrale de l'appareil floral persiste après la libération des spores.

Esta enfermedad se identifica por el temprano aparecimiento de panojas enfermas y achaparramiento de las plantas de la mayoría de variedades. Generalmente todas las flores en la panoja están infectadas. Las partes florales infectadas tienden a largarse y proliferar. La membrana se rompe temprano, dejando salir la masa polvorienta de esporas negras. La porción central de la estructura floral persiste después de que las esporas han salido.

DISTRIBUTION: Af., A., Eur., N.A., C.A., S.A.

Sorghum vulgare + Sphacelotheca sorghi (Lk.) Clint. =

ENGLISH Covered Kernel Smut
FRENCH Charbon Couvert
SPANISH Carbón cubierto, Carbón duro, Carbón del grano
PORTUGUESE
ITALIAN Carbone coperto, Carbone della saggina
RUSSIAN **головня завязей покрытая**
SCANDINAVIAN
GERMAN Kornbrand, Gedeckter Brand
DUTCH
HUNGARIAN
BULGARIAN **покрита главня**
TURKISH Darı kapalı rastığı
ARABIC
PERSIAN Sēyàhàkê zoráte khōōshêē سیاهک ذرت خوشه‌ای
BURMESE Haing, Aing roga ဝ ပြေ စိလိုသံကား ဆေ5ဂုင်
THAI
NEPALI Junelo Ko kalogede rog जुनेलीको कालो गेडे रोग
HINDI Dana Kand Rog दाना कंद रोग
VIETNAMESE
CHINESE Gao líang lì hei swài bìng 高梁粒黑穗病
JAPANESE Tubukuroho-byô 粒黒穂病

Usually all, but occasionally only a part, of the kernels on a diseased plant are destroyed. In diseased spikes, enlarged cylindrical or cone-shaped galls are formed instead of the kernels. At first these galls are covered with a light gray or brown membrane that later may break and release the dark brown spores. Plants affected by this disease appear normal except for the diseased spikes.

Usualmente tote le granos (ma a vices solmente un portion de illos) in un planta morbide es destructe. In spicas morbide gallas ampliate de forma cylindric o conic se forma in loco de granos. Al initio, iste gallas es coperte de un membrana gris o brun clar le qual plus tarde se rumpe e discarga le sporas brun obscur. Plantas afflicte per iste maladia pare esser normal, salvo le spicas morbide.

D'ordinaire, tous les grains d'une plante malade sont détruits, mais quelquefois un certain nombre seulement le sont. Sur les épis malades, de grosses galles cylindriques ou en forme de cône se forment au lieu des grains. Au début, ces galles sont recouvertes d'une membrane gris pâle qui, plus tard, peut se briser et libérer les spores brun foncé. Les plantes atteintes de cette maladie paraissent normales, exception faite des épis malades.

Generalmente todo el grano, pero ocasionalmente sólo una parte de éste es destruída en una planta enferma. En panojas enfermas, en vez del grano se forman agallas grandes, cilíndricas o en forma de cono. Al principio estas agallas están cubiertas con una membrana de color gris claro o café que después puede abrirse dejando salir las esporas de color café obscuro. Las plantas afectadas por esta enfermedad parecen normales, excepto por las panojas enfermas.

DISTRIBUTION: Af., A., Aust., Eur., N.A., C.A., S.A.

274

Sorghum spp. + Xanthomonas holcicola (Elliott) Starr & Burkholder =

ENGLISH	Bacterial Streak
FRENCH	Bactériose Striée
SPANISH	Estría bacteriana, Xanthomoniasis, Tizón bacteriano, Raya bacterial
PORTUGUESE	
ITALIAN	
RUSSIAN	**штриховатость бактериальная**
SCANDINAVIAN	
GERMAN	Blattfleckenbakteriose
DUTCH	
HUNGARIAN	
BULGARIAN	
TURKISH	
ARABIC	
PERSIAN	
BURMESE	
THAI	
NEPALI	
HNDI	
VIETNAMESE	
CHINESE	Gao líang tíao ku bìng 高粱條枯病
JAPANESE	

This disease causes narrow, water-soaked streaks on the leaves. These streaks may occur on plants from the seedling stage to near maturity. At first no color is evident except the light yellow drops of exudate standing out on the young streaks. Later, narrow red-brown margins or blotches of color appear in the streaks and after a few weeks, the streaks are red throughout and no longer appear water-soaked. Numerous streaks may join to form large, dead tissue areas.

Iste maladia produce strias stricte e aquose in le folios. Iste strias pote formar se in plantulas o in stadios subsequente, quasi usque a maturitate del plantas. Al initio, nulle coloration es presente, salvo le guttas jalne clar de exsudato visibile in le strias juvene. Plus tarde, margines stricte brun-rubie o pustulas colorate se manifesta in le strias, e post qualque septimanas le strias es totalmente rubie e non aquose plus. Strias numerose a vices coalesce usque a formar grande areas de texito morte.

Cette maladie produit des bandes étroites et d'aspect detrémpé sur les feuilles. Ces bandes peuvent apparaître sur les plantes depuis le stade plantule jusqu'à la maturité. Au début, aucune couleur n'apparaît excepté celle des gouttelettes jaune pâle d'un exsudat suintant des nouvelles bandes. Plus tard, des marges étroites, brun rouge, ou des éclaboussures de couleurs apparaissent dans les bandes puis au bout de quelques semaines, les bandes sont rouge d'un bout à l'autre et n'ont plus cet aspect détrempé. Plusieurs bandes peuvent s'unir pour former de grandes plages nécrotiques.

Esta enfermedad causa rayas angostas y empapadas de agua en las hojas. Estas rayas pueden aparecer en las plantas desde muy pequeñas hasta cerca de la madurez. Al principio no se nota ningún color, excepto las gotas de exudación de color amarillo claro que aparecen sobre las rayas jóvenes. Más tarde aparecen márgenes angostos, de color rojo-café o manchones de color aparecen en las rayas y después de unas semanas, las rayas adquieren un color rojo total y ya no parecen empapadas de agua. Numerosas rayas pueden unirse para formar grandes áreas de tejido muerto.

DISTRIBUTION: Af., Aust., N.A., S.A.

<u>Thea</u> <u>sinensis</u> + <u>Guignardia camelliae</u> (Cooke) Butler =

ENGLISH   Brown Blight, Copper Blight
FRENCH
SPANISH   Tizón café, Tizón de cobre
PORTUGUESE
ITALIAN
RUSSIAN
SCANDINAVIAN
GERMAN
DUTCH
HUNGARIAN
BULGARIAN
TURKISH
ARABIC
PERSIAN
BURMESE
THAI
NEPALI
HINDI
VIETNAMESE
CHINESE   Chá shù ku bìng   茶樹枯病
JAPANESE   Akahagare-byō   赤葉枯病

This disease first appears as small grayish spots on the uppersurface of older leaves which increase in size and finally coalesce, forming large blotches. These are gray, bounded by a dark line. The leaves finally assume a blistered or swollen appearance and eventually the epidermis over the blisters bursts, usually forming a triangular figure.

Iste maladia se manifesta al initio como minute maculas gris al superfacie superior de folios plus vetule; iste maculas cresce e al fin coalesce usque a formar grande pustulas. Iste pustulas es gris, limitate per un linea obscur. Le folios pare tumide o pustulose, e al fin le epidermis super le pustulas se finde,usualmente formante un designo triangular.

Cette maladie se manifeste d'abord, à la face supérieure des vieilles feuilles, par de petites taches grisâtres qui s'agrandissent et finalement se joignent pour former de grandes éclaboussures. Celles-ci sont grises et entourées d'une ligne noire. Les feuilles finissent par se boursoufler ou s'enfler, et l'épiderme qui recouvre les vésicules éclate éventuellement pour former une sorte de triangle.

Esta enfermedad aparece primero como pequeñas manchas grisáceas sobre el haz de las hojas más viejas, que aumentan en tamaño y finalmente se unen formando grandes manchones. Estos son de color gris, rodeadas por una línea obscura. Las hojas finalmente adquieren una apariencia ampollada o hinchada y eventualmente la epidermis sobre las ampollas se revienta, generalmente formando una figura triangular.

DISTRIBUTION: A., Eur., N.A.

Thea sinensis + Pestalotiopsis theae (Sawada) Steyaert (Pestalotia theae Sawada) =

ENGLISH	Gray Blight
FRENCH	
SPANISH	Mancha morena de la hoja, Tizón gris
PORTUGUESE	
ITALIAN	
RUSSIAN	
SCANDINAVIAN	
GERMAN	
DUTCH	
HUNGARIAN	
BULGARIAN	
TURKISH	Yaprak gri lekesi
ARABIC	
PERSIAN	Pestaloozeyaye chay پستالوزای چای
BURMESE	Phyu-hmaung-ywet-pok roga ပင်စတလိုဇ္ဇား သိ ေဆး
THAI	
NEPALI	
HINDI	
VIETNAMESE	Cháy lá
CHINESE	Chá lúen ban bìng 茶輪斑病
JAPANESE	Rinhan-byô 輪斑病

This disease begins as small brown spots on the leaves. Later they show on the uppersurface as small brown spots with greyish centers with light to dark brown margins. Lesions are usually circular to oval with concentric zonations marked out on the uppersurface by the dark fruiting bodies. On young leaves these zonations are often absent. Coalescence of the spots may occur.

Iste maladia comencia como minute maculas brun in le folios; plus tarde illos se manifesta como minute maculas brun in le superfacie superior con centros gris e con margines brun clar o obscur. Le lesiones usualmente es circular o oval con zonationes concentric marcate in le superfacie superior per le structuras sporal obscur. In folios juvene, iste zonationes frequentemente es absente. Le maculas pote coalescer.

Cette maladie débute par de petites taches brunes sur les feuilles et puis, à la face supérieure, par de petites taches brunes avec un centre grisâtre et des bords brun pâle à brun foncé. D' ordinaire, les lésions sont circulaires ou ovales avec des ceintures concentriques marquées, à la face supérieure, par les fructifications noires. Sur les jeunes feuilles, ces ceintures sont souvent absentes. Il peut y avoir fusion des taches.

Esta enfermedad comienza como pequeñas manchas de color café sobre las hojas; luego, éstas aparecen sobre el haz como pequeñas manchas de color café con centros grisáceos y márgenes claros u obscuros. Las lesiones son generalmente circulares u ovaladas, con zonificaciones concéntricas marcadas sobre el haz por los cuerpos fructíferos obscuros. Sobre las hojas jóvenes estas zonificaciones frecuentemente no aparecen. Puede suceder que las manchas se unan.

DISTRIBUTION: Af., A., Aust., N.A., S.A.

Thea sinensis + Pseudomonas theae Okabe et Goto (Bacillus theae Hori et Bokura) =

ENGLISH	Red Blight
FRENCH	
SPANISH	Tizón rojo
PORTUGUESE	
ITALIAN	
RUSSIAN	
SCANDINAVIAN	
GERMAN	
DUTCH	
HUNGARIAN	
BULGARIAN	
TURKISH	
ARABIC	
PERSIAN	
BURMESE	
THAI	
NEPALI	
HINDI	
VIETNAMESE	
CHINESE	Chá chìr shaw bìng 茶 赤 燒 病
JAPANESE	Akayake-byð 赤燒病

This disease is characterized by small, circular spots which are pale brown and have sharply distinct borders. These spots occur on the upper surface of the leaves. Later these spots coalesce and their surface becomes a dark color or sometimes black with concentric broken lines, during moist weather. During dry weather, they appear reddish-brown. These same symptoms appear on the stems and twigs. In addition, the tissue under the bark turns black.

Iste maladia se characterisa per minute maculas circular pallide e brun e con bordos acutemente distincte in le superfacies superior del folios. Plus tarde iste maculas coalesce e lor superfacie deveni de color obscur o a vices nigre con lineas interrupte concentric in tempore humide. In tempore sic, illos es brun-rubiette. Iste mesme symptomas se manifesta in le pedunculos e in le ramettos. In plus, le texito sub le cortice deveni nigre.

Cette maladie est caractérisée par de petites taches circulaires brun pâle, à marges nettement distinctes. Ces taches se présentent à la face supérieure des feuilles. Plus tard, les taches se fusionnent et leur surface devient foncée ou parfois noire, avec, par temps humide, des lignes concentriques brisées. Par temps sec, elles paraissent brun rougeâtre. Des symptômes semblables se montrent sur les tiges et les rameaux. En plus, les tissus noircissent sous l'écorce.

Esta enfermedad se caracteriza por pequeñas manchas circulares, de color café claro y con bordes marcadamente distinguibles. Estas manchas aparecen sobre el haz de las hojas. Luego estas manchas se unen y su superficie se vuelve de color obscuro o algunas veces negro, con líneas quebradas concéntricas, durante el tiempo húmedo. Durante el tiempo seco, aparecen de color café-rojizo. Estos mismos síntomas aparecen sobre los tallos y ramillas. Además, el tejido debajo de la corteza se vuelve negro.

DISTRIBUTION: A.

278

Thea sinensis (Theobroma cacao, Coffea spp., Hevea brasiliensis etc.) + Ustulina zonata (Lév.)
Sacc. (Ustulina deusta (Hoffm. ex Fr.) Petr.) =

ENGLISH	Root Disease, Collar Rot
FRENCH	
SPANISH	Enfermedad radicular, Podredumbre del cuello
PORTUGUESE	
ITALIAN	
RUSSIAN	
SCANDINAVIAN	
GERMAN	
DUTCH	
HUNGARIAN	
BULGARIAN	
TURKISH	
ARABIC	
PERSIAN	
BURMESE	
THAI	
NEPALI	
HINDI	Charcoal viglan चारकोल विगलान
VIETNAMESE	
CHINESE	Chá shù tàn gen bìng 茶樹炭根病, Chá shù tàn pí bìng 茶樹炭皮病
JAPANESE	

The fungus causing this disease works between the wood and the bark, forming white or yellowish fan-shaped patches which become black on the edge when they come in contact with a crack
in the bark. A cross section of the root shows numerous irregular black lines. The large fungus
bodies protrude from cracks in the bark. They show as white swollen cushions and spread over
the surface as flattened plates. When mature, these plates are gray, concentrically zoned and
marked with black dots.

Le fungo que causa iste maladia se activa inter le ligno e le cortice, producente areas blanc
o jalne flabelliforme cuje margine deveni nigre in contacto con un fissura in le cortice. In
section transversal, le radice monstra numerose lineas nigre irregular. Le grande structuras
sporal protrude ab le fissuras in le cortice. Illos se manifesta como blanc cossinos tumide e
se extende super le superfacie como plattas applanate. Iste plattas, quando illos es maturate,
es gris, concentricamente zonate, e marcate per punctos nigre.

Le champignon qui cause cette maladie oeuvre entre l'écorce et l'arbre, formant des plaques
blanches ou jaunâtres en forme d'éventail et qui noircissent à la marge quand elles viennent
en contact avec une fissure de l'écorce. Une coupe transversale de la racine laisse voir de
nombreuses lignes noires irrégulières. De grosses masses fongueuses sortent des fissures
de l'écorce. Elles apparaissent comme des coussins blancs renflés et s'étendent à la surface
comme des plaques aplaties. A maturité, ces plaques sont grises, ceinturées de façon concen-
trique et marquées de points noirs.

El hongo causante de esta enfermedad trabaja entre la corteza y la madera, formando manchones blancos o amarillentos en forma de abanico, los cuáles se vuelven negros en la orilla
cuando entran en contacto con alguna rajadura en la corteza. Una sección cruzada de la raíz
muestra numerosas líneas negras irregulares. Los grandes cuerpos del hongo salen fuera
a través de las rajaduras en la corteza. Estos se muestran como cojines blancos hinchados
y se diseminan sobre la superficie como placas planas. Cuando se maduran, estos placas son
grises, concéntricamente zonadas y marcadas con puntos negros.

DISTRIBUTION: Af., A., Aust., Eur., N.A., C.A., S.A.

Theobroma cacao + Calonectria rigidiuscula (Berk. & Br.) Sacc. =

ENGLISH Die-back, Canker
FRENCH
SPANISH Muerte regresiva, Cancro
PORTUGUESE
ITALIAN
RUSSIAN
SCANDINAVIAN
GERMAN
DUTCH
HUNGARIAN
BULGARIAN
TURKISH
ARABIC
PERSIAN
BURMESE
THAI
NEPALI
HINDI
VIETNAMESE Sưng lở trên cành và trái
CHINESE
JAPANESE

This fungus causes the bark to become dry and grayish-brown. When cut, the tissue is found to be discolored. The bark splits and allows a brownish-red gummy fluid to ooze out. This gum dries and gives a dark rusty appearance. Under some weather conditions, an abnormal number of flowers are produced which never set fruit. Cankers may occur on either the branches or the main stem and cause girdling. The leaves are reduced in size and are yellowish in color.

Iste fungo face le cortice desiccar se e devenir brun-gris. Si on lo talia, on vide que le texito es discolorate. Le cortice se finde e lassa un liquor gummose rubie-brunette exsudar se. Iste gumma se desicca e ha un aspecto ferruginose obscur. Sub alicun conditiones de tempore, un numero anormal de flores se forma, le quales non produce fructos. Canceres occurre o in le ramos o in le trunco principal, le quales illos imbracia. Le folios es de dimension diminute e de color jalne.

Le champignon cause un desséchement de l'écorce et la rend brun grisâtre. Quand on sectionne les tissus, on voit qu'ils sont décolorés. L'écorce se fend et laisse exsuder un liquide gommeux. La gomme s'assèche et donne une apparence rouillée et foncée. Dans certaines conditions de climat, il se forme un nombre anormal de fleurs qui ne portent jamais fruit. Des chancres peuvent se former soit sur les branches ou sur la tige principale et causer l'encerclement. Les feuilles deviennent jaunâtres et n'atteignent pas leur grosseur normale.

Este hongo causa que la corteza se vuelve seca y de color café grisáceo. Cuando se corta, el tejido se ve que está descolorido. La corteza se raja y permite que un fluído gomoso rojo-cafesoso salga afuera. Esta goma se seca, dándole una apariencia herrumbrosa. Bajo algunas condiciones de tiempo, se produce un número anormal de flores, las cuáles nunca fijan fruto. Pueden aparecer cancros ya sea en las ramas o en el tallo principal y causar una atadura. Las hojas se reducen en tamaño y son de color amarilloso.

DISTRIBUTION: Af., A ., Aust., N.A., C.A, W.I.

280

Trifolium spp. + Kabatiella caulivora (Kirchn.) Kirchn. =

ENGLISH	Northern Antracnose
FRENCH	**Anthracnose** Septentrionale, Anthracnose "du Nord"
SPANISH	Antracnosis del norte
PORTUGUESE	
ITALIAN	Antracnosi
RUSSIAN	**антракоз "северный"**
SCANDINAVIAN	Stængelsyge
GERMAN	Stengelbrenner
DUTCH	Stengelbrand
HUNGARIAN	
BULGARIAN	
TURKISH	
ARABIC	
PERSIAN	
BURMESE	
THAI	
NEPALI	
HINDI	
VIETNAMESE	
CHINESE	
JAPANESE	Kukiware-byô 茎割病

The first symptoms to appear on the leaves and stems are medium brown to black lesions, which may or may not develop gray or light brown centers. The elongated, slightly sunken stem lesions crack in the center. Severely affected plants appear as if scorched by fire. The stalk lesions cause leaves to wilt. Affected parts dry out rapidly and become brittle.

Le prime symptomas in le folios e in le pedunculos es lesiones brun o nigre, le quales pote monstrar centros gris o brun clar o non. Le elongate lesiones legiermente depresse in le pedunculos se finde in le centro. Plantas severmente infectate pare esser tostate per igne. Le lesiones in le pedunculos face le folios marcescer. Partes afflicte se desicca rapidemente e deveni fragile.

Les premiers symptômes à apparaître sur les feuilles et les tiges sont des lésions brun moyen à noires, dont le centre peut être gris ou brun pâle. Les lésions sur les tiges, allongées et quelque peu déprimées, se fendillent en leur centre. Les plantes gravement atteintes semblent avoir été touchées par le feu. Les lésions sur les tiges provoquent une flétrissure des feuilles. Les parties atteintes se dessèchent rapidement et deviennent cassantes.

Los primeros síntomas aparecen sobre las hojas y tallos y son lesiones de color café mediano a negras, que pueden o no desarrollar centros grises o café claro. Las lesiones alargadas, un poco hundidas en el tallo se rajan en el centro. Las plantas severamente afectadas parecen como si hubieran sido quemadas por el fuego. Las lesiones de los tallos causan que las hojas se marchiten. Las partes afectadas se secan rápidamente y se vuelven quebradizas.

DISTRIBUTION: Af., A., Aust., Eur., N.A.

Trifolium spp. (Medicago spp., and others) + Peronospora trifoliorum De Bary =

ENGLISH	Downy Mildew
FRENCH	Mildiou, Maladie du Trèfle
SPANISH	Mildiú lanoso
PORTUGUESE	
ITALIAN	Peronospora del trifoglio, Muffa
RUSSIAN	пероноспороз
SCANDINAVIAN	
GERMAN	Falscher Mehltau
DUTCH	
HUNGARIAN	
BULGARIAN	
TURKISH	
ARABIC	
PERSIAN	
BURMESE	
THAI	
NEPALI	
HINDI	
VIETNAMESE	
CHINESE	
JAPANESE	Beto-byô　べと病

The characteristic symptoms of this disease are light green leaves, especially at the base of the stem, and the grayish-white fungus on the surface of the leaves. The stems are shorter and smaller and the leaflets are twisted and rolled with severe infection. When infection occurs throughout the plant, the stems are swollen and the foliage chlorotic and the leaf tissue collapses.

Le symptomas characteristic de iste maladia es folios verde clar, specialmente al pede del pedunculo, e le fungo blanc-gris al superfacie del folios. Le pedunculos es plus curte e plus minute que normal e le foliettos es distorte e inrolate in infection sever. Si le infection se extende per tote le planta, le pedunculos es tumide le foliage es chlorotic, e le mesophyllo mori.

Les symptômes caractéristiques de cette maladie sont des feuilles vert pâle, particulièrement à la base de la tige, ainsi que la présence du champignon gris-blanc à la surface des feuilles. Lorsque l'infection est grave, les tiges sont plus courtes et plus chétives et les folioles, tordues et enroulées. Lorsque toute la plante est infectée, les tiges se gonflent, le feuillage devient chlorotique, et les tissus des feuilles s'affaissent.

El síntoma característico de esta enfermedad son las hojas de color verde claro, especialmente en la base del tallo y el hongo blanco-grisáceo sobre la superficie de las hojas. Los tallos son más cortos y pequeños y las hojillas están retorcidas y enrolladas con infección severa. Cuando la infección aparece en toda la planta, los tallos se hinchan, el follaje es clorótico y el tejido foliar sufre un colapso.

DISTRIBUTION: Af., A., Aust., Eur., N.A., C.A., S.A.

282

Trifolium spp. (Medicago spp. and others) + Sclerotinia trifoliorum Erikss. =

ENGLISH	Sclerotinia Wilt, Root Rot
FRENCH	Flétrissure Sclérotique, Maladie à Sclérotes, Chancre
SPANISH	Pudrición radicular, Marchitez de sclerotinia, Podredumbre radicular
PORTUGUESE	
ITALIAN	Cancro, Mal dello sclerozio
RUSSIAN	рак корнай и основания стеблей, склеротиниоз
SCANDINAVIAN	Kløverens knoldbægersvamp, Kløverråte
GERMAN	Kleekrebs, Krebs
DUTCH	Kanker, Klaverkanker
HUNGARIAN	
BULGARIAN	
TURKISH	
ARABIC	
PERSIAN	
BURMESE	
THAI	
NEPALI	
HINDI	
VIETNAMESE	
CHINESE	
JAPANESE	Kinkaku-byõ　菌核病

The symptoms of this disease vary with season, weather conditions and tissues invaded. Brown leaf spots occur in wet periods late in the growing season. The heavily infected leaflets drop off. Infection spreads downward into the roots. The new growth of infected plants wilts and the dead tissue becomes overgrown by the white fungus growth, especially under high moisture conditions.

Le symptomas de iste maladia se varia secundo saison, conditiones de tempore, e le texito invase. Maculas foliar brun occurre durante periodos humide tarde in le saison. Le foliettos severmente infectate se separa e cade. Le infection se extende in basso in le radices. Le partes emergente de plantas infectate marcesce e le texito morte es coperte del fungo blanc, specialmente sub conditiones multo humide.

Les symptômes varient avec les saisons, les conditions atmosphériques et les tissus atteints. Des taches brunes apparaissent durant les périodes pluvieuses de la fin de la saison de végétation. Les folioles gravement atteintes tombent. L'infection descend vers les racines. La nouvelle pousse des plantes atteintes se flétrit, et les tissus morts sont envahis par le mycélium blanc du champignon, particulièrement dans des conditions d'humidité élevée.

Los síntomas de esta enfermedad varían con la estación, las condiciones del tiempo y los tejidos invadidos. En períodas húmedos aparecen manchas foliares de color café, tarde durante la época de crecimiento. Las hojitas severamente infectadas se caen. La infección se disemina hacia abajo dentro de las raíces. El nuevo crecimiento de plantas infectadas se marchita y el tejido muerto se cubre con el crecimiento blanco del hongo, especialmente bajo condiciones de alta humedad.

DISTRIBUTION: A., Aust., Eur., N.A., S.A.

Trifolium spp. + Uromyces trifolii (DC.) Lév. =

ENGLISH	Rust
FRENCH	Rouille
SPANISH	Chahuixtle, Roya, Polvillo
PORTUGUESE	
ITALIAN	Ruggine
RUSSIAN	
SCANDINAVIAN	
GERMAN	
DUTCH	
HUNGARIAN	
BULGARIAN	
TURKISH	
ARABIC	
PERSIAN	Zàngê shàbdàr زنگ شبدر
BURMESE	
THAI	
NEPALI	
HINDI	
VIETNAMESE	
CHINESE	
JAPANESE	Hasabi-byô 葉さび病

One stage of the fungus in the life cycle of this disease causes light-yellow to orange-yellow lesions on the stems, leaf stalks and leaves. The other stage causes small brown lesions on any portion of the green plant.

Durante un prime periodo de su vita, iste fungo produce lesiones jalne clar o jalne-orange in le pedunculos, petiolos, e folios. Durante le secunde periodo, illo produce minute lesiones brun per tote le partes del planta verde.

Durant le premier stade de son développement, au cours du cycle d'évolution de cette maladie, le champignon produit des lésions jaune pâle à jaune orangé sur les tiges, les pétioles et les feuilles. Au cours du second stade, le champignon produit de petites lésions brunes sur n'importe quelle partie de la plante.

Un estado del hongo en el ciclo de vida de esta enfermedad causa lesiones amarillo claro a amarillo-anaranjado en los tallos, pédunculos y hojas. El otro estado causa pequeñas lesiones de color café en cualquier porción de la planta verde.

DISTRIBUTION: Af., A., Aust., Eur., N.A., C.A., S.A.

284

Triticum aestivum (Gramineae) + Cercosporella herpotrichoides Fron =

ENGLISH	Eyespot Foot Rot
FRENCH	Tache Ocellée, Pietin-verse, Cercosporellose
SPANISH	Podredumbre de pié, mancha de ojo
PORTUGUESE	
ITALIAN	Spezzamento dei culmi, Mal del piede dei cereali
RUSSIAN	ломкость стеблей, гниль корневой шейки, церкоспореллез
SCANDINAVIAN	Fotsyke, Knækkefodsyge
GERMAN	Lagerfusskrankheit, Halmbruchkrankheit, Augenflekken, Medaillonflecken
DUTCH	Oogvlekkenziekte
HUNGARIAN	Szártőbetegség
BULGARIAN	паразитно полягане
TURKISH	Çim boğaz yanıklığı
ARABIC	
PERSIAN	
BURMESE	
THAI	
NEPALI	
HINDI	
VIETNAMESE	
CHINESE	
JAPANESE	

The most conspicuous symptom of this disease is the falling over of diseased plants near the end of the growing season. The lesions are first evident on the leaves as elliptical to ovate spots with light brown centers and dark brown margins. Similar spots occur on other parts of the plant. Necrosis also occurs on the roots near the groundline. When infection occurs early, the plants are killed before the grain matures.

Le symptoma le plus characteristic de iste maladia es le cader de plantas morbide presso le fin del saison. Le lesiones se manifesta in le folios como elliptic o oval maculas con centros brun clar e margines brun obscur. Maculas similar occurre in altere partes del planta. Necrosis occurre anque in le radices presso le plano del solo. Si infection occurre in le prime parte del saison, le plantas mori ante que le granos se matura.

Le symptôme le plus voyant de cette maladie est le renversement des plantes atteintes, vers la fin de la saison de végétation. Les lésions apparaissent d'abord sur les feuilles sous forme de taches elliptiques à ovées, avec un centre brun pâle et des bords brun foncé. Des taches semblables se forment sur d'autres parties de la plante. La nécrose des racines se produit aussi au ras du sol. Quand l'infection est précoce, les plantes meurent avant que le grain soit mûr.

El síntoma más conspicuo de esta enfermedad es la caída de las plantas enfermas cerca del final de la época de cultivo. Las lesiones son evidentes primero sobre las hojas como manchas elípticas a ovaladas, con centros café claro y márgenes café obscuro. Manchas similares aparecen en otras partes de la planta. También ocurre necrosis en las raíces cerca del nivel del suelo. Cuando la infección aparece temprano, las plantas mueren antes de que el grano madure.

DISTRIBUTION: Af., Aust., Eur., N.A.

Triticum spp. + Corynebacterium tritici (Hutchinson) Burkholder =

ENGLISH	Yellow Slime
FRENCH	
SPANISH	Limo amarillo
PORTUGUESE	
ITALIAN	
RUSSIAN	
SCANDINAVIAN	
GERMAN	
DUTCH	
HUNGARIAN	
BULGARIAN	
TURKISH	
ARABIC	
PERSIAN	Khōoshèhĕ sámghĕyĕh gàndŏm خوشهٔ صمغی گندم
BURMESE	
THAI	
NEPALI	
HINDI	
VIETNAMESE	
CHINESE	
JAPANESE	

Infection of this disease causes all plant parts to bear yellow slimy bacterial exudate. When the spikes are covered with this slime, they become distorted or fail to emerge properly and produce little or no grain. Long, slimy yellow spots occur on the leaves. In some countries this disease is associated with nematode disease which produces a gall and in these cases the symptoms are a combination of those caused by the two organisms.

In iste maladia ìnfection face tote le partes del planta portar exsudato bacterial viscose jalne. Quando le spicas es coperte de iste fango, illos deveni distorte o non emerge correctement e illos produce pauc o nulle granos. Longe maculas jalne viscose se manifesta in le folios. In alicun paises iste maladia es associate con un morbo causate per nematodos, e illo produce un gallo. In   tal casos, le symptomas es le symptomas combinate del duo organismos.

L'infection par cette maladie provoque l'apparition d'un exsudat bactérien jaune, visqueux, sur tous les organes de la plante. Quand les épis sont couverts par cet exsudat, ils se déforment ou n'émergent pas normalement et produisent peu ou pas de grains. De longues taches jaunes, visqueuses apparaissent sur les feuilles. Dans certains pays, cette maladie accompagne une maladie causée par des nématodes, qui produit une cécidie. Dans ces cas, les symptômes sont une combinaison de ceux produits par les deux organismes.

La infección de esta enfermedad causa que todas la partes de la planta tengan una exudación bacterial limosa. Cuando las espigas están cubiertas con este limo, se vuelven deformes o no emergen apropiadamente y producen poco o ningún grano.   En las hojas aparecen manchas largas, limoso-amarillas. En algunos países esta enfermedad está asociada con enfermedad de nemátodos, la cuál produce agallas y en estos casos los síntomas son una combinación de éstos causados por ambos organismos.

DISTRIBUTION: Af., A., Aust., Eur.

286

Triticum spp. (Hordeum spp. and other Gramineae) + Erysiphe graminis DC. ex Mérat =

ENGLISH	Powdery Mildew
FRENCH	Blanc des Graminées, Oïdium
SPANISH	Mildiu polvoriento, Oidio
PORTUGUESE	
ITALIAN	Mal bianco, Albugine dei cereali, Nebbia
RUSSIAN	мучнистая роса
SCANDINAVIAN	Græssernesmeldug, Gräsmjöldagg, Grasmeldugg, Hvedens meldug
GERMAN	Echter Getreidemehltau, Mehltau
DUTCH	Meeldauw
HUNGARIAN	Buzalisztharmat
BULGARIAN	брашнеста мана
TURKISH	Tahıl küllemesi
ARABIC	
PERSIAN	Sefedàke haghegheye gàndom سفیدک حقیقی گندم
BURMESE	
THAI	
NEPALI	Gahoon Ko pitho-chhare rog गहुँको पिठो-छरे रोग , Gahoon Ko sete-pat गहुँको सेतेपात
HINDI	Choorni Phaphundi चूरनी फाफुन्दी
VIETNAMESE	
CHINESE	Shǐao mài bór fěn bìng 小麥白粉病
JAPANESE	Udonko-byð うどんこ病

This disease develops on the leaves and floral parts of the plant. The fungus fruiting bodies appear on the surface of these portions and at first are light gray in color, but later dark dots appear. With severe infection, chlorosis and browning of the tissue are apparent, followed by a gradual drying out of the tissue.

Iste maladia se forma in le folios e in le partes floral del planta. Le structuras sporal del fungo se manifesta al superfacie de iste portiones e es, al initio, de color gris clar, ma plus tarde punctos obscur se manifesta. In infection sever, chlorosis e brunimento del texito es evidente e, plus tarde, desiccation gradual del texito.

Cette maladie se manifeste sur les feuilles et les parties florales de la plante. Les organes de fructification du champignon apparaissent à la surface de ces parties atteintes, et sont d'abord gris pâle. Plus tard, ce sont des points sombres qui apparaissent. Dans le cas d'une infection grave, on peut observer la chlorose et le brunissement des tissus atteints, suivis d'un dessèchement graduel des tissus.

Esta enfermedad se desarrolla en las hojas y partes florales de la planta. Los cuerpos fructíferos del hongo aparecen sobre la superficie de estas porciones y al principio son de color gris, pero leugo aparecen manchas más grandes. Cuando la infección es severa, es aparente una clorosis y color café de los tejidos, seguido por un secamiento gradual del tejido.

DISTRIBUTION: Af., A., Aust., Eur., N.A., C.A., S.A.

Triticum aestivum (Gramineae) + Fusarium culmorum (W. G. Smith) Sacc. =

ENGLISH	Root and Culm Rot, Seedling Blight
FRENCH	Fusariose, Pourriture Fusarienne des Racines
SPANISH	Pudrición radicular, Podredumbre de la raíz y del tallo, Tizón de la plántula
PORTUGUESE	
ITALIAN	
RUSSIAN	гниль корней фузариозная, также фузариоз
SCANDINAVIAN	
GERMAN	Fusarium-Fusskrankheit
DUTCH	
HUNGARIAN	Szártőbetegség
BULGARIAN	фузариоза
TURKISH	
ARABIC	
PERSIAN	
BURMESE	
THAI	
NEPALI	
HINDI	
VIETNAMESE	
CHINESE	
JAPANESE	

The disease occurs as a seedling blight, foot rot and head blight. The blighted seedlings are characterized by a light brown to reddish-brown water-soaked rot and blight either before or after emergence. The spike blight causes a water-soaked appearance, followed by the loss of green color. During warm humid weather, the fungus causes the spiklets to appear pink in color. The kernels are shriveled with a scabby appearance.

Le morbo se manifesta como necrosis de plantulas, putrefaction radical, e necrosis de inflores-centias. Le plantulas necrotic se characterisa per un putrefaction e necrosis aquose brun clar o brun-rubiette o ante o post emergentia. Le necrosis del spica ha un aspecto aquose e plus tarde perdita del color verde. In tempore calide e humide, le fungo face le spiculas semblar de color rosee. Le granos es crispate e ha un aspecto crustose.

La maladie apparaît comme une brûlure des semis, une pourriture des racines et une brûlure de l'épi. Les semis atteints sont caractérisés par une pourriture détrempée, brun pâle à brun rougeâtre, et se flétrissent avant ou après l'émergence. La brûlure de l'épi produit un aspect détrempé, suivi de la perte de la couleur verte. Par temps chaud et humide, le champignon donne aux épillets une couleur rosée. Les grains sont ratatinés et ont une apparence galeuse.

La enfermedad aparece como un tizón de la plántula, podredumbre del pié y tizón de la parte superior. Las plántulas con tizón se caracterizan por la podredumbre de color café claro a café-rojizo, empapada de agua y el tizón ya sea antes o después de la emergencia. El tizón en la espiga provoca una apariencia empapada de agua, seguida por la pérdida del color verde. Durante épocas cálidas y húmedas, el hongo causa que las espículas tengan un color rosado. Los granos se arrugan y tienen una apariencia costrosa.

DISTRIBUTION: Af., A., Aust., Eur., N.A., C.A., S.A.

288

<u>Triticum</u> spp. (<u>Secale</u> spp. and other Gramineae) + <u>Fusarium</u> <u>nivale</u> (Fr.) Ces. (Calonectria <u>nivalis</u> Schaf.) =

ENGLISH	Snow Mold, Foot Rot, Head Blight
FRENCH	Moisissure de Neige, Moisissure Nivale des Céréales
SPANISH	Pudrición radicular, Moho de nieve, Podredumbre del pié, Tizón
PORTUGUESE	
ITALIAN	Mal del piede, Muffa della neve
RUSSIAN	плесень снежная
SCANDINAVIAN	Sneskimmel, Snömögel, Snømugg
GERMAN	Schneeschimmel Getreide
DUTCH	Kiemschimmel, Sneeuwschimmel
HUNGARIAN	
BULGARIAN	снежна плесен
TURKISH	
ARABIC	
PERSIAN	
BURMESE	
THAI	
NEPALI	
HINDI	
VIETNAMESE	
CHINESE	
JAPANESE	Kosyoku-yukigusare-byô 紅色雪腐病

The fungus is conspicuous on the leaves because of the white superficial growth which is very abundant under moist conditions. Later the leaves become bleached and dried out. The disease appears on all plants in large areas with healthy plants adjoining. The disease also causes a spike blight where individual grain are blighted. The kernels are shriveled and light brown in color.

Le fungo es remarcabile in le folios a causa del strato superficial blanc le qual es copiose sub conditiones humide. Plus tarde le planta deveni blanchite e desiccate. Le maladia afflige tote le plantas intra grande areas, ma plantas salubre es contigue. Le morbo anque attacca le spicas e singule granos es necrotic. Le granos es crispate e de color brun clar.

Le champignon est visible sur les feuilles à cause de la croissance superficielle d'un mycélium blanc, très abondant par temps humide. Plus tard, les feuilles se décolorent et se dessèchent. La maladie apparaît sur toutes les plantes sur de grandes étendues à proximité de plantes saines. La maladie cause aussi une brûlure de l'épi dont chacun des grains est atteint. Les grains sont ratatinés et brun pâle.

El hongo es conspicuo en las hojas debido al crecimiento blanco superficial que es muy abundante bajo condiciones húmedas. Más tarde las hojas se destiñen y se secan. Esta enfermedad aparece en todas las plantas en grandes áreas, con plantas sanas alrededor. La enfermedad también causa un tizón de la espiga donde granos individuales son atacados. Los granos se arrugan y se vuelven de color café claro.

DISTRIBUTION: Af., A., Aust., Eur., N.A.

Triticum aestivum (Gramineae) + Gaeumannomyces graminis (Sacc.) von Arx & Olivier
(Ophiobolus graminis Sacc.) =

ENGLISH	Take-all, Foot Rot, White-heads
FRENCH	Piétin-échaudage, Epis Blanc
SPANISH	Pietin, Mal del pie, Podredumbre del pie, Tómalo todo, Cabezas blancas
PORTUGUESE	Mal do Pé
ITALIAN	Mal del piede, Diradamento del grano
RUSSIAN	офиоболез, гниль корней офиоболезная
SCANDINAVIAN	Fotsyke, Goldfodsyge, Rotdödare
GERMAN	Schwarzbeinigkeit, Weissährigkeit, Ophiobolus-Fusskrankheit
DUTCH	Tarwehalmdooder, Halmdooder
HUNGARIAN	Szártőbetegség
BULGARIAN	черно кореново гниене
TURKISH	Tahıl kök boğazı hastalığı
ARABIC	
PERSIAN	
BURMESE	
THAI	
NEPALI	
HINDI	
VIETNAMESE	
CHINESE	
JAPANESE	Tatigare-byō 立枯病

The symptoms of this disease vary greatly under different environmental conditions. Under moist conditions, the disease appears at the time the plants are producing spikes. The green color fades and rapid bleaching of the leaves, stems and spikes occurs. The main roots and stem tissue show a dry rot accompanied by a dark brown to black surface mat of fungus material. Under dry conditions the plants are short, and few plants show the dead bleached condition.

Le symptomas de iste maladia se varia grandemente secundo varie conditiones de ambiente. Sub conditiones humide, le morbo se manifesta quando le spicas emerge. Le color verde se attenua e le folios, pedunculos, e spicas es rapidemente blanchite. Le radices principal e le texito del pedunculo monstra un putrefaction sic con un strato superficial del fungo brun obscur o nigre. Sub conditiones sic, le plantas es curte, e pauc plantas monstra le aspecto morte e blanchite.

Les symptômes de cette maladie varient beaucoup selon les conditions du milieu. Par temps humide, la maladie apparaît au moment où la plante forme son épi. La couleur verte disparaît, et la décoloration rapide des feuilles, des tiges et des épis survient. Les racines principales et les tissus de la tige laissent voir une pourriture sèche accompagnée en surface d'un tapis mycélien du champignon brun foncé à noir. Par temps sec, les plantes sont courtes, et peu d'entre elles montrent cette décoloration morbide.

Los síntomas de esta enfermedad varían mucho bajo diferentes condiciones ambientales. Bajo condiciones húmedas, la enfermedad aparece cuando la planta está produciendo espigas. El color verde desaparece y ocurre un blanquimiento de las hojas, tallos y espigas. El tejido de la raíz principal y del tallo muestra una podredumbre seca acompañada de un colchón superficial café obscuro a negro de material fungoso. Bajo condiciones secas las plantas son bajas y unas pocas muestran la condición de blanquimiento muerto.

DISTRIBUTION : Af., A., Aust., Eur., N.A., S.A.

Triticum aestivum (Gramineae) + Helminthosporium gramineum Rabh. (Pyrenophora gramineae S. Ito & Kuribay) =

ENGLISH	Stripe
FRENCH	
SPANISH	Raya
PORTUGUESE	Helmintosporiose
ITALIAN	
RUSSIAN	
SCANDINAVIAN	
GERMAN	Streifenkrankheit Gerste
DUTCH	
HUNGARIAN	
BULGARIAN	
TURKISH	
ARABIC	
PERSIAN	Làkèhè ghàhvèēyè gàndòm لكهٔ قهوه ای گندم
BURMESE	
THAI	
NEPALI	
HINDI	
VIETNAMESE	
CHINESE	
JAPANESE	

The first symptom of this disease is yellow striping of the older leaf blades. Some seedling blight occurs, but only on very susceptible varieties. The yellow stripes soon turn brown as tissue necrosis progresses and finally the tissues dry out and the leaves split.

Le prime symptoma de iste maladia es striation jalne in le laminas foliar plus vetule. Necrosis del plantulas occurre, ma solmente in le varietates multo susceptibile. Le strias jalne rapidemente deveni brun, quando le necrosis del texito se avantia, e al fin le texitos se desicca e le folios se finde.

Les premiers symptômes de cette maladie sont des rayures jaunes sur le limbe des vieilles feuilles. Il se produit de la brûlure des semis, mais seulement chez les variétés très susceptibiles. Les rayures jaunes deviennent bientôt brunes à mesure que progresse la nécrose des tissus. Finalement, les tissus se dessèchent et les feuilles se déchirent.

El primer síntoma de esta enfermedad son las rayas amarillas que aparecen sobre las hojas maduras. Ocurre algo de tizón en las plántulas, pero solamente en variedades muy susceptibles. Las rayas amarillas pronto se vuelven de color café a medida que el tejido necrótico progresa y finalmente el tejido se seca y las hojas se rajan.

DISTRIBUTION: Af., A., Eur., N.A., S.A.

Triticum spp. (Hordeum spp., Secale spp. and other Gramineae) + Leptosphaeria nodorum
Muller (Septoria nodorum Berk.) =

ENGLISH	Glume Blotch, Node Canker
FRENCH	Tache des Glumes
SPANISH	Septoriosis, Mancha de la gluma y nudo, Manchón de gluma, Cancro del nudo
PORTUGUESE	Mancha da gluma
ITALIAN	Imbrunimento delle spighe
RUSSIAN	
SCANDINAVIAN	
GERMAN	
DUTCH	Kafjesbruin
HUNGARIAN	
BULGARIAN	петносване
TURKISH	
ARABIC	
PERSIAN	
BURMESE	
THAI	
NEPALI	
HINDI	
VIETNAMESE	
CHINESE	Shǐao mài fú ku bìng 小麥秄枯病
JAPANESE	Hugare-byð ふ枯病

This disease first appears as light green to yellow spots between the veins of the leaves. The lesions spread rapidly to form light brown irregular blotches with a speckled appearance as the fungus fruiting bodies develop. Under very favorable weather conditions for the disease, defoliation or infection at the groundline occurs, resulting in weakened or dead plants. The disease occurs on the floral parts as small, linear to oblong, light brown to dark brown in color.

Iste maladia se manifesta al initio como maculas verde clar o jalne inter le venas foliar. Le lesiones se extende rapidemente usque a formar pustulas irregular brun clar con un aspecto maculate quando le structuras sporal del fungo se developpa. Sub conditiones de tempore multo favorabile pro le maladia, defoliation o infection presso le solo occurre, causante plantas morte o infirmate. Le maladia occurre in le partes floral como minute maculas linear o oblonge, de color brun clar o brun obscur.

Cette maladie se manifeste d'abord sous forme de taches vert pâle à jaunes entre les nervures des feuilles. Les lésions s'étendent rapidement pour former des éclaboussures brun pâle, à contours irréguliers, lesquelles prennent une apparence tachetée à mesure que se développent les organes de fructification du champignon. Dans des conditions atmosphériques très favorables à la maladie, la défoliation ou l'infection se produisent au niveau du sol avec, comme résultat, l'affaiblissement ou la mort des plantes. La maladie se manifeste sur les parties florales sous forme de petites lésions linéaires à oblongues, de couleur brun pâle à brun foncé.

Esta enfermedad aprece primero como manchas de color verde claro a amarillas entre las venas de las hojas. Las lesiones se diseminan rápidamente hasta formar manchones irregulares de color café claro con uns apariencia moteada a medida que se desarrollan los cuerpos fructíferos del hongo. Bajo condiciones muy favorables de tiempo para la enfermedad, ocurre una defoliación o infección al nivel del suelo, resultando en plantas débiles o muertas. La enfermedad aparece en las partes florales como pequeñas manchas lineales a oblongas, de color café claro a café obscuro.

DISTRIBUTION: Af., A., Aust., Eur., N.A., S.A.

292

Triticum aestivum (Hordeum vulgare, Solanum tuberosum, Lycopersicon esculentum) + Pseudomonas atrofaciens (McCulloch) Stevens =

ENGLISH	Basal Glume Rot of Wheat and Barley, Stunt of Potatoes and Tomatoes
FRENCH	Bactériose des Bases des Glumes
SPANISH	Podredumbre basal de gluma
PORTUGUESE	
ITALIAN	
RUSSIAN	почернение оснований чешуй, бактериоз базальный
SCANDINAVIAN	
GERMAN	Basale Bakteriose, Basale Spelzenfäule, Bakterienblattfleckenkrankheit
DUTCH	
HUNGARIAN	
BULGARIAN	базална бактериоза
TURKISH	
ARABIC	
PERSIAN	
BURMESE	
THAI	
NEPALI	
HINDI	
VIETNAMESE	
CHINESE	
JAPANESE	

This disease appears first on the leaves as small, light green, oval to oblong spots or lesions. The spots then enlarge and sometimes coalesce. As the center of the lesion dies, the color changes to a gray-brown. The wide, pale green to straw-colored halo-like border around the dead center is very seldom observed. When the disease is severe, the infection extends into the spikes and grain.

Iste maladia se manifesta al initio in e folios como minute maculas o lesiones oval o oblonge verde clar. Alora le maculas cresce e a vices coalesce. Quando le centro del lesion mori, le color deveni brun-gris. Rarmente, on vide un large bordo similar a un halo verde pallide o paleate circum le centro morte. Quando le morbo es sever, le infection se extende in le spicas e in le granos.

Cette maladie apparaît d'abord sur les feuilles sous forme de petites taches ou lésions vert pâle, ovales à oblongues. Les taches s'agrandissent ensuite et parfois se fusionnent. Lorsque le centre de la lésion meurt, sa couleur vire au gris brun. La bordure, sorte d'aréole entourant le centre mort, large et de couleur vert pâle à paille, est très rarement remarquée. Lorsque a maladie est grave, l'infection s'étend aux épis et aux grains.

Esta enfermedad aparece primero en las hojas como pequeñas manchas de color verde claro, ovaladas a oblongas. Las manchas luego crecen y algunas veces se juntan. A medida que el centro de las lesiones muere, el color cambia a café-grisáceo. La orilla en forma de halo alrededor del centro muerto, de color verde pálido a paja, rarmente se observa. Cuando la enfermedad es severa, la infección se extiende dentro de las espigas y grano.

DISTRIBUTION: Af., A. Aust., Eur., N.A.,

Triticum spp. (Hordeum spp. and other Gramineae) + Puccinia glumarum (Schm.) Erikss. & Henn. (Puccinia striiformis Westend.) =

ENGLISH   Yellow Rust, Stripe Rust
FRENCH   Rouille Jaune, Rouille Jaune Striée
SPANISH   Roya estriada, Roya de las glumas, Polvillo de las glumes, Roya amarilla
PORTUGUESE   Ferrugem linear, Amarela
ITALIAN   Ruggine gialla, Ruggine striata del grano
RUSSIAN   ржавчина желтая
SCANDINAVIAN   Gulrust, Gulrost, Hvedens gulrust
GERMAN   Gelbrost Getreide
DUTCH   Geleroest
HUNGARIAN
BULGARIAN   жълта ръжда
TURKISH   Sarı pas
ARABIC
PERSIAN   Zàngè zàrdè gàndòm کله زردگند م
BURMESE   Thet-turnt-na-nwin roga ဝင်ဒီ:နိ:ဃၥ:ဓ္ချိင်ကိ ၄ၥၢ၀၆၆
THAI   Ra-Snim-Lek ราสนิมเหล็ก
NEPALI Gahoon Ko paheline rog गहुँको पहेलिने राेग, Gehoon Ka harda kal गेहुँका हर्दा काल
HINDI   Peela Ratua पीला रतुवा
VIETNAMESE
CHINESE   Shǐao mài húang shìou bìng 小麥黃銹病, Shǐao mài tiao shìou bìng 小麥條銹病
JAPANESE   Kisabi-byǒ 黃さび病

This disease usually appears early in the season and is characterized by linear, narrow bands of the fungus fruiting bodies on the leaves when the plants are small or again when the plants are maturing. Another phase of the fungus produces long stripes between the veins on the leaves and small lesions on the floral parts.

Iste maladia usualmente se manifesta in le prime parte del saison e se characterisa per bandas stricte linear de structuras sporal del fungo in le folios o quando le plantas es juvene o quando illos se approxima a maturation. Un altere stadio del fungo produce strias elongate intervenal in le folios e minute lesiones in le partes floral.

Cette maladie se manifeste habituellement au début de la saison et est caractérisée par des bandes linéaires, étroites, formées des organes de fructification du champignon sur les feuilles, lorsque les plantes sont jeunes, ou encore lorsqu'elles mûrissent. Durant une autre phase de son développement, le champignon produit de longues rayures entre les nervures des feuilles et de petites lésions sur les parties florales.

Esta enfermedad generalmente aparece temprano en la estación y se caracteriza por bandas angostas y lineales de los cuerpos fructíferos del hongo sobre las hojas cuando las plantas están pequeñas o de nuevo cuando las plantas están madurando. Otra fase de este hongo produce rayas largas entre las venas en las hojas y pequeñas lesiones en las partes florales.

DISTRIBUTION: Af., A., Eur., N.A., C.A., S.A.

Triticum aestivum + *Puccinia graminis* Pers. f. sp. *tritici* Eriks. & E. Henn. =

ENGLISH	Stem Rust
FRENCH	Rouille Noire des Céréales
SPANISH	Roya negra, Roya del tallo, Roya lineal, Chahuixtle del tallo, Polvillo de la caña
PORTUGUESE	Ferrugem negra, Ferrugem do colmo, Ferrugem preta
ITALIAN	Ruggine lineare del grano, Ruggine comune
RUSSIAN	ржавчина стеблевая, ржавчина линейная
SCANDINAVIAN	Sortrust, Svartrost, Svartrust
GERMAN	Schwarzrost
DUTCH	Zwarteroest
HUNGARIAN	
BULGARIAN	черна стъблена ръжда
TURKISH	Kara pas
ARABIC	Sadaa صدأ
PERSIAN	Zàngē sēyāhē gàndōm زنگ سیاه گندم
BURMESE	Pin-se-na-nwin roga ပင်စည်နီးဝါးရောဂါ
THAI	
NEPALI	Gahoon Ko sindure rog गहुँको सिन्दुरे रोग , Harda kal हर्दाकाल
HINDI	Kala Ratua काला रतुवा
VIETNAMESE	
CHINESE	Shiao mài gan shìou bìng 小麥稈銹病 , Shiao mài hei shìou bìng 小麥黑銹病
JAPANESE	Kurosabi-byō 黒さび病

This disease is characterized by reddish-brown pustules on the stems, leaf blades, floral parts and sometimes the grain. Soon after the pustules appear, they rupture the epidermis, exposing a powdery mass of spores. The elongated shape of the pustules gives them a ragged appearance. If the disease is abundant, two or more pustules may run together to form a streak. As the crop matures, the reddish-brown pustules gradually turn black.

Iste maladia se characterisa per pustulas brun-rubiette in le folios, laminas foliar, partes floral, e, a vices, le granos. Post que le pustulas se manifesta, illos rapidemente rumpe le epidermis, discoperiente un massa pulverose de sporas. Le contorno elongate del pustulas causa que illos ha un aspecto zigzag. Si le maladia es copiose, duo o plures del pustulas pote coalescer usque a formar un stria. Quando le plantas se matura, le pustulas brun-rubiette gradualmente deveni nigre.

Cette maladie est caractérisée par des pustules brun rougeâtre sur les tiges, les feuilles, les parties florales, et parfois les grains. Peu après leur apparition, les pustules percent l'épiderme et exposent une masse de spores poudreuse. La forme allongée des pustules leur donne une apparence chiffonnée. Si la maladie est grave, deux pustules ou plus peuvent se fusionner pour former une rayure. A mesure que la récolte mûrit, le brun rougeâtre des pustules vire graduellement au noir.

Esta enfermedad se caracteriza por las pústulas café rojizas en los tallos, hojas, partes florales y algunas veces en el grano. Pronto después que aparecen las pústulas, éstas rompen la epidermis dejando expuesta una masa polvorienta de esporas. La forma alargada de las pústulas les da una apariencia rasgada. Si la enfermedad es abundante, dos o más pústulas pueden aparecer juntas para formar una raya. A medida que el cultivo madura, las pústulas café rojizo gradualmente se vuelven negras.

DISTRIBUTION: Af., A., Aust., Eur., C.A., S.A.

Triticum spp. (Avena sativa, Oryza sativa and other Gramineae) + Sclerospora macrospora
Sacc. =

ENGLISH	Downy Mildew
FRENCH	Mildiou, Sclérosporose
SPANISH	Mildiú lanoso
PORTUGUESE	
ITALIAN	Peronospora del frumento
RUSSIAN	склероспороз, пожная мучнистая роса
SCANDINAVIAN	
GERMAN	Sklerosporose
DUTCH	
HUNGARIAN	
BULGARIAN	мана
TURKISH	
ARABIC	
PERSIAN	
BURMESE	
THAI	
NEPALI	
HINDI	
VIETNAMESE	
CHINESE	
JAPANESE	ôka-isyuku-byô 黄化萎縮病

The infected plants are erect, yellowish-green, somewhat dwarfed, and they branch excessively. The leaves are thickened, remain erect and develop in a close whorl. Many of the branches turn brown and die. The large brown fruiting bodies of the fungus that appear between the veins of the leaf blade constitute the most important diagnostic symptom. The floral parts look more like leaf structures.

Le plantas infectate es erecte, verde-jalne, aliquanto nanos, e illos monstra ramification excessive. Le folios es inspissate, resta erecte, e se developpa verticillate. Multe ramos deveni brun e mori. Le symptoma diagnostic le plus importante es le grande structuras sporal brun del fungo, le quales se manifesta inter le venas del lamina foliar. Le partes floral es multo similar a structuras foliar.

Les plantes atteintes sont dressées, vert jaunâtre, quelque peu rabougries, et tallent de façon excessive. Les feuilles épaississent, restent dressées et se développent en un verticille compact. Plusieurs des talles brunissent et meurent. Les gros organes de fructification du champignon, qui apparaissent entre les nervures du limbe des feuilles, constituent le symptôme le plus important du diagnostic. Les parties florales ressemblent davantage à la partie feuillue de la plante.

Las plantas infectadas son erectas, verde amarilloso, algo achaparradas y tienen ramas excesivas. Las hojas se engruesan, permanecen erectas y se desarrollan en un verticilo cerrado. Muchas de las ramas se vuelven de color café y mueren. Los grandes cuerpos fructíferos del hongo que aparecen entre las venas de las hojas constituyen el síntoma más importante de diagnóstico. Las partes florales parecen más como estructuras foliares.

DISTRIBUTION: Af., A., Aust., Eur., N.A.

Triticum spp. (other Gramineae) + Septoria tritici Rob. & Desm. =

ENGLISH	Speckled Leaf Blotch
FRENCH	Tache Septorienne, Nuile des Céréales, Taches Foliaires et sur les Glumes
SPANISH	Septoriosis, Mancha de la hoja, Manchón foliar moteado
PORTUGUESE	Mancha da folha
ITALIAN	Seccume delle foglie
RUSSIAN	септориоз
SCANDINAVIAN	Graapletsyge
GERMAN	Schwarzfleckigkeit, Weizenblattflecken, Braunfleckigkeit, Spelzenfleckigkeit
DUTCH	Bladvlekken
HUNGARIAN	
BULGARIAN	ранен листен пригор
TURKISH	Buğday septoriası
ARABIC	
PERSIAN	Septōrēyāyē gàndōm سپتوریای گندم
BURMESE	
THAI	
NEPALI	
HINDI	Parn Daag परण दाग
VIETNAMESE	
CHINESE	Shǐao mài yèh ku bìng 小麥葉枯病
JAPANESE	Hagare-byō 葉枯病

This disease occurs cheifly on leaves, forming light green to yellow spots between the veins. A speckled appearance is later produced by the fruiting bodies in the leaf. These lesions are various shades of brown, elongated, linear and sometimes vein-limited with a diffuse margin. Where infection is severe, shriveling of leaves and defoliation may result in dead plants. Seedlings may also be killed.

Iste maladia occurre principalmente in folios, producente maculas verde clar o jalne intervenal. Plus tarde le folio ha un aspecto maculate a causa del structuras sporal. Iste lesiones es de varie tintas de brun, elongate, linear, e, a vices, limitate per le venas con un margine diffuse. Si le infection es sever, crispation de folios e defoliation occurre, lassante plantas morte. Anque plantulas pote morir.

Cette maladie se manifeste principalement sur les feuilles, formant des. taches vert pâle à jaunes entre les nervures. Plus tard, les organes de fructification du champignon donnent aux feuilles une apparence mouchetée. Ces lésions présentent divers tons de brun, sont allongées, linéaires et parfois limitées aux nervures, avec un contour diffus. Lorsque l'infection est grave, le ratatinement des feuilles et la défoliation peuvent causer la mort des plantes. Les semis peuvent aussi être tués.

Esta enfermedad ataca principalmente las hojas, formando manchas verde claro a amarillas entre las venas. Más tarde se produce una apariencia moteada causada por los cuerpos fructíferos en la hoja. Estas lesiones tienen varios tonos de café, son alargadas, lineales y algunas veces limitadas por las venas con un margen difuso. Cuando la infección es severa, las hojas se arrugan, hay defoliación y las plantas pueden morir. Las plantulas también pueden morir.

DISTRIBUTION: Af., A., Aust., Eur., N.A., C.A., S.A.

<u>Triticum</u> <u>aestivum</u> + <u>Tilletia</u> <u>caries</u> (DC.) Tul. =

ENGLISH	Rough-spored Bunt
FRENCH	Carie du Blé
SPANISH	Carie, Carbón hediondo, Carbón de espora rugosa
PORTUGUESE	
ITALIAN	Carie, Golpe, Carbone fetido, Carie del frumento
RUSSIAN	**головня мокрая, головня вонючая, головня твердая**
SCANDINAVIAN	Hvedens stinkbrand
GERMAN	Stinkbrand Weizen, Weizensteinbrand, Steinbrand, Schmierbrand, Hartbrand
DUTCH	Steenbrand
HUNGARIAN	
BULGARIAN	**обикновена мазна главня**
TURKISH	Buğday sürmesı
ARABIC	
PERSIAN	Sēyahàkē pēnhànē gàndōm سیاهک پنهان گندم
BURMESE	
THAI	
NEPALI	
HINDI	Bunt Rog बंट रोग
VIETNAMESE	
CHINESE	Shǐao mài sing hei sùai bìng 小麥腥黑穗病
JAPANESE	Ami-namagusa-kuroho-byō 網なまぐさ黒穂病

Generally the symptoms of this disease are not apparent until the crop is starting to mature. However, some varieties show dwarfing of the plants, small light-colored spots on the leaves and a grayish foliage during early growth. The diseased plants may appear bluish-green to grayish-green in color. The smut balls that replace the grain are conspicuous in the diseased spikes. The smut balls are nearly the shape of the grain and are grayish-green, changing to brown as the grain ripens.

Generalmente le symptomas de iste maladia non se manifesta ante que le plantas comencia maturar se. In alicun varietates, tamen, nanismo del planta, minute maculas de color clar in le folios, e foliage gris in le prime stadios de crescimento occurre. Le plantas morbide pote esser de color verde-blau o verde-gris. Le massas de carbon que reimplacia le granos es remarcabile in le spicas morbide. Le massas carbonose es de contorno quasi indentic con le grano e de color verde-gris, deveniente brun quando le grano se matura.

En général, les symptômes de cette maladie n'apparaissent que lorsque la récolte commence à mûrir. Cependant, certaines variétés exhibent un nanisme de la plante, de petites taches pâles sur les feuilles et un feuillage grisâtre au début de la croissance. Les plantes malades peuvent prendre une coloration vert bleuâtre à vert grisâtre. Les masses charbonneuses qui remplacent les grains sont visibles dans les épis malades. Les masses charbonneuses ont presque la forme des grains, et sont vert grisâtre virant au brun à mesure que les grains mûrissent.

Generalmente los síntomas de esta enfermedad no son aparentes hasta que el cultivo está comenzando a madurar. Sinembargo, algunas variedades muestran achaparramiento de la planta, pequeñas manchas de color claro en las hojas y follaje grisáceo durante el primer crecimiento. Las plantas enfermas pueden tener un color verde azuloso a verde grisáceo. Las bolas de carbón que aparecen en lugar de los granos son conspicuas en las espigas enfermas. Estas bolas de carbón son casi de la misma forma que los granos y son de color verde grisáceo, cambiando a café cuando el grano madura.
DISTRIBUTION: Af., A., Aust., Eur., N.A., S.A.

<u>Triticum</u> aestivum + <u>Tilletia</u> foetida (Wallr.) Liro =

ENGLISH	Smooth-spored Bunt
FRENCH	Carie
SPANISH	Carie, Carbón aspestoso, Carbón de espora suave
PORTUGUESE	
ITALIAN	Carie, Carbone puzzolente del grano
RUSSIAN	
SCANDINAVIAN	
GERMAN	
DUTCH	
HUNGARIAN	
BULGARIAN	обикновена мазна главня
TURKISH	
ARABIC	
PERSIAN	Sēyāhàkē pēnhānē gàndōm سیاهک پنهان گندم
BURMESE	
THAI	
NEPALI	Gahoon Ko pudke rog गहुँको पुड्के रोग, Gehun Ka Jhakhara kal गहुंका फखराकाल
HINDI	
VIETNAMESE	
CHINESE	Shǐao mài sing hei sùai bǐng 小麥腥黑穗病
JAPANESE	Maru-namagusa-kuroho-byð 丸なまぐさ黒穂病

This disease is not easily identified until the time the spikes are formed, although infected plants may be slightly stunted. The infected heads have a bluish cast, in contrast to the green of healthy spikes. At maturity the diseased spikes appear plumper than normal spikes. During the development of the spike, the grain is replaced by a short, plump smut ball. These are round in contrast to grain-shaped as those described in the previous disease.

Iste maladia non se indentifica facilemente ante que le spicas se forma, ben que un pauc de nanismo pote occurrer in plantas morbide. Le capites infectate ha un tinta blau, in contradistinction al verde normal de spicas salubre. A maturation le spicas morbide appare plus inspissate que le spicas normal. Durante le developpamento del spica, le grano se reimplacia per un curte, grasse massa carbonose. Isto es rotunde e non del mesme contorno que le grano -- como in le maladia precedente.

Cette maladie n'est pas facilement reconnaissable avant la formation des épis, bien que les plantes atteintes soient légèrement rabougries. Les épis infectés ont une teinte bleutée contrastant avec le vert des épis sains. A maturité, les épis malades paraissent plus dodus que les épis normaux. Durant le développement de l'épi, les grains sont remplacés par de courtes masses charbonneuses dodues. Celles-ci sont rondes plutôt que de la forme des grains, comme celles décrites dans la maladie précédente.

Esta enfermedad no se identifica fácilmente sino hasta la época en que las espigas están formadas, aunque las plantas infectadas pueden estar algo achaparradas. Las espigas infectadas tienen un tono azuloso, en contraste con el verde de las sanas. En la madurez las espigas enfermas parecen más gruesas que las normales. Durante el desarrollo de la espiga, el grano es substituído por una pelota de carbón corta y gruesa. Estas son redondas en contraste con la forma del grano como aquellas descritas en las anteriores enfermedades.

DISTRIBUTION: Af., A., Aust., Eur., N.A., S.A.

Triticum aestivum + Urocystis tritici Kornicke (Urocystis agropyri (Preuss) Schroet.) =

ENGLISH	Flag Smut
FRENCH	Charbon des Tiges et Feuilles du Blé
SPANISH	Carbón de la espigo(volador), Carbón de bandera
PORTUGUESE	
ITALIAN	Carbone del culmo e delle foglie del grano, Golpe nera
RUSSIAN	головня стеблей
SCANDINAVIAN	
GERMAN	Streifenstengelbrand
DUTCH	
HUNGARIAN	
BULGARIAN	стъблена главня
TURKISH	
ARABIC	Taphahom Lewaii تفحم لوائي
PERSIAN	
BURMESE	
THAI	
NEPALI	
HINDI	Dhvaj Kand धवाज कंद
VIETNAMESE	
CHINESE	
JAPANESE	

This disease is evident from the seedling stage until maturing of the crop. The early symptoms are the gray to grayish-black linear lesions in the older leaf blades. The fungus fruiting bodies form in the leaf tissue between the veins and are covered by the epidermis during early stages of development. Later the epidermis ruptures, releasing the black spore mass and finally the leaf tissue becomes frayed. In most susceptible varieties the plants are dwarfed and spike development is stopped prior to its emergence from the leaf whorl.

Iste maladia se manifesta ab le plantulas usque a maturation de plantas. Le prime symptomas es le lesiones gris o nigre-gris linear in le laminas foliar plus vetule. Le structuras sporal del fungo se forma in le mesophyllo intervenal e durante le prime stadios de developpamento illos es coperte del epidermis. Plus tarde, le epidermis se rumpe, discargante le massa nigre de sporas e al fin le mesophyllo deveni lacerate. In le varietates le plus susceptibile, le plantas suffre nanismo e le developpamento de spicas se arresta ante que illos emerge ab le verticillo foliar.

Cette maladie se manifeste à partir du stade plantule jusqu'à la maturité de la récolte. Les premiers symptômes sont des lésions linéaires, grises à noir grisâtre, sur le limbe des vieilles feuilles. Les organes de fructification du champignon se forment dans les tissus de la feuille, entre les nervures, et sont couverts par l'épiderme durant les premiers stades de leur développement. Plus tard, l'épiderme se perce, libérant la masse de spores noire, et finalement les tissus de la feuille deviennent lacérés. Chez la plupart des variétés susceptibles, les plantes sont rabougries, et le développement de l'épi s'arrête avant que celui-ci n'émerge du verticille foliaire.

Esta enfermedad es evidente desde el estado de plántula hasta la madurez del cultivo. Los primero síntomas son las lesiones grises a negras grisáceas sobre las hojas maduras. Los cuerpos fructíferos del hongo se forman en ell tejido foliar entre las venas y están cubiertos por la epidermis durante las primeras etapas del desarrollo. Más tarde la epidermis se rompe, dejando salir la masa negra de esporas y finalmente el tejido foliar se vuelve deshilachado. En las varidades más susceptibles las plantas se achaparran y el desarrollo de la espiga se suspende antes de emerger del verticilo foliar.

DISTRIBUTION: Af., A., Aust., Eur., N.A., S.A.

300

Triticum aestivum (Secale cereale, Hordeum vulgare) + Xanthomonas translucens var. undulosa (Smith, Johnson & Reddy) Dowson ex Elliott =

ENGLISH	Black Chaff
FRENCH	Glume Noire, Bactériose Noire
SPANISH	Espiga negra, Granza negra
PORTUGUESE	
ITALIAN	
RUSSIAN	бактериоз черный, чернопленчатость
SCANDINAVIAN	
GERMAN	Schwarze Bakteriose, Schwarzspelzigkeit
DUTCH	
HUNGARIAN	
BULGARIAN	черна бактериоза
TURKISH	
ARABIC	
PERSIAN	
BURMESE	
THAI	
NEPALI	
HINDI	
VIETNAMESE	
CHINESE	
JAPANESE	Karakuroho-byô　から黒穂病

This disease occurs on the floral parts, stems and leaves. On the floral structures, the lesions appear as small, linear to striated brown to black spots, frequently coalescing to blacken the infected areas completely. The leaf lesions arc light brown, irregularly linear and coalesce to form blotches. The young lesions on all tissue are water-soaked and droplets of exudate appear on the surface.

Iste maladia se manifesta in le partes floral, pedunculos, e folios. In le structuras floral, le lesiones se manifesta como minute maculas brun o nigre, linear o striate, frequentement coalescente usque a obscurar le areas infectate completemente. Le lesiones foliar es brun clar, irregularmente linear, e illos coalesce usque a formar pustulas. Le juvene lesiones in tote le texitos es saturate de aqua e minute guttas de exsudato se manifesta al superfacie.

Cette maladie se manifeste sur les parties florales, les tiges et les feuilles. Les lésions sur les parties florales sont de petites taches linéaires ou striées, brunes à noires, qui,en s'unissant fréquemment, noircissent complètement les parties atteintes. Les lésions sur les feuilles sont brun pâle, irrégulièrement linéaires, et s'unissent pour former des éclaboussures. Les jeunes lésions sur tous les tissus sont détrempées, et des gouttelettes d'exsudat apparaissent à la surface.

Esta enfermedad aparece en las partes florales, tallos y hojas. En las estructuras florales las lesiones aparecen como pequeñas manchas lineales a estriadas, de color café a negras, que frecuentemente se juntan y ennegrecen completamente el área infectada. Las lesiones en la hoja son de color café claro, irregularmente lineales y se juntan para formar manchones. Las lesiones jóvenes en todos los tejidos están empapadas de agua y en la superficie aparecen gotitas de exudación.

DISTRIBUTION: Af., A., Aust., Eur., N.A.

Ulmus spp. + Ceratostomella ulmi Buis (Ceratocystis ulmi (Buism.) Moreau) =

ENGLISH	Dutch Elm Disease
FRENCH	Dépérissement, Thyllose Parasitaire de l'orme, Maladie de l'orme
SPANISH	Enfermedad del olmo holandés
PORTUGUESE	
ITALIAN	Moria degli olmi, Grafiosi dell'olmo
RUSSIAN	
SCANDINAVIAN	
GERMAN	Höllandische Ulmenkrankheit, Ulmensterben
DUTCH	Iepenziekte
HUNGARIAN	
BULGARIAN	
TURKISH	
ARABIC	
PERSIAN	
BURMESE	
THAI	
NEPALI	
HINDI	
VIETNAMESE	
CHINESE	
JAPANESE	Oranda-byö　オランダ病

The first symptom of this disease is a yellowing and wilting of foliage of affected branches, followed by a leaf cast. The extremities of the twigs die. Necrosis may proceed and result in death of the whole tree within a year. However, this is rare as the spread of the fungus from an annual ring to the next seldom occurs. Final death, therefore, depends on two or three re-infections and this may take several years.

Le prime symptomas de iste maladia es jalnessa e marcescentia de foliage in le ramos afflicte, postea cadita de folios. Le apices del ramettos mori. Necrosis a vices se avantia e le resultado pote esser le morte del arbore integre intra un sol anno. Isto, tamen, occurre rarmente perque le fungo rarmente se extende ab un anello annual a un altere. Morte total, alora, es le resultato de duo o tres reinfectiones e iste processo dura per plure annos.

Le premier symptôme de cette maladie est le jaunissement et le flétrissement du feuillage des branches atteintes, suivis d'une dégringolade des feuilles. L'extrémité des rameaux meurt. La nécrose peut s'installer et entraîner la mort de l'arbre entier en deçà d'un an. Cela est rare, toutefois, car il n'arrive pas souvent que le champignon se propage d'un anneau annuel au suivant. Il s'ensuit que la mort de tout l'arbre dépend de deux ou trois réinfections, ce qui peut prendre plusieurs années à se produire.

El primer síntoma de esta enfermedad es un amarillamiento y marchitez del follaje de las ramas afectadas, seguido por una caída de las hojas. Los extremos de las ramillas se mueren. Puede seguir una necrosis y resultar en la muerte de todo el árbol en el lapso de un año. Sin embargo ésto es raro, ya que el hongo casi nunca se disemina de un anillo anual de la corteza a otro. La muerte final, por lo tanto, depende de dos o tres re-infecciones y esto puede tomar varios años.

DISTRIBUTION: A., Eur., N.A.

302

Vicia faba + Botrytis fabae Sardiña (Botrytis cinerea Pers. ex Fr.) =

ENGLISH	Gray-Mold Shoot Blight
FRENCH	Maladie des Taches, Taches Brunes, Moisissure Grise
SPANISH	Mancha chocolate, Tizón de moho gris de los brotes
PORTUGUESE	
ITALIAN	Macculatura delle fava
RUSSIAN	**пятнистость шоколадная, гниль серая, ботрититз**
SCANDINAVIAN	
GERMAN	Schokoladefleckenkrankheit, Graufäule, Braun Blattfleckenkrankheit
DUTCH	Grauwe schimmel Meiziekte
HUNGARIAN	
BULGARIAN	
TURKISH	Kurşuni küf
ARABIC	Tabbakoo chikolati تبغ شيكولاتى
PERSIAN	
BURMESE	
THAI	
NEPALI	
HINDI	
VIETNAMESE	
CHINESE	Tsán dou hwei sèh méi bìng 蠶豆灰色黴病
JAPANESE	Sekisyoku-hanten-byô 赤色斑点病

This disease produces symptoms on the pods and stems of young plants, producing a water-soaked condition, followed by wilt and later death. Pods resting on the ground often become infected and develop a slime which results in heavy loss during transit of the crop. On above-ground plant parts, infection usually occurs where thc old blossoms have fallen on the plant parts or have been retained at the tip of the pod.

Iste maladia produce symptomas in le siliquas e in le pedunculos de plantas juvene, causante un aspecto aquose, marcescentia, e, al fin, morte. Siliquas que reposa super le solo es sovente infectate e developpa un secretion viscose, le qual causa perdita sever durante le transportation del production. In partes epigee infection usualmente occurre ubi le flores vetule ha cadite al partes inferior o in le siliquas si le flores has essite retenite.

Cette maladie produit des symptômes sur les gousses et les tiges des jeunes plantes auxquelles elle donne un aspect délavé, suivi du flétrissement et puis de la mort. Les gousses qui séjournent sur le sol deviennent souvent infectées et produisent un mucus qui cause de lourdes pertes durant le transport de la récolte. Sur les parties aérienes, l'infection se produit généralement aux endroits de la plante où les vieilles parties florales sont tombées, ou au bout de la gousse quand des débris de fleurs y adhèrent.

Esta enfermedad produce síntomas en las vainas y tallos de las plantas jóvenes, produciendo una condición de empapamiento de agua, seguida por marchitez y más tarde muerte. Las vainas que topan al suelo frecuentemente se infectan, y desarrollan un limo que resulta en una fuerte pérdida durante el tránsito de la cosecha. En las partes aéreas de la planta, la infección generalmente ocurre donde las flores viejas han caído sobre diferentes partes de la planta, o han sido retenidas al extremo de la vaina.

DISTRIBUTION: Af., A., Eur., S.A.

Vicia faba (Legume hosts) + Uromyces viciae-fabae (Pers.) Schroet. =

ENGLISH	Rust
FRENCH	Rouille
SPANISH	Roya, Polvillo
PORTUGUESE	
ITALIAN	Ruggine della fava
RUSSIAN	
SCANDINAVIAN	Sjokoladeflekk
GERMAN	Ackerbohne, Saubohnenrost
DUTCH	Roest
HUNGARIAN	
BULGARIAN	
TURKISH	Bakla pası
ARABIC	Sadaa صـدا
PERSIAN	Zángê bäghälä زنگ باقلا
BURMESE	Na-nwin, Sa-nwin roga ဃ ရှ ၃ ၵ ၝ ၆ ၀ ၆၀း
THAI	
NEPALI	Bakula Ko sindure rog वकुलाका सिन्दुरे रोग
HINDI	Ratua रतुवा
VIETNAMESE	
CHINESE	Tsán dòu shìou bìng 蠶豆銹病
JAPANESE	Sabi-byô さび病

This disease causes pustules to appear on any part of the plant aboveground, being most numerous on the underneath side of the leaves, less abundant on the pods and stems. Infection is evident at first as minute, almost white, slightly raised pustules which later become reddishbrown. The many pustules per leaf give a rusty color which rubs off if touched. Occasionally a yellow halo surrounds the lesions. With severe infection, the leaves shrivel and fall from the plant.

Iste maladia produce pustulas in tote le partes epigee del planta, plus numerose in le superfacie inferior del folios e minus copiose in le siliquas e in le pedunculos. Al initio le infection se manifesta como minute pustulas legiermente elevate, quasi blanc, le quales plus tarde deveni brun-rubiette. Le multe pustulas in cata folio da un color ferruginose, le qual se remove per un legier friction. A vices un halo jalne imbracia le lesiones. In infection sever le folios se crispa e se separa e cade ab le planta.

Cette maladie fait apparaître des pustules sur toutes les parties aériennes de la plante; les pustules sont plus nombreuses à la face inférieure des feuilles, et moins abondantes sur les gousses et les tiges. L'infection se manifeste d'abord sous forme de minuscules pustules légèrement surélevées, presque blanches, et qui plus tard deviennent brun rougeâtre. La profusion de pustules donne à la feuille une couleur de rouille qui s'efface au toucher. Occasionnellement, un halo jaune entoure les lésions. Dans les cas graves, les feuilles se ratatinent et tombent.

Esta enfermedad causa el aparecimiento de pústulas en cualquier parte aérea de la planta, siendo más numerosas en el envés de las hojas y menos abundantes en las vainas y tallos. La infección es evidente primero como pústulas diminutas, casi blancas algo levantadas, que más tarde se vuelven de color café rojizo. Cuando hay muchas pústulas en una hoja dan un color herrumbroso que se desprende al tocarlo. Ocasionalmente un halo amarillo rodea las lesiones. Con infección severa, las hojas se arrugan y caen.

DISTRIBUTION: Af., A., Aust., Eur., N.A., C.A., S.A.

304

Vitis spp. + Coniella diplodiella (Speg.) Petr. & Syd. (Coniothyrium diplodiella (Speg.) Sacc.) =

ENGLISH	White Rot, Dieback
FRENCH	Rot Blanc, Rot Livide, Rot Pâle, Coitre, Maladie de la Grêle
SPANISH	Pudrición blanca, muerte regresiva
PORTUGUESE	
ITALIAN	
RUSSIAN	гниль белая на побегах и листьях
SCANDINAVIAN	
GERMAN	Weissfäule, Hagelpilz
DUTCH	
HUNGARIAN	
BULGARIAN	бяло гниене
TURKISH	Beyaz çürüklük
ARABIC	
PERSIAN	
BURMESE	
THAI	
NEPALI	
HINDI	
VIETNAMESE	
CHINESE	
JAPANESE	Sirogusare-byô 白腐病

This disease occurs chiefly on the fruit; however, it also infects the fruit stalks and injures shoots, stems and leaves. Lesions on the leaves are marginal and irregular with the centers various shades of brown, becoming lighter towards the edge.

Iste maladia occurre principalmente in le fructos; nonobstante, illo infecta anque le pedunculos e causa damno a ramos e folios. Lesiones foliar es marginal, irregular, con le centros de varie tintas de brun, deveniente plus clar verso le bordo.

Cette maladie s'attaque surtout au fruit; elle peut toutefois infecter les coursons et endommager les pousses, les sarments et les feuilles. Les lésions sur les feuilles sont marginales et irrégulières, leur centre étale plusieurs nuances de brun et devient plus pâle vers la périphérie.

Esta enfermedad ataca principalmente los frutos; sin embargo, también ataca el ápice del fruto y daña los brotes, tallos y hojas. Las lesiones en las hojas son marginalmente irregulares, con los centros de varios tonos de café, que se vuelven más claros hacia la orilla.

DISTRIBUTION: Af., A., Aust., Eur., N.A., S.A.

<u>Vitis</u> <u>vinifera</u> + <u>Cryptosporella</u> <u>viticola</u> Shear =

ENGLISH	Dead Arm, Black Knot, Necrosis, Side Arm
FRENCH	Nécrose des Sarments
SPANISH	Brazo muerto, Nudo negro, Necrosis, Brazo lateral
PORTUGUESE	Morte dos ramos
ITALIAN	Necrosi delle viti
RUSSIAN	отмирание побегов, усыхание побегов, пятнистость черная
SCANDINAVIAN	
GERMAN	Phomopsis-Beerenfäule, Triebnekrose, Schwarzfleckenkrankheit
DUTCH	
HUNGARIAN	
BULGARIAN	
TURKISH	
ARABIC	
PERSIAN	
BURMESE	
THAI	
NEPALI	
HINDI	
VIETNAMESE	
CHINESE	
JAPANESE	Turuware-byô つる割病

This fungus attacks the trunk and main branches of the vines primarily, although it can also attack young canes and leaves. On the canes and leaves, the lesions first appear as small dark spots. As the canes grow, the spotted tissue may split. On the trunk, infection produces a canker. The canker enlarges each year until it finally girdles the trunk, killing the vine above the canker. Sometimes new shoots may develop near the canker; they are usually weak and have yellow and cupped leaves.

Iste fungo attacca primarimente le trunco e le ramos principal del vites, ma illo pote attaccar anque cannas juvene e folios. In le cannas e in le folios, le lesiones se manifesta al initio como minute maculas obscur. Quando le cannas cresce, le texito maculate a vices se finde. In le trunco, infection produce un cancere. Le cancere cresce cata anno usque a imbraciar le trunco al fin, faciente le vite supra le cancere morir. A vices ramettos nove se forma presso le cancere; illos es usualmente infirme e ha folios jalne cuppiforme.

Ce champignon s''attaque principalement au cep et aux branches principales de la vigne, mais il peut aussi attaquer les jeunes sarments et les feuilles. Sur les sarments et les feuilles, les lésions se présentent d'abord sous forme de petits points noirs. Pendant que le sarment grandit, les tissus tachetés peuvent se fendre. Sur le cep, l'infection produit un chancre. Le chancre grossit chaque année jusqu'à encercler le tronc, ce qui entraîne la mort de la partie située au-dessus du chancre. De nouvelles pousses peuvent parfois se former près du chancre. Ces pousses sont généralement faibles et portent des feuilles jaunes en forme de coupe.

Este hongo ataca el tronco y ramas principales de las guías principalmente, aunque también puede atacar las ramas nuevas y las hojas. En las ramas y hojas las lesiones aparecen primero como pequeñas manchas obscuras. A medida que las ramas crecen, el tejido manchado puede rajarse. En el tronco, la infección produce un cancro. El cancro crece cada año hasta que finalmente rodea el tronco, matando la guía arriba del cancro. Algunas veces pueden desarrollarse nuevos brotes cerca del cancro; éstos son generalmente débiles y tienen hojas amarillas y enroscadas.

DISTRIBUTION: Af., A., Eur., N.A., S.A.

Vitis spp. + Elsinoe ampelina (d By.) Shear (Sphaceloma ampelinum De Bary, Gloeosporium ampelophagum (Pass.) Sacc.) =

ENGLISH	Anthracnose, Bird's-eye Rot
FRENCH	Anthracnose, Anthracnose Maculée
SPANISH	Antracnosis, Podredumbre ojo de pájaro
PORTUGUESE	Antracnose, Variola
ITALIAN	Antracnosi della vite, Vaiolo, Picchiola, Morbiglione, Petecchia
RUSSIAN	антракноз пятнистый
SCANDINAVIAN	
GERMAN	Fleckenanthraknose, Schwarzer Brenner, Vogelaugenkrankheit, Schwindpocken
DUTCH	
HUNGARIAN	
BULGARIAN	антракноза
TURKISH	Çelik marazı
ARABIC	
PERSIAN	Änträknôze mô آنتراکنوز مو
BURMESE	Ywet-pyauk roga ၀ လ်ယ်ဝ ၀ ၄ ၄ိ၀ ၆ ဆ ၆ ၁ ၇ ၆ ၇, ၈ ၆
THAI	Anthracnose แอนแทรกโนส
NEPALI	
HINDI	Anthracnose ऐनथ्रेक्नोज
VIETNAMESE	Đen lá - Thối quả
CHINESE	
JAPANESE	Kokutô-byô 黒とう病

The symptoms of this disease on the berries and other parts of the vine are rather striking. The berries, young shoots, petioles, leaf veins and fruit stems are attacked. Numerous spots occur on the shoots, which sometimes combine and cause girdling which kills the vine tips. Similar spots occur on the petioles and leaves. Spots on the berries are circular, sunken and ashy-gray. In later stages of the disease, the spots have dark borders.

Le symptomas de iste maladia in le baccas e in altere partes del vite es remarcabile. Le maladia attacca baccas, ramos juvene, petiolos, venas foliar, e pedunculos. Maculas numerose se forma in le partes juvene, le quales a vices coalesce e imbracia le vites, faciente le apices morir. Maculas similar occurre in le petiolos e folios. Maculas in le baccas es circular, depresse, e cinerose. Plus tarde le maculas ha bordos obscur.

Les symptômes de cette maladie sur les raisins et les autres parties de la vigne sont plutôt frappants. Les raisins, les jeunes pousses, les pétioles, les nervures de la feuille et les coursons sont tous sujets à l'attaque. Sur les pousses, on voit apparaître de nombreuses taches qui parfois s'unissent pour causer un encerclement qui tue les bouts de la vigne. Des taches semblables se forment sur les pétioles et les feuilles. Sur les fruits, les taches sont circulaires, déprimées et gris cendré. Durant les dernières phases de la maladie, les taches ont des bords foncés.

Los síntomas de este enfermedad en los frutos y otras partes de la planta son bastante notorios. Los frutos, brotes nuevos, pecíolos, venas de la hoja y ápices del fruto son atacados. En los brotes aparecen numerosas manchas, las cuáles a veces se combinan y causan un cinto que rodea y mata los extremos de las guías. Manchas similares aparecen en los pecíolos y hojas. Las manchas en los frutos son circulares, hundidas y de color gris ceniza. En estados avanzados de la enfermedad, las manchas tienen orillas obscuras.

DISTRIBUTION: Af., A., Aust., Eur., N.A., C.A., S.A.

Vitis spp. + Erwinia vitivora (Bacc.) du Pless. =

ENGLISH	Bacterial Blight
FRENCH	Gommose Bacillaire de la Vigne, Gélivure
SPANISH	Tizón bacterial
PORTUGUESE	
ITALIAN	Mal nero, Spaccatura, Mal dello spacco, Gommosi bacillare
RUSSIAN	
SCANDINAVIAN	
GERMAN	
DUTCH	
HUNGARIAN	
BULGARIAN	чернилка
TURKISH	
ARABIC	
PERSIAN	
BURMESE	
THAI	
NEPALI	
HINDI	
VIETNAMESE	
CHINESE	
JAPANESE	

The disease develops first on the young shoots as pale, yellowish-green, water-soaked spots which darken to grayish-black. These extend as stripes along one side of the shoot. The stripes widen around the stems including both fruit and leaf petioles. Longitudinal cracks appear on the surface of the lesions. The affected shoots are severely malformed by the end of the season and the leaves become reddish-yellow or bright brown.

Iste maladia se manifesta al initio in le partes juvene como pallide maculas aquose verde-jalnette, le quales se obscura verso nigre-gris. Istos se extende in strias per un latere del canna. Le strias se allarga circum le pedunculos e le petiolos. Fissuras longitudinal se manifesta al superfacie del lesiones. Le partes afflicte es severmente deforme al fin del saison e le folios deveni jalne-rubiette o brun brillante.

La maladie se développe d'abord sur les jeunes sarments sous forme de taches détrempées, pâles, vert jaunâtre, qui deviennent noir grisâtre. Celles-ci se prolongent en bandes sur un côté du sarment. Les bandes s'élargissent autour des tiges, des fruits et des pétioles. Des fentes longitudinales apparaissent à la surface des lésions. Les sarments affectés deviennent très difformes à la fin de la saison et les feuilles deviennent jaune rougeâtre ou brun clair.

La enfermedad se desarrolla primero en los brotes jóvenes como manchas pálidas, verde amarilloso, empapadas de agua que se obscurecen y llegan a ser negro grisáceo. Estas se extienden como rayas a lo largo de un lado del brote. Las rayas se ensanchan alrededor de los tallos abarcando ambos el fruto y los pecíolos de la hoja. Rajaduras longitudinales aparecen en la superficie de las lesiones. Los brotes afectados están severamente malformados al final de la estación y las hojas se vuelven de color amarillo rojizo o café brillante.

DISTRIBUTION: Af., Eur.

Vitis vinifera + Guignardia bidwellii (Ellis) Viala & Ravaz =

ENGLISH	Black Rot
FRENCH	Pourriture Noire
SPANISH	Pudrición negra seca
PORTUGUESE	
ITALIAN	Black-rot delle viti
RUSSIAN	**гниль ягод и листьев черная фомозная**
SCANDINAVIAN	
GERMAN	Phoma-Blatt-und Beerenschwarzfäule
DUTCH	
HUNGARIAN	
BULGARIAN	**черно гниене**
TURKISH	Siyah çürüklük
ARABIC	
PERSIAN	
BURMESE	
THAI	
NEPALI	
HINDI	Kala viglan काला विगलान
VIETNAMESE	
CHINESE	
JAPANESE	Kurogusare-byō 黒腐病

This fungus attacks all the new growth of the vine. The first symptoms to be observed are necrotic lesions on the leaves. The lesions are scattered or grouped, circular and at first red in color. Later, when they enlarge, sharp black lines appear at the margins and the centers become reddish-brown to grayish-tan. Lesions on the shoots, leaf stems and veins are purple to black, slightly depressed and elongated. Lesions appear on the berries when they are half grown causing them to shrink, mummify and turn black.

Iste fungo attacca tote le partes nove del vite. Le prime symptomas evidente es lesiones necrotic foliar. Le lesiones es disperse e in gruppamentos, circular, e al initio de color rubie. Plus tarde, quando illos cresce, distincte lineas nigre se manifesta al margines e le centros deveni brun-rubiette o bronzate-gris. Lesiones in le ramos, petiolos, e venas es purpuree o nigre, legiermente depresse, e elongate. Lesiones se forma in le baccas in le medie del saison, faciente los crispar se, mumificar se, e devenir nigre.

Ce champignon s'attaque à toutes les nouvelles pousses de la vigne. Les premiers symptômes observés sont des lésions nécrotiques sur les feuilles. Les lésions sont soit groupées, soit disséminées, rondes et rouges au début. Plus tard, quand elles s'accroissent, des lignes noires bien définies apparaissent à la marge et le centre devient brun rougeâtre à gris sale. Les lésions sur les sarments, les tiges et les nervures sont pourprées à noires, légèrement affaissées et allongées. Des lésions apparaissent sur les fruits à demi formés qui se ratatinent, se momifient et noircissent.

Este hongo ataca todo el crecimiento nuevo de la planta. Los primeros síntomas que se observan son lesiones necróticas en las hojas. Las lesiones están diseminadas o agrupadas, circulares y al principio son de color rojo. Más tarde, cuando se agrandan, aparecen líneas negras bien marcadas en los márgenes y los centros se vuelven café rojizo a crema grisáceo. Las lesiones en los brotes, hojas, tallos y venas son de color púrpura a negro, algo hundidas y alargadas. Las lesiones aparecen sobre los frutos a la mitad del desarrollo, causando que se encojan, se momifiquen y se vuelvan negros.
DISTRIBUTION: A., Eur., N.A., W.I., S.A.

309

Vitis spp. (Ampelopsis) + Physopella ampelopsidis (Diet. & P. Syd.) Cumm. (Physopella vitis (Thum.) Arth.) =

ENGLISH   Leaf Rust
FRENCH
SPANISH   Roya de la hoja
PORTUGUESE
ITALIAN
RUSSIAN
SCANDINAVIAN
GERMAN
DUTCH
HUNGARIAN
BULGARIAN
TURKISH
ARABIC
PERSIAN
BURMESE   Na-nwin, Than-chi roga ဖုက်ဆို့ဝလ အ၆ဂ် လျှ၆ဆ၆၃၆၆
THAI
NEPALI
HINDI
VIETNAMESE
CHINESE
JAPANESE

This fungus attacks chiefly the leaves, causing premature defoliation and inducing the plants to produce short shoots. The symptoms on the leaves are typical of many of those caused by the rust fungi. The fruiting structures are scattered and yellowish in color.

Iste fungo attacca principalmente le folios, causante defoliation prematur e le production de pedunculos curte. Le symptomas foliar es typic pro le fungos de (fer)rugine. Le structuras sporal es disperse e de color jalnette.

Ce champignon s'attaque surtout aux feuilles, dont il cause la chute prématurée et fait que les plantes ne produisent que des tiges courtes. Les symptômes sur les feuilles sont typiques des rouilles, en général. Les fructifications sont parsemées et jaunâtres.

Este hongo ataca principalmente las hojas, causando defoliación prematura, induciendo a la planta a producir brotes cortos. Los síntomas en las hojas son típicos de muchos de aquéllos causados por el hongo de la roya. Las estructuras fructíferas están diseminadas y son de color amarillento.

DISTRIBUTION: A., N.A., C.A., S.A.

<u>Vitis</u> spp. + <u>Plasmopara viticola</u> (Berk. & Curt. ex De Bary) Berl. & de Toni =

ENGLISH	Downy Mildew
FRENCH	Mildiou
SPANISH	Mildiú lanoso, Peronóspora, Mildiu felpudo, Mildiú velloso
PORTUGUESE	Mildio, Peronospora
ITALIAN	Peronospora della vite
RUSSIAN	**мильдью, ложная мучнистая роса**
SCANDINAVIAN	Vinskimmel, Vinbladmögel
GERMAN	Falscher Mehltau, Blattfall-krankheit
DUTCH	Valse meeldauw
HINGARIAN	Szőlőperonoszpóra
BULGARIAN	**мана (пероноспора)**
TURKISH	Mildiyö (Pronos)
ARABIC	Bayad Zaghaby بياض زغبي
PERSIAN	Sefedáke döroogheye mö سفیدک دروغی مو
BURMESE	Ywet-chauk roga ဝလ်ဒစ ၇ပိုး၊ ၇ဘိုက်ဝိ၇ိုလ
THAI	Ra-Num-Karng ราน้ำค้าง
NEPALI	
HINDI	Mriduromil phaphundi मृदुरोमिल फाफुन्दी
VIETNAMESE	Đốm phấn làm cháy lá
CHINESE	
JAPANESE	Beto-byō べと病

This fungus primarily attacks the leaves. The first sign of infection is the appearance of light yellow spots. Then a white moldy growth of the fungus forms on the undersides of the leaves. The spots may be few or many. When they merge, they can cover most of the leaf. The affected leaves eventually turn brown, become dry and crumpled and fall. The disease may also cause severe malformation of the shoots or berries early in the season. Early symptoms on these parts appear as water-soaked depressions.

Iste fungo attacca principalmente le folios. Le prime symptoma de infection es maculas jalne clar. Postea, un blanc strato mucide del fungo se forma in le superfacies inferior del folios. Le maculas pote esser o numerose o pauc. Quando illos coalesce, illos coperi le major parte del folio. Le folios afflicte al fin deveni brun, desiccate, e corrugate, e illos se separa e cade. Le maladia causa anque malformation sever del partes nove o del baccas in le prime parte del saison. Symptomas initial in iste partes se manifesta como depressiones aquose.

Ce champignon s'attaque surtout aux feuilles. Le premier signe d'infection est l'apparition de taches jaune pâle. Ensuite, une moisissure blanche se forme à la face inférieure des feuilles. Les taches peuvent être clairsemées ou nombreuses. Quand elles se réunissent, elles peuvent couvrir presque toute la feuille. Eventuellement, les feuilles affectées tournent au brun, se dessèchent, s'émiettent et tombent. La maladie peut aussi causer de graves déformations des tiges ou des fruits tôt dans la saison. Les premiers symptômes sur ces parties se manifestent sous forme de dépressions délavées.

Este hongo ataca principalmente las hojas. El primer signo de la infección es el aparecimiento de manchas amarillo claro. Luego, un crecimiento mohoso blanco del hongo se forma en el envés de las hojas. Las manchas pueden ser pocas o muchas. Cuando éstas emergen pueden cubrir casi toda la hoja. Las hojas afectadas eventualmente se vuelven de color café, se secan, se arrugan y caen. La enfermedad puede también causar severa malformación de los brotes o frutos al principio de la estación. Los primeros síntomas en estas partes aparecen como depresiones empapadas de agua.

DISTRIBUTION: Af., A., Aust., Eur., N.A., C.A., S.A.

Vitis spp. (Fragaria chiloensis, Malus pumila = Malus sylvestris)+ Rosellinia necatrix
(Prill.) Berl. =

ENGLISH	Root Rot, White Root Rot
FRENCH	Pourridié de la Vigne, Pourriture des Racines
SPANISH	Pudrición radicular, Pudrición radicular blanca
PORTUGUESE	Podridão das raízes
ITALIAN	Mal bianco
RUSSIAN	гниль корней розеллиниозная
SCANDINAVIAN	
GERMAN	Rosellinia-Wurzelschimmel
DUTCH	
HUNGARIAN	
BULGARIAN	
TURKISH	Kök cürüklüğü
ARABIC	
PERSIAN	Pōōsēdēgēyē rēshēhē mō پوسیدگی ریشه مو
BURMESE	
THAI	
NEPALI	
HINDI	
VIETNAMESE	
CHINESE	
JAPANESE	Siromonpa-byō 白紋羽病

The above-ground symptoms of this disease are not distinctive in that other root troubles show similar effects. Sparse foliage, slow growth and terminal die-back of twigs are some features usually observed. On the roots, the symptoms are more distinctive. The surface of the root may be covered with a layer of cottony-white growth, or this may appear when the outer layer of dead bark is removed. Then white strands of the fungus which later turn brown, and finally black fruiting bodies of the fungus appear on the dead bark. Rotting of the roots results and often final death of the plants.

Le symptomas epigee de iste maladia non es distinctive perque altere afflictiones radical da aspectos similar. Foliage sparse, crescimento lente, e morte descendente del ramettos es usualmente remarcate. In le radices, le symptomas es plus distinctive. Le superfacie del radice es a vices coperte de un strato blanc cotonose, o a vices iste symptoma se manifesta quando on remove le cortice morte exterior. Alora, hyphas blanc del fungo, le quales plus tarde deveni brun, e al fin nigre structuras sporal del fungo se manifesta in le cortice morte. Putrefaction del radices occurre e sovente final del plantas.

Les symptômes extérieurs de cette maladie ne sont pas caractéristiques, puisque d'autres désordres radiculaires ont des effects semblables. Un feuillage peu abondant, une croissance lente et le dessèchement du bout des rameaux sont des points couramment observés. Sur les racines, les symptômes sont plus distinctifs. La surface de la racine peut être recouverte d'une couche blanche et cotonneuse, ce qui peut aussi se présenter quand la couche extérieure de l'écorce morte est enlevée. Alors apparaissent sur l'écorce morte des filaments blancs du champignon qui plus tard tournent au brun pour finalement laisser place aux fructifications noires du champignon. Alors les racines pourrissent et souvent les plantes succombent.

Los síntomas en las partes áreas de la planta causados por esta enfermedad no se distinguen de otros problemas radiculares que muestran efectos similares. Algunos signos que generalmente se observan son el follaje ralo, crecimiento lento y muerte regresiva de los extremos de las ramillas. En las raíces los síntomas son más distinctivos. La superficie de la raíz puede estar cubierta con una capa de un crecimiento algodonoso blanco, o éste puede aparecer al quitar la capa exterior de corteza muerta. Entonces aparecen en la corteza muerta hebras blancas del hongo, las cuáles después se vuelven de color café, y finalmente cuerpos fructíferos negros del hongo. La pudrición de las raíces sucede y frecuentemente la muerte final de las plantas.

DISTRIBUTION: Af., A., Aust., Eur., N.A., S.A.

Vitis spp. + Uncinula necator (Schwein.) Burrill (Oidium tuckerii Berkeley) =

ENGLISH Powdery Mildew
FRENCH Blanc, Oïdium de la Vigne
SPANISH Oidium, Oidio, Ceniza, Quintal, Cenicilla, Mildiú polvoriento
PORTUGUESE Oidio
ITALIAN Oidio delle vite, Crittogama
RUSSIAN
SCANDINAVIAN Vinmeldug, Vinmjöldagg
GERMAN Echter Mehltau, Oidium Äscher
DUTCH Meeldauw van de druivelaar
HUNGARIAN Szőlő-lisztharmat
BULGARIAN оидиум
TURKISH Bağ küllemesi
ARABIC Bayad Dakiki بيـاض دقيقي
PERSIAN Sēyāhākê hághēghēgê mô سفيدک حقيقی مو
BURMESE Far-oo-hmo roga ဆဉ်ဓိနှုလာဉ်ကောတာ
THAI Ra-Pang-Khaow ราแป้งขาว
NEPALI
HINDI Choorni phaphundi चूरनी फाफुन्दी
VIETNAMESE
CHINESE
JAPANESE Udonko-byô うどんこ病

This disease may occur on any new growth, but is most conspicuous on the upper surface of leaves. Here small, rather indistinct, white patches appear which later leave a powdery appearance. Usually these enlarge so that the entire upper surface of the leaf has a dusty coating. Severely attacked leaves curl upward during hot, dry weather. On the canes, the spots are limited, and these become brown or black later. The berries are often misshapen or show rusty spots.

Iste maladia occurre in tote le partes nove, ma es le plus remarcabile in le superfacies superior del folios. Minute areas blanc, aliquanto indistincte, se manifesta e plus tarde resulta in un aspecto pulverose. Usualmente, iste areas cresce usque a coperir tote le superfacie foliar superior con pulvere. Folios severmente afflicte se inrola in alto in tempore calide e sic. In le cannas, le maculas es limitate e illos deveni brun o nigre plus tarde. Sovente le baccas es malformate o monstra maculas (fer)ruginose.

Cette maladie peut se trouver sur toutes les pousses nouvelles, mais on la remarque surtout à la face supérieure des feuilles. De petites taches blanches, d'abord presque imperceptibles, prennent bientôt l'aspect d'une poudre. D'ordinaire, ces taches s'agrandissent de façon que toute la face supérieure de la feuille devient une couche de poussière. Les feuilles gravement atteintes s'enroulent vers le haut par temps chaud et sec. Sur les tiges, les taches sont limitées et, par la suite, deviennent brunes ou noires. Les fruits sont souvent déformés ou montrent des taches de couleur rouille.

Esta enfermedad puede ocurrir en cualquier nuevo crecimiento, pero es más conspicua en el haz de las hojas. Aquí aparecen manchones blancos, pequeños, poco distinguibles, que más tarde adquieren una apariencia polvorienta. Generalmente estos se agrandan, por lo que toda la superficie de la hoja tiene una capa polvorienta. Las hojas severamente atacadas se enroscan hacia arriba durante el tiempo cálido y seco. En los tallos, las manchas son limitadas y éstas mas tarde se vuelven de color café o negras. Los frutos frecuentemente están malformados y muestran manchas herrumbrosas.

DISTRIBUTION: Af., A., Aust., Eur., N.A., C.A., S.A.

Zea mays + Bacterium stewartii (E. F. Smith) Dowson =

ENGLISH   Bacterial Wilt
FRENCH   Flétrissure Bactérienne, Flétrissement Bactérien de Stewart
SPANISH   Marchitez bacteriana
PORTUGUESE   Bacteriose do milho, Murcha bacteriana
ITALIAN   Batteriosi del mais
RUSSIAN   чвядание бактериальное, бактериоз початков
SCANDINAVIAN
GERMAN   Bakterielle Kolbenkrankheit, Bakterielle Welke, Stewart'sche Krankheit
DUTCH
HUNGARIAN
BULGARIAN
TURKISH
ARABIC
PERSIAN
BURMESE
THAI   Hiaw-Chaow เหี่ยวเฉา
NEPALI
HINDI
VIETNAMESE   Chết héo
CHINESE   Yùh mǐh shìh jǐunn wěi tiao bìng 玉米細菌萎凋病
JAPANESE

The symptoms produced by this disease vary with the type of host affected. If sweet maize varieties are infected, the leaves show linear, pale green to yellow streaks with irregular or wavy margins which parallel the veins and may extend the length of the leaf. The streaks soon become dry and brown. Infected plants may produce premature, bleached and dead tassels. Field maize symptoms are short to long, irregular, pale green to yellow streaks in the leaves which appear when the plants are maturing. Sometimes the entire leaves die and dry up.

Le symptomas de iste maladia se varia secundo le planta-hospite afflicte. Si varietates de mais dulce es infectate, le folios monstra strias verde pallide o jalne, linear, con margines irregular o undulate, le quales es parallel al venas e a vices se extende per tote le longitude del folio. Le strias rapidemente deveni desiccate e brun. Plantas infectate pote producer spiculas prematur, blanchite, e morte. Symptomas in le varietates agreste es strias curte o elongate irregular verde pallide o jalne in le folios le quales se manifesta quando le plantas se approxima a maturation. A vices folios integre mori e se desicca.

Les symptômes de cette maladie varient selon l'hôte. Dans le cas du maïs sucré, les feuilles montrent des raies linéaires, de vert pâle à jaunes, avec des marges irrégulières et ondoyantes qui sont parallèles aux nervures et qui peuvent se prolonger tout le long de la feuille. Les raies deviennent aussitôt sèches et brunes. Les plantes atteintes peuvent produire des panicules prématurées, blanchies et mort-nées. Dans le cas du maïs fourrager, les symptômes prennent la forme de raies irrégulières vert pâle à jaunes, de longueur variable, apparaissant sur les feuilles au moment de la maturation. Quelquefois, les feuilles tout entières meurent et sèchent.

Los síntomas producidos por esta enfermedad varían con el tipo de hosperdera afectada. Si variedades de maíz dulce son afectadas, las hojas muestran rayas lineales, verde pálido a amarillas, con márgenes irregulares y ondulados que aparecen paralelas a las venas y pueden extenderse a lo largo de la hoja. Las rayas pronto se vuelven secas y de color café. Las plantas infectadas pueden producir espigas prematuras, blanqueadas y muertas. Los síntomas de campo en el maíz son rayas cortas a largas, irregulares, verde pálido a amarillas en las hojas, las cuáles aparecen cuando la planta está madurando. Algunas veces la hoja entera muere y se seca.

DISTRIBUTION: Eur., N.A., W.I.

314

Zea mays + Cochliobolus lunatus Nelson & Haasis (Curvularia lunata (Wakk.) Boed.) =

ENGLISH   Glume Mold,   Leaf Spot, False Blast, Kernel Discoloration
FRENCH
SPANISH   Moho de la gluma, Mancha foliar, falsa ráfaga, decoloración del grano
PORTUGUESE
ITALIAN
RUSSIAN
SCANDINAVIAN
GERMAN
DUTCH
HUNGARIAN
BULGARIAN
TURKISH
ARABIC
PERSIAN
BURMESE
THAI   Bi-Jut   ใบจุด
NEPALI
HINDI
VIETNAMESE
CHINESE
JAPANESE

This pathogen causes a seedling blight which occurs before or after emergence. Brown leaf spots develop on the less severely infected seedlings. Circular to elongate brown leaf spots are first small without marked water-soaking and later spread with reddish-brown margins with gray centers. On severely infected plants, the leaves dry out before the plants are mature.

Iste pathogeno causa un necrosis de plantulas que occurre o ante o post emergentia. Maculas foliar brun se forma in le plantulas minus severmente infectate. Maculas foliar brun circular o elongate es al initio minute e non visibilemente aquose; plus tarde illos se extende con brun-rubiette margines con centros gris. In plantas severmente infectate, le folios se desicca ante que le plantas se matura.

Ce pathogène cause une brûlure des semis qui survient avant ou après la levée. Des taches brunes se forment sur les feuilles des plantules les moins gravement infectées. Les taches brunes, circulaires à allongées, sont d'abord petites, sans aspect détrempé bien marqué, et s'accroissent par la suite, avec un centre gris et des marges brun rougeâtre. Sur les plantes gravement infectées, les feuilles sèchent avant que la plante atteigne la maturité.

Este patógeno causa un tizón de las plántulas que ocurre antes o después de la emergencia. Manchas foliares de color café se desarrollan en las plántulas menos severamente infectadas. Manchas circulares o alargadas, de color café aparecen en las hojas, siendo primero pequeñas sin empapamiento de agua marcado y luego se diseminan con márgenes café rojizos con centros grises. En plantas severamente infectadas, las hojas se secan antes de que las plantas están maduras.

DISTRIBUTION: Af., A., Aust., Eur., N.A., C.A., S.A.

Zea mays + Diplodia macrospora Earle (Diplodia zeae (Schw.) Lév.) =

ENGLISH	Dry Rot of Ears and Stalks, Leaf Spot
FRENCH	Maladie du "dry-rot", Pourriture Sèche des Epis
SPANISH	Podredumbre de la mazorca, Pudrición seca, Pudrición negra de la caña y mazorca
PORTUGUESE	Podridão sêca da espiga, Podridão branca da espiga
ITALIAN	Antracnosi fogliare
RUSSIAN	диплодиоз, гниль сухая (початков и стеблей)
SCANDINAVIAN	
GERMAN	Trockenfäule
DUTCH	
HUNGARIAN	
BULGARIAN	
TURKISH	
ARABIC	
PERSIAN	
BURMESE	
THAI	Fak-Nao ฝักเน่า
NEPALI	Makai Ko danth kuhine rog मकैको दाढ कुहिने रोग
HINDI	
VIETNAMESE	Bệnh thối ngang thân - Thối trái; Pokkah boeng
CHINESE	Yùh mǐh mǐao fǔ bìng 玉米苗腐病 , Yùh mǐh jing ku bìng 玉米莖枯病
JAPANESE	

Husks of early-infected ears appear bleached or straw-colored. Sometimes the entire ear turns grayish-brown, shrunken and completely rotted. Lightweight ears usually stand upright with the inner husks matted together by the fungus growth. Black fruiting bodies may be scattered on husks and sides of kernels. Ears infected late in the season show no external symptoms, but when broken and kernels are removed, a white mold can be seed growing between the kernels.

Le vaginas de spicas infectate in le prime parte del saison appare blanchite o paleate. A vices tote le spica deveni brun-gris, contracte, e totalmente putrefacte. Spicas legier usualmente resta erecte con le vaginas interior copulate con le fungo. Structuras sporal nigre es a vices disperse al vaginas e al lateres de granos. Spicas infectate in le ultime parte del saison monstra nulle symptomas externe, ma si on los rumpe e remove le granos, un mucor blanc se maifesta inter le granos.

Les enveloppes des épis attaqués tôt paraissent blanchies ou couleur paille. Quelquefois, tout l'épi devient brun grisâtre, ratatiné et complètement pourri. Les épis légers sont généralement dressés, et leurs enveloppes intérieures deviennent enchevêtrées parmi le mycélium. Des fructifications noires du champignon peuvent être dispersées sur les enveloppes et les côtés des grains. Les épis infectés tard dans la saison ne présentent aucun symptôme externe, mais lorsqu'on les brise et qu'on enlève les grains, on peut voir végéter, entre ceux-ci, une moisissure blanche.

Las tusas de las mazorcas infectadas temprano aparecen blanqueadas o de color paja. Algunas veces la mazorca entera se vuelve de color café grisáceo, encogida y completamente podrida. Las mazorcas livianas generalmente se mantienen erectas con las tusas del interior enredadas entre sí debido al crecimiento del hongo. Cuerpos fructíferos negros pueden estar diseminados sobre las tusas y lados de los granos. Las mazorcas infectadas tarde en la estación no muestran síntomas externos, pero al pelarlas y sacarles los granos, puede verse un moho blanco que crece entre los granos.

DISTRIBUTION: Af., A., Aust., N.A., C.A., S.A.

Zea mays (Oryza sativa) + Fusarium moniliforme (Sheld.) Snyder & Hansen (Gibberella fijikuroi (Saw.) Wr.) =

ENGLISH Gibberella Blight of Corn, Bakanae Disease of Rice
FRENCH   Pourriture Fusarienne, Fusariose Enivrante
SPANISH Marchitez, Necrosis foliar, Fusariosis, Enfermedad de Bakanae
PORTUGUESE Podridão rosada da espiga
ITALIAN
RUSSIAN фузариоз, гиббереллез, "пьяный хлеб", гниль красная фузариозная
SCANDINAVIAN
GERMAN Gibberella-Taumelkrankheit
DUTCH
HUNGARIAN fuzáriumos tőszáradás és csöpenészedés
BULGARIAN фузариоза
TURKISH
ARABIC
PERSIAN
BURMESE
THAI
NEPALI
HINDI
VIETNAMESE
CHINESE Yùh mǐh huong swài bǐng 玉米紅穗病
JAPANESE Naetatigare-byô 苗立枯病

Symptoms of the disease on stalks are usually not seen until the crop is maturing. At this time the stalks may break near the groundline. If the stalks are split, the rotted areas may be extensive. Ear rot is often associated with stalk rot because the fungus grows rapidly from one to the other. A salmon-pink to reddish-brown discoloration appears on individual kernels. As the disease progresses, a cottony-pink mold growth develops on infected kernels.

Symptomas de iste maladia in le pedunculos usualmente non se manifesta ante le maturation de plantas. Alora, le pedunculos pote rumper se presso le solo. Si le pedunculos se finde, le areas putrefacte pote esser extensive. Putrefaction del spicas sovente se associa con putrefaction de pedunculos perque le fungo se extende rapidemente ab le uno al altere. Un discoloration rosee salmon o brun-rubiette se manifesta in alicun granos. A mesura que le maladia se avantia, un strato mucide rosee e cotonose se forma in le granos infectate.

Les symptômes de cette maladie ne sont généralement pas perçus sur les tiges avant que la récolte soit en train de mûrir. A ce moment, les tiges peuvent se briser au ras du sol. Si les tiges se fendent, les plages pourries peuvent être considérables. La pourriture de l'épi accompagne souvent la pourriture des tiges, car le champignon pousse rapidement de l'un à l'autre. Une couleur rose saumon à brun rougeâtre apparaît sur certains grains. Pendant que la maladie évolue, une moisissure rose cotonneuse se développe sur les grains infectés.

Los síntomas de esta enfermedad en los tallos generalmente no se ven hasta que el cultivo está madurando. Para esta época los tallos pueden quebrarse cerca de la superficie del suelo. Si los tallos se rajan, las áreas podridas pueden ser extensivas. La pudrición de las mazorcas está frecuentemente asociada con la pudrición del tallo, ya que el hongo crece rápidamente de uno a otro. Una decoloración rosado salmón a café rojiza aparece en mazorcas individuales. A medida que la enfermedad progresa, un crecimiento de moho algondonoso rosado se desarrolla en los granos infectados.

DISTRIBUTION: Af., A., Aust., N.A., C.A., S.A.

Zea mays (and others) + Helminthosporium carbonum Ullstrup =

ENGLISH	Leaf Spot, Charred Ear, Seedling Blight
FRENCH	
SPANISH	Tizón de la hoja, Helminthosporiosis, Mancha foliar, Mazorca chamuscada
PORTUGUESE	
ITALIAN	
RUSSIAN	гельминтоспориоз початков, стеблей и листьев
SCANDINAVIAN	
GERMAN	
DUTCH	
HUNGARIAN	
BULGARIAN	
TURKISH	
ARABIC	
PERSIAN	
BURMESE	
THAI	Bi-Jut ใบจุด
NEPALI	
HINDI	
VIETNAMESE	
CHINESE	Yǔh mǐh yúan ban bìng　玉 米 圓 斑 病
JAPANESE	Gomahagare-byō　ごま葉枯病

This fungus produces lesions on the leaves which are characterized by numerous elongated spots between the veins with limited and parallel margins. These spots are chocolate-brown in color. The disease also occurs on the kernels as a black moldy growth, resulting in a charred appearance on the infected ear.

Iste fungo produce lesiones foliar, le quales se characterisa per numerose maculas elongate inter le venas con margines limitate e parallel. Iste maculas es brun como chocolate. Le morbo occurre anque in le granos como un strato mucide nigre e produce un aspecto carbonisate in le spica infectate.

Ce champignon produit sur les feuilles des lésions caractérisées par de nombreuses taches allongées entre les nervures, avec des marges limitées et parallèles. Ces taches sont brun chocolat. La maladie se présente également sur les grains sous la form d'une moisissure noire qui rend l'épi infecté comme carbonisé.

Este hongo produce lesiones en las hojas, las cuáles se caracterizan por numerosas manchas alargadas entre las venas con márgenes limitados y paralelos. Estas manchas son de color café chocolate. La enfermedad también aparece en los granos como un crecimiento mohoso negro, que da como resultado una apariencia chamuscada de la mazorca.

DISTRIBUTION: Af., A., Aust., Eur., N.A.

318

Zea <u>mays</u> (Gramineae) + <u>Helminthosporium</u> <u>maydis</u> Nisik. & Miyake (<u>Cochliobolus</u> <u>heterostrophus</u> (Drechsl.) Drechsl.) =

ENGLISH	Leaf Blotch, Seedling Blight, Southern Leaf Blight
FRENCH	Helminthosporiose, Taches Foliaires
SPANISH	Mancha café, Manchón de la hoja, tizón de la plántula, tizón foliar sureño
PORTUGUESE	
ITALIAN	
RUSSIAN	
SCANDINAVIAN	
GERMAN	Helminthosporiose
DUTCH	
HUNGARIAN	
BULGARIAN	
TURKISH	
ARABIC	
PERSIAN	
BURMESE	
THAI	Bi-Mhai ใบไหม้
NEPALI	
HINDI	
VIETNAMESE	Đốm lá gốc
CHINESE	Yǔh mǐh hu má yèh ku bìng 玉米胡蔴葉枯病
JAPANESE	

The lesions caused by this fungus are long, with parallel sides and buff to brown borders. One race of this pathogen produces lesions which are spindle-shaped or elliptical, with greenish-yellow or chlorotic halos. Seedlings from infected kernels may wilt and die within three weeks after planting. Cob rot can occur with heavy losses in harvesting and shelling.

Le lesiones producte per iste fungo es elongate con bordos parallel e castanie o brun. Un linea genealogic de iste pathogeno produce lesiones le quales es fusiforme o elliptic, con halos jalne-verde o chlorotic. Plantulas ab granos infectate pote marcescer e morir intra tres septimanas post que on los planta. Putrefaction de spicas a vices occurre con perdita sever in le recolta-mento e in le disgranamento.

Les lésions causées par ce champignon sont longues, avec des côtes parallèles et des bords jaune clair à bruns. Une race de ce pathogène produit des lésions fusiformes ou elliptiques, avec un halo jaune verdâtre ou chloritique. Les semis issus de grains infectés peuvent se flé-trir et mourir durant les trois semaines qui suivent la mise en terre.Il peut y avoir pourriture de l'épi, suivie de lourdes pertes durant la récolte et l'égrenage.

Las lesiones causadas por este hongo son largas, con lados paralelos y orillas de color ante a café. Una raza de este patógeno produce lesiones en forma de huso o elíptica, con halos amarillo verdosos o cloróticos. Las plántulas provenientes de granos infectados pueden marchi-tarse y morir tres semanas después de la siembra. La podredumbre de las mazorcas puede ocurrir causando fuertes pérdidas en la cosecha y desgrane.

DISTRIBUTION: Af., A., Aust., Eur., N.A., C.A., S.A.

Zea mays (Gramineae) + Helminthosporium turcicum Pass.   =

ENGLISH	Leaf Blight
FRENCH	Brûlure des Feuilles, Maladie des Stries, Taches Foliaires
SPANISH	Helminthosporiosis, Tizón de la hoja, Mancha de la hoja, Quemazón foliar
PORTUGUESE	Helmintosporiose
ITALIAN	Nebbia del granturco
RUSSIAN	**пятнистость гельмнтоспорозная**
SCANDINAVIAN	
GERMAN	Helminthosporium-Blattkrankheit
DUTCH	
HUNGARIAN	
BULGARIAN	**листен пригор**
TURKISH	
ARABIC	
PERSIAN	
BURMESE	Twet-nar-chauk roga ၐ တၐ်ၐ ၑ ၉ၾ ၜ ၐ ၐ်ၝ ၐ်
THAI	Bi-Mhai ใบไหม้
NEPALI	Makai Ko dadhuwa rog मकैको डढुवा रोग
HINDI	Parn Angmaari पर्ण अंगमारी
VIETNAMESE	Đốm loang trên lá
CHINESE	Yǔh mǐh méi wén bìng 玉米煤紋病
JAPANESE	Susumon-byô すす紋病

The symptoms of this disease may be seen on all aerial parts of the plant, but the most conspicuous lesions are found on the leaves. Individual lesions are elliptical and are large. The edges are well-defined and the infected area is tan or brown. As infection progresses the fungus sporulates in the centers of the lesions, giving them a black dusty appearance. Severe infection causes premature death of the leaves and consequent yield reduction.

Le symptomas de iste maladia se manifesta in tote le partes aeree del planta, ma le lesiones le plus remarcabile se trova in le folios. Lesiones es elliptic e grande. Le margines es ben definite e le area infectate es bronzate o brun. Quando le infection se avantia, le fungo sporula in le centros del lesiones, causante un aspecto pulverose nigre. Infection sever causa morte prematur del folios e diminution del rendimento.

Les symptômes de cette maladie peuvent s'observer sur toutes les parties aériennes de la plante, mais les lésions les plus voyantes se trouvent sur les feuilles. Les lésions individuelles sont elliptiques et considérables. Leurs bords sont bien définis et la partie infectée est de couleur chamois ou brune. A mesure que l'infection progresse, le champignon produit des spores au centre des lésions, ce qui leur donne une apparence poussiéreuse et noire. Une infection grave cause la mort prématurée des feuilles et réduit par le fait même le rendement.

Los síntomas de esta enfermedad pueden verse en todas las partes aéreas de la planta, pero las lesiones más conspicuas se encuentran en las hojas. Las lesiones individuales son elípticas y grandes. Las orillas están bien definidas y el área infectada es de color crema o café. A medida que la infección progresa, el hongo esporula en los centros de las lesiones, dándoles una apariencia polvorienta negra. La infección severa causa la muerte prematura de las hojas y la consecuente reducción en rendimiento.

DISTRIBUTION: Af., A., Aust., Eur., N.A., C.A., S.A.

320

Zea mays + Physoderma maydis Miyabe =

ENGLISH	Brown Spot
FRENCH	Physodermose
SPANISH	Mancha de la vaina, Mancha carmelita, Manchas negras, Mancha café
PORTUGUESE	Mancha parda
ITALIAN	
RUSSIAN	физодермоз, пятнистость бурая
SCANDINAVIAN	
GERMAN	Physodermose, Physoderma-Blattkrankheit
DUTCH	
HUNGARIAN	
BULGARIAN	
TURKISH	
ARABIC	
PERSIAN	
BURMESE	
THAI	Bi-Jut-Si-Numtan ใบจุดสีน้ำตาล
NEPALI	Makai Ko chakke thople rog मकैको चक्के थोप्ले रोग
HINDI	Bhura Dhabba भूरा धब्वा
VIETNAMESE	
CHINESE	
JAPANESE	Hanten-byō 斑点病

Lesions first appear as very small, oblong to round, yellowish spots on the leaves, stalks and sometimes the outer ear husks. Lesions may appear in bands across the leaf blades. Infected tissues turn brown to reddish-brown and coalesce to form large irregular blotches. Stalks infected at nodes beneath the sheaths frequently break at infection centers.

Lesiones se manifesta al initio como multe minute maculas jalne oblonge o rotunde in le folios, pedunculos, e a vices le vaginas exterior del spicas. Lesiones pote formar se in bandas a traverso le laminas foliar. Texitos infectate deveni brun o brun-rubiette e coalesce usque a formar grande pustulas irregular. Pedunculos le quales es infectate in nodos sub le vaginas frequentemente se rumpe al centro del infection.

Les lésions apparaissent d'abord sous forme de très petites taches allongées à rondes, de couleur jaunâtre, sur les feuilles, les tiges et parfois sur les enveloppes. Des lésions peuvent se former en lisières en travers du limbe. Les tissus infectés deviennent bruns à brun rougeâtre et se fusionnent pour former de grosses éclaboussures irrégulières. Les tiges, infectées aux noeuds sous les gaines, se brisent souvent vis-à-vis de ces foyers d'infection.

Las lesiones aparecen primero como manchas muy pequeñas, oblongas a redondas, amarillosas, sobre las hojas, tallos y algunas veces en el exterior de las tusas de la mazorca. Las lesiones pueden aparecer en bandas a través de la hoja. Los tejidos infectados se vuelven de color café a café rojizo y se juntan para formar grandes manchones irregulares. Los tallos infectados en los nudos debajo de las envolturas frecuentemente se quiebran en los centros de infección.

DISTRIBUTION: Af., A., Aust., N.A., C.A., S.A.

Zea mays + Physopella zeae (Mains) Cummins & Ramachar =

ENGLISH   Rust
FRENCH
SPANISH   Roya blanca de Guatemala, Roya
PORTUGUESE
ITALIAN
RUSSIAN
SCANDINAVIAN
GERMAN
DUTCH
HUNGARIAN
BULGARIAN
TURKISH
ARABIC
PERSIAN
BURMESE
THAI
NEPALI
HINDI
VIETNAMESE
CHINESE
JAPANESE

This fungus produces pustules which are cream-colored and occur in groups on the upper leaf surface. These are covered by the epidermis except for a small pore or slit. Later these pustules develop purplish, circular or oblong blotches with creamy centers. The dark chocolate-brown fungus fruiting structures are usually scattered in groups and appear late in the growing season.

Iste fungo produce pustulas le quales es de color de crema e occurre in gruppamentos in le superfacie superior del folios. Iste pustulas es coperte del epidermis, salvo un minute poro o fissura. Plus tarde iste pustulas produce ampullas purpuree, circular o oblonge, con centros de color de crema. Le structuras sporal obscur brun como chocolate usualmente es disperse in gruppamentos e se manifesta in le ultime parte del saison.

Ce champignon produit des pustules de couleur crème qui se présentent en groupes à la face supérieure de la feuille. Ces pustules sont recouvertes de l'épiderme, ne laissant à découvert qu'un pore ou une petite fissure. Plus tard, ces pustules deviennent des éclaboussures pourprées, rondes ou oblongues, avec un centre crémeux. Les fructifications brun chocolat foncé du champignon sont généralement disséminées en groupes et apparaissent vers la fin de la saison de végétation.

Este hongo produce pústulas que son de color crema y aparecen en grupos sobre el haz de las hojas. Estas están cubiertas por la epidermis excepto por un pequeño poro o abertura. Más tarde estas pústulas desarrollan manchones de color púrpura, circulares o alargados, con centros color crema. Las estructúras fructíferas del hongo, de color café chocolate obscuro están generalmente diseminadas en grupos y aparecen tarde en la época de cultivo.

DISTRIBUTION: C.A., S.A., W.I.

322

Zea mays + Puccinia maydis Béreng. (Puccinia sorghi Schw.) =

ENGLISH	Rust
FRENCH	Rouille
SPANISH	Roya, Roya común de la hoja
PORTUGUESE	Ferrugem
ITALIAN	Ruggine del granturco
RUSSIAN	ржавчина бурая
SCANDINAVIAN	
GERMAN	Braunrost
DUTCH	Roest
HUNGARIAN	
BULGARIAN	листна ръжда
TURKISH	Mısir pası
ARABIC	
PERSIAN	
BURMESE	
THAI	
NEPALI	Makai Ko sindure rog मकैको सिन्दुरे रोग
HINDI	Parn Ratua परण रतुवा
VIETNAMESE	Ri lá
CHINESE	Yúh mǐh shǐou bìng 玉米銹病
JAPANESE	Sabi-byð さび病

This disease is characterized by the appearance of pustules on any above-ground plant part,
being most abundant on leaves. The circular to elongate, golden-brown to cinnamon-brown
pustules are sparsely scattered over both leaf surfaces becoming brownish-black as the plant
matures and the fungus fruiting structures appear. When severe, chlorosis and death of leaves
may occur.

Iste maladia se characterisa per le emergentia de pustulas in tote le partes epigee del planta,
ma le plus frequentemente in le folios. Le pustulas circular o elongate, bronzate o brun como
cannella, es disperse passim per tote le duo superfacies foliar. Illos deveni nigre-brunette
quando le planta se matura e le structuras sporal del fungo se manifesta. Si le infection es
sever, chlorosis e morte del folios occurre.

On reconnaît cette maladie à l'apparition de pustules sur les parties aériennes, surtout sur
les feuilles. Les pustules, de forme ronde à allongée et de couleur brun doré à brun cannelle,
sont dispersées çà et là sur les deux faces des feuilles et deviennent brun foncé à l'approche
de la maturité de la plante. A ce moment, les fructifications apparaissent. Quand l'attaque
est grave, la décoloration et la mort des feuilles peuvent survenir.

Esta enfermedad se caracteriza por la aparición de pústulas en cualquier parte aérea de la
planta, siendo más abundantes en las hojas. Las pústulas circulares o alargadas, de color
café dorado a café canela están diseminadas a grandes trechos sobre ambas superficies de
las hojas volviéndose negras-cafesosas a medida que la planta madura y aparecen las estructuras fructíferas del hongo. Cuando la enfermedad es severa, puede ocurrir clorosis y muerte
de las hojas.

DISTRIBUTION: Af., A., Aust., Eur., N.A., C.A., S.A.

Zea mays + Puccinia polysora Underw. =

ENGLISH	Leaf Rust
FRENCH	Rouille du Sud
SPANISH	Roya, Roya tropical de la hoja
PORTUGUESE	
ITALIAN	
RUSSIAN	ржавчина "южная", ржавчина американская
SCANDINAVIAN	
GERMAN	Südlicher Rost
DUTCH	
HUNGARIAN	
BULGARIAN	
TURKISH	
ARABIC	
PERSIAN	
BURMESE	Than-chi roga ပတ်စီး နှး ယ၁း ေဆ၅ရိုင
THAI	Ra-Snim-Leck ราสนิมเหล็ก
NEPALI	
HINDI	
VIETNAMESE	
CHINESE	
JAPANESE	

This fungus produces symptoms which resemble those two previous diseases which have been described on the same host. The pustules are light cinnamon-brown, circular to oval and densely scattered on both leaf surfaces.

Iste fungo produce symptomas que es similar al duo morbos precedente le quales se associa con le mesme planta-hospite. Le pustulas es brun clar como cannella, circular o oval, e densemente disperse in tote le duo superfacies foliar.

Ce champignon produit des symptômes qui ressemblent à ceux des deux maladies ci-devant décrites sur ce même hôte. Les pustules sont brun cannelle clair, de rondes à ovales, et disséminées densément sur les deux faces des feuilles.

Este hongo produce síntomas que se asemejan a los de las dos enfermedades previamente descritas en la misma hospedera. Las pústulas son de color café canela claro, circulares a ovaladas y densamente diseminadas en ambas superficies de las hojas.

DISTRIBUTION: Af., A., Aust., N.A., C.A., S.A.

324

Zea mays (Sorghum vulgare) + Sphacelotheca reiliana (Kühn) Clint (Ustilago reiliana f. zeae (Kuehn) Pass.) =

ENGLISH	Head Smut
FRENCH	Charbon des inflorescences, Charbon
SPANISH	Carbón de la espiga, Carbón de la cabeza
PORTUGUESE	
ITALIAN	Carbone della pannocchia
RUSSIAN	головня пыльная
SCANDINAVIAN	
GERMAN	Kopfbrand, Rispenbrand
DUTCH	
HUNGARIAN	
BULGARIAN	
TURKISH	
ARABIC	
PERSIAN	
BURMESE	
THAI	
NEPALI	Makai Ko kalopoke मकैको कालोपोके
HINDI	Sheersh Kand शीरश कंद
VIETNAMESE	
CHINESE	Yùh mǐh sih hei swài bìng 玉米絲黑穗病
JAPANESE	Ilokuroho-byð 糸黒穗病

This disease first appears when ears and tassels are formed. The floral structures may be partially or completely converted to fungus masses containing the brownish-black spores. Tassel infection may be confined to individual spikelets, causing shoot-like growth or the entire tassel may proliferate forming a distorted leafy structure. Ears of infected plants may be diseased with leafy buds replacing normal ears. Occasionally the fungus structures develop as long, thin stripes in the leaves.

Iste maladia se manifesta al initio quando le spicas e le inflorescentias masculin se forma. Le structuras floral es a vices partialmente o totalmente reimplaciate per massas fungal que contine le sporas nigre-brunette. Infection del inflorescentias masculin a vices se limita a spiculas individue, causante excrescentia, o a vices le inflorescentia integre prolifera, producente un verticillo distorte. Spicas de plantas infectate es infirme e buttones de foliage reimplacia le spicas normal. A vices, le structuras fungal se manifesta como strias elongate e svelte in le folios.

Cette maladie apparaît d'abord quand les épis et les panicules se forment. Les parties florales sont partiellement ou complètement transformées en masses fongueuses contenant les spores noir brunâtre. L'infection de la panicule peut se confiner à des épillets individuels, causant des excroissances ressemblant à des tiges, ou la panicule tout entière peut proliférer pour former une structure foliare tordue. Les épis malades peuvent se transformer en bourgeons foliaires. Occasionnellement, le champignon forme de longues lisières minces sur les feuilles.

Esta enfermedad aparece primero cuando las mazorcas y espigas están formadas. Las estructuras florales peuden estar parcial o completamente cubiertas con las masas del hongo que contienen las esporas negro-cafesosas. La infección de la espiga puede estar confinada a espículas individuales, causando un crecimiento como de brote o toda la espiga puede proliferar formando una estructura foliar deforme. Las mazorcas de plantas infectadas pueden estar enfermas y tener capullos de hojas en vez de las mazorcas normales. Ocasionalmente las estructuras del hongo se desarrollan como rayas largas, delgadas en las hojas.

DISTRIBUTION: Af., A., Aust., Eur., N.A., C.A., S.A.

<u>Zea</u> <u>mays</u> + <u>Ustilago</u> <u>maydis</u> (DC.) Corda =

ENGLISH	Smut
FRENCH	Charbon
SPANISH	Carbón cubierto, Huitlacoche, Carbón del maíz, Carbón volador
PORTUGUESE	Carvão comun, Bouba da espiga
ITALIAN	Carbone del granturco
RUSSIAN	головня пузырчатая
SCANDINAVIAN	Majsbrand, Majssot
GERMAN	Maisbeulenbrand, Maisbrand, Beulenbrand, Maisflugbrand
DUTCH	Builenbrand, Maïsbrand
HUNGARIAN	Kukorica golyvásuzögje
BULGARIAN	обикновена главня
TURKISH	Mısır parı, Mısır rastığı
ARABIC	Taphahom تفخم
PERSIAN	
BURMESE	
THAI	Ra-Khamouw-Dum ราเขมาดำ
NEPALI	Makai Ko kalogeda rog मकैको कालो गेडा रोग
HINDI	Kand कंद
VIETNAMESE	Bệnh than trên trái
CHINESE	Yùh mǐh hei swài bìng 玉米黑穗病
JAPANESE	Kuroho-byô 黒穂病

This disease occurs as white or grayish-white galls of varying size on any aerial part of the plant. The individual galls vary greatly as to size. They may be seen along the midrib of a leaf, at the base of the leaf, on the stalk or in the tassels or ears. When the galls are forming, they are light-colored throughout but as they mature, masses of black spores are formed. The galls rupture and release the fungus spores. Seedling infection results in distorted and dwarfed plants.

Iste maladia occurre como gallas blanc o blanc-gris de varie dimensiones in tote le partes aeree del planta. Le gallas se varia grandemente in dimension. Illos se manifesta in le nervatura foliar, al pede foliar, in le pedunculos, o in le spicas e spiculas. Le gallas comencia como maculas de color clar, ma quando illos se matura, massas de sporas nigre se developpa. Le gallas se rumpe e discarga le sporas fungal. Infection de plantulas causa plantas nanos e distorte.

Cette maladie se présente sous forme de galles blanches ou grisâtres de grosseur variable sur les parties aériennes de la plante. La grosseur des galles varie beaucoup. On les rencontre le long des nervures des feuilles, à la base de la feuille, sur les tiges ou sur les panicules ou les épis. Quand les galles se forment, elles sont pâles d'un bout à l'autre, mais en vieillissant, elles se transforment en masses de spores noires. Les galles se rompent et libèrent les spores du champignon. L'infection des plantules se traduit par des plants tordus et rabougris.

Esta enfermadad aparece como agallas blancas o blanco-grisáceas de diferentes tamaños en cualquier parte áerea de la planta. Las agallas indivuales varían mucho en cuanto al tamaño. Pueden verse en la vena central de la hoja, en la base de la hoja, en el tallo o en las espigas o mazorcas. Cuando se están formando las agallas, son de color claro todas, pero a medida que maduran se forman masas de esporas negras. Las agallas se rompen dejando salir las esporas del hongo. La infección de las plántulas da como resultado plantas achaparradas y deformes.

DISTRIBUTION: Af., A., Aust., Eur., N.A., C.A., S.A

330

332

334

338

Mycosphaerella tabifica   Seedling Root Rot, Leaf Spot, Black Rot, Heart Rot   28
Uromyces betae   Rust   29

Brassica oleracea and other Brassica spp.   Cabbage

Alternaria brassicae   Gray Leaf Spot   30
Alternaria brassicicola   Black Spot, Brown Rot, Head Browning   31
Cercosporella brassicae   White Spot   32
Erysiphe polygoni   Powdery Mildew   33
Fusarium conglutinans   Yellows   34
Mycosphaerella brassicicola   Ring Spot   35
Olpidium brassicae   Seedling Disease, Big Vein   36
Peronospora brassicae =
   Peronospora parasitica   Downy Mildew   37
Phoma lingam   Black Leg, Leaf Spot   38
Plasmodiophora brassicae   Club Root   39
Pseudomonas maculicola   Leaf Spot   40
Xanthomonas campestris =
   Pseudomonas campestris   Black Rot   41

Capsicum spp.   Pepper

Phythophthora capsici   Phythophthora Blight, Fruit Rot   42
Xanthomonas vesicatoria   Bacterial Spot of Tomato, Stem Canker   135

Carthamus tinctorius   Safflower

Alternaria carthami   Leaf Spot   43
Puccinia carthami   Rust   44

Castanea dentata   Chestnut

Endothia parasitica   Blight, Canker   45
Phytophthora cambivora   Ink Disease   46
Phytophthora cinnamomi   Heart Rot of Stems and Buds of Pineapple, Root Rot   10

Citrus sinensis and other Citrus spp.   Orange

Alternaria citri =
   Alternaria mali   Black Rot of Oranges, Fruit Rot, Leaf Spot   47
Botryodiplodia theobromae =
   Diplodia natalensis   Die-back, Stem End Rot, Botryodiplodia Rot, Charcoal Rot   48
Corticium salmonicolor   Pink Disease of Rubber and Citrus   101
Diaporthe citri =
   Phomopsis citri   Melanose of Fruit and Foliage, Gummosis, Phomopsis Rot   49
Elsinoe fawcettii =
   Sphaceloma fawcettii   Scab   50
Gloeosporium limetticola   Anthracnose, Withertip   51
Guignardia citricarpa =
   Phoma citricarpa   Black Spot   52
Oospora citri-aurantii   Sour Rot   53
Penicillium italicum   Blue Mold   54
Phytophthora cactorum   Collar Rot of Apple and Pear, Die-back, Root Rot   144
Phytophthora citrophthora =
   Phytophthora citricola   Brown Rot of Fruit, Brown Rot Gummosis   55
Pseudomonas syringae   Citrus Blast, Black Pit of Lemon   56
Rosellinia bunodes   Secondary Root Rot of Coffee and Citrus , Black Root   61
Sphaeropsis tumefaciens   Branch Knot   57
Xanthomonas citri =
   Phytomonas citri   Citrus Canker, Bacterial Canker   58

344

346

348

350

363

374

376

378

380

382

386

400

402

418

422

424

433

436

440

446

450

Glossario de Parolas Frequente in Interlingua
(Abbreviationes: A anglese, E espaniol, F francese)

a vices A at times E a veces F quelquefois
ab A from, since E de, desde F de, depuis
alicun A any, some E algún F quelque
alte A high E alto F haut
ancora A still, yet E todavía F encore
anello A ring E anillo F anneau
anque A also, too E también F aussi
aspere A rough E áspero F rêche
attaccar A to attack E atacar F attaquer
avantiar A to progress, advance E avanzar F avancer
banda A stripe E raya F bande
ben que A although E bien que F bien que
bulliente A boiling E herviendo F ébullition
button A bud E yema F bourgeon
cader A to drop, fall E caer F tomber
campo A field E campo F champ
cannellate A grooved, fluted E acanalado F cannelé
carnose A fleshy E gordo F charnu
castanie A chestnut-colored E color de castaña F châtain
circumcinger A to girdle E cercar F cordelière
clar A light, clear E claro F clair
collar A crown E corona F couronne
collina A hill E colina F colline
collo A neck E cuello F collet
comestibile A edible E comestibile F comestible
componer A to compound, put together E componer F combiner
coperir A to cover E cubrir F couvrir
corda A cord, rope E cuerda F cordelette
corrugate A wrinkled E arrugado F ridé
corrumper A to rot E pudrir F pourrir
crescer A to grow E cresciento F grossir
crispar A to shrivel E arrugar F ratatiner
depost A after E después de, detrás de F après
desiccar A to dry out E secarse F dessécher
devenir A to turn into, become E hacer F devenir
disligar A to unbind E desligar F délier
dum A while, during E durante F pendant que
dur A hard E duro F dur
esser A to be E ser, estar F être
extender A to develop, extend E desarrollar F développer
ferruginose A rusty-brown E rufosa F brun-roux
folios A leaves E hojas F feuilles
forsan A perhaps, maybe E quizá F peut-être
fracturar A to break E romper F briser
fructo A fruit E fruto F fruit
gris A gray E gris F gris
gutta A drop E gotita, gota F goutte
humiditate A moisture, dampness E humedad F humidité
immagasinage A warehousing, storing up E almacenar F emmagasinage
inaperte A closed E no abierto F n'ouvert pas
inrolar A to curl, to roll up E rizar F friser
iste A this, these E este F ceci
jalnette A yellowish E amarillo F jaunâtre
junctura A joint, juncture E juntura F jointure

juvene A young E joven F jeune
lassar A to leave E dejar F laisser
lignose A woody E leñose F liégeux
macula A spot, fleck E mancha F tache
maltractamento A maltreatment, rough handling E dañoso F mauvais traitement
maniera A manner, way E manera F manière
marcescer A to wilt E marchitar F Se flétrir
mat A dull E sin brillo F mat
micre A small E pequeño, menudo F petit
molle A soft E blando F mou
monstrar A to show E mostrar F montrer
morir A to die E morir F mourir
mucor A mold E moho F moisi
musca A fly E mosca F mouche
nanismo A stunting E impedimento de crecimiento F rabougrissement
obscur A dark E obscuro F foncé
pacchetteria de conservas A packing house E frigorífico F entrepôt
pauc A slightly, few E poco F peu
pender A to hang E pender F pendre
postea A afterwards E después F ensuite
poter A to be able to E poder F être capable de
precoce A early E temprano F précoce
presso A near, close E cerca de F près de
punctillate A speckled E manchedo F tacheture
punctos A spots E puntos, manchas F taches
purpuree A purple E purpurino, morado F violet
putrefaction A damping-off E putrefacción F pourriture
qualque A some, any E cualquier F quelque
radice A root E raiz F racine
ramettos A twigs E ramitas F branchettes
rendimento A yield E cosecha F rendement
restar A to remain E restar F rester
rumper A to break E romper F casser
semines A seeds E semillas F graines
sequente A following, next E seguiente F suivant
septimana A week E semana F semaine
siliqua A seed pod E silicua, vaina F silique
sin A without E sin F sans
spicula A spikelet E espiguilla F épillet
stadio A stage (of development) E etapas F étape
stirpe A stem E tallo F tige
sub A under, below E debajo de F sous
tenere A tender E tierno F tendre
texito A tissue E tejido F tissus
tumide A swollen, inflated E hinchado F enflé
turgide A turgid E turgente, hinchado F turgide, enflé
usque A (all the way) to, until E hasta F jusque
vagina A sheath E vaina F manchon
venas A veins, streaks E venas F nervures
verdiero A orchard E huerto F verger
vetule A old E viejo F vieux

ANNOTATED BIBLIOGRAPHY

Balashev, L. L. (Ed.) СЛОВАРЬ ПОЛЕЗНЫХ РАСТЕНИЙ НА ДВАДЦАТИ ЕВРОПЕЙСКИХ ЯЗЫКАХ,
(Dictionaries of Useful Plants in Twenty Languages) 1970. Publishing House "Nauka." Moscow.
368 p.
476 species of cultivated plants, trees and shrubs, 54 terms from plant morphology in twenty
European languages. Purpose is   to simplify the reading, translation and abstracting of literature in agrobotany. Three sections: dictionary of names of cultivated plants, trees and shrubs;
dictionary of morphological terms; indexes to the dictionaries. Languages are   Bulgarian,
Czech, Danish, German, English, Spanish, French, Italian, Hungarian,   Dutch, Norwegian,
Polish, Portuguese,   Rumanian, Russian, Albanian, Slovakian, Serbo-croation, Finnish
and Swedish.

Conners,   I. L. An Annotated Index of Plant Diseases in Canada and fungi recorded on
plants in Alaska, Canada and Greenland. 1967. Canada Department of Agriculture. Publication 1251. 381 p.
Summary of recorded occurences of diseases of cultivated Canadian plants and fungi on wild and
cultivated plants in Canada, Alaska   and   Greenland. Entries listed according to Latin
host name with brief description of host in English. Pathogens listed alphabetically under descriptions with English and French common names and geographical information. Index arranged
by Latin host and pathogen names and common disease names. Extensive bibliography.

Commonwealth Mycological Institute.   CMI Descriptions of Pathogenic Fungi and Bacteria.
1974. Kew, Surrey, England.
Series of looseleaf sheets, 10 to a folder, issued four times each year. Illustrations and descriptions   of pathogens, information on resulting diseases, their hosts, geographical distribution, transmission,   location of type description and synonyms of pathogen. Literature
citations. Information based on CMI Distribution Maps of Plant Diseases, Review of Plant
Pathology and Review of Medical and Veterinary Mycology.

Commonwealth Mycological Institute. Distribution Maps of Plant Diseases. 1974, Kew, Surrey,
England.
Group of world maps with locations of   diseases circled. Information includes Latin pathogen
names, synonyms, host names in English and Latin and distribution by continent and country.
References and indications of herbarium specimens available given on reverse side. Maps con-
tinually revised and updated.

Diakova, G. A. (Ed.)   ФИТОПАТОЛОГИЧЕСКИЙ СЛОВАРЬ-СПРАВОЧНИК   РУССКО-АНГЛО-НЕМЕЦКО-
ФРАНЦУЗСКИЙ. (Phytopathological Dictionary in Russian, English, German and French.) 1969.
Academy of Sciences of   the USSR. Institute of Scientific and Technical Information. "Nauka."
Mowcow. 478 p.
More than 5000 names of fungal, bacterial and virus diseases of cultivated plants, general
terms in phytopathology, mycology and plant protection, names of most common cultivated plants
Forest trees, shrubs and ornamentals not included. Material arranged according to plant types.
Synonyms for causal agents given. Indexes in four languages, Latin pathogen index, list of
authors and their abbreviations and Table of Contents.

Kwizda, Richard. Vocabularium Nicentium Florae, Wörterbuch der Wichtigsten Pflanzenschäd-
lange und Unkräuter. (Dictionary of Plant Pests and Diseases.) 1963. Wien, Springer Verlag.
128 p.
500 names of plant pests, diseases, weeds, forest diseases, pests in storage and wood parasites on field crops, vegetables, fruit and forest trees. The languages included are German,
Latin, Danish, English, French, Italian, Dutch, Russian, Swedish and Spanish. Each language
has own separate index. Literature citations given.

Merino-Rodriguez, Manuel (Compiler and Arranger.) <u>Lexicon of Plant Pests and Diseases.</u>
1966. Elsevier Publication Company. Amsterdam, London, New York. 351 p.
Each entry numbered and arranged alphabetically within sections according to Latin name of
organism. English is key language in appendices (1. Symptoms of diseases, 2. Non-parasitic
diseases and 3. Unclassified virus diseases.) Names are those recorded in current usage.
Synonyms of pathogens are given in many entries. Indexes in each language: Latin, English,
French, Italian, Spanish and German. Sections labeled Zooparasites, Bacteria, Ascomycetes,
Basidiomycetes, Phycomycetes and Fungi Imperfecti.

Nijdam, J. (Ed.) Tuinbouwkundig Woordenboek in Acht Talen <u>(Horticultural Dictionary in Eight</u>
<u>Languages.</u>) 1961. Plv. Directeur Rijks Hogere Tuinbouwschool Te Utecht. Ministerie van
Landbouw En Visseriu-Directir Tuinbouw.
4000 words and expressions included are related to the practice of horticulture (production of
vegetables, fruits, flowers, bulbs, trees and shrubs, herbs and seeds, and apiculture.) Words
arranged alphabetically according to Dutch spellings. Alphabetical indexes for each language
(Dutch, English, French, German, Danish, Swedish, Spanish and Latin) refer to the entry
numbers and letters.

Plant Pathology Committee, The British Mycological Society. <u>List of Common British Plant</u>
<u>Diseases.</u> 1944. Cambridge University Press. 61 p.
Entries by English common name of host, followed by Latin in parentheses. Common
names of diseases under each entry in English, Swedish, Spanish plus Italian, Danish, French,
Russian, German, American and Dutch when known, along with Latin pathogen name. Effort
made to use only one common name per entry even though others known to exist. One index
combining all common names in all languages plus scientific names of hosts and parasites
and synonyms. List of authors' names and abbreviations.

United States Department of Agriculture. <u>Index of Plant Diseases in the United States.</u>
Agriculture Handbook No. 165. 1960. 531 p.
A compilation of more than 1200 host genera and some 50,000 parasitic and non-parasitic
diseases. Listed in alphabetical order by Family name, Genus of host, followed by short
description of host, with pathogens listed alphabetically under descriptons. Indexes by Family
and Genera and alphabetical English common names. Appendix material: list of authors of
plant parasite names and recommended abbreviations. Index has been for many years the accepted standard for pathogen names, authorities and synonyms.